Introducing Einstein's Relativity

Albert Einstein (1879–1955)

Introducing
Einstein's Relativity

Ray d'Inverno

Faculty of Mathematical Studies, University of Southampton

CLARENDON PRESS · OXFORD
1992

Oxford University Press, Walton Street, Oxford OX2 6DP

Oxford New York Toronto
Delhi Bombay Calcutta Madras Karachi
Petaling Jaya Singapore Hong Kong Tokyo
Nairobi Dar es Salaam Cape Town
Melbourne Auckland
and associated companies in
Berlin Ibadan

Oxford is a trade mark of Oxford University Press

Published in the United States
by Oxford University Press, New York

A catalogue record for this book is
available from the British Library

Library of Congress Cataloging in Publication Data
d'Inverno, R. A.
Introducing Einstein's relativity/R. A. d'Inverno.
Includes bibliographical references and index.
1. Relativity (Physics) 2. Black holes (Astronomy)
3. Gravitation. 4. Cosmology. 5. Calculus of tensors. I. Title.
QC173.55.I58   1992   530.1'1—dc20   91-24894
ISBN 0-19-859653-7 (hardback)
ISBN 0-19-859686-3 (pbk)

Typeset by Macmillan India Ltd, Bangalore 25
Printed in the United States of America

This book is dedicated to my beloved wife Pauline

Contents

Overview

The organization of the book 1

1.1 Notes for the student

There is little doubt that relativity theory captures the imagination. Nor is it surprising: the anti-intuitive properties of special relativity, the bizarre characteristics of black holes, the exciting prospect of gravitational wave detection and with it the advent of gravitational wave astronomy, and the sheer scope and nature of cosmology and its posing of ultimate questions; these and other issues combine to excite the minds of the inquisitive. Yet, if we are to look at these issues meaningfully, then we really require both physical insight and a sound mathematical foundation. The aim of this book is to help provide these.

The book grew out of some notes I wrote in the mid-1970s to accompany a UK course on general relativity. Originally, the course was a third-year undergraduate option aimed at mathematicians and physicists. It subsequently grew to include M.Sc. students and some first-year Ph.D. students. Consequently, the notes, and with it the book, are pitched principally at the undergraduate level, but they contain sufficient depth and coverage to interest many students at the first-year graduate level. To help fulfil this dual purpose, I have indicated the more advanced sections (level-two material) by a grey shaded bar alongside the appropriate section. Level-one material is essential to the understanding of the book, whereas level two is enrichment material included for the more advanced student. To help put a bit more light and shade into the book, the more important equations and results are given in tint panels.

In designing the course, I set myself two main objectives. First of all, I wanted the student to gain insight into, and confidence in handling, the basic equations of the theory. From the mathematical viewpoint, this requires good manipulative ability with tensors. Part B is devoted to developing the necessary expertise in tensors for the rest of the book. It is essentially written as a self-study unit. Students are urged to attempt all the exercises which accompany the various sections. Experience has shown that this is the only real way to be in a position to deal confidently with the ensuing material. From the physical viewpoint, I think the best route to understanding relativity theory is to follow the one taken by Einstein. Thus the second chapter of Part C is devoted to discussing the principles which guided Einstein in his search for a relativistic theory of gravitation. The field equations are approached first from a largely physical viewpoint using these principles and subsequently from a purely mathematical viewpoint using the variational principle approach. After a chapter devoted to investigating the quantity which goes on the 'right-hand side' of the equations, the structure of the equations is discussed as a prelude to solving them in the simplest case. This part of the course ends by considering the experimental status of general relativity. The course originally assumed that the student had some reasonable knowledge of special relativity. In fact, over the years, a growing number of students have taken the course without this background, and so, for completeness, I eventually added Part A. This is designed to provide an introduction to special relativity sufficient for the needs of the rest of the book.

The second main objective of the course was to develop it in such a way that it would be possible to reach three major topics of current interest, namely, black holes, gravitational waves, and cosmology. These topics form the subject matter of Parts D, E, and F respectively.

Each of the chapters is supported by exercises, numbering some 300 in total. The bulk of these are straightforward calculations used to fill in parts omitted in the text. The numbers in parentheses indicate the sections to which the exercises refer. Although the exercises in general are important in aiding understanding, their status is different from those in Part B. I see the exercises in Part B as being absolutely essential for understanding the rest of the book and they should not be omitted. The remaining exercises are desirable. The book is neither exhaustive nor complete, since there are topics in the theory which we do not cover or only meet briefly. However, it is hoped that it provides the student with a sound understanding of the basics of the theory.

A few words of advice if you find studying from a book hard going. Remember that understanding is not an all or nothing process. One understands things at deeper and deeper levels, as various connections are made or ideas are seen in different contexts or from a different perspective. So do not simply attempt to study a section by going through it line by line and expect it all to make sense at the first go. It is better to begin by reading through a few sections quickly — skimming — thereby trying to get a general feel for the scope, level, and coverage of the subject matter. A second reading should be more thorough, but should not stop if ideas are met which are not clear straightaway. In a final pass, the sections should be studied in depth with the exercises attempted at the end of each section. However, if you get stuck, do not stop there, press on. You will often find that the penny will drop later, sometimes on its own, or that subsequent work will produce the necessary understanding. Many exercises (and exam questions) are hierarchical in nature. They require you to establish a result at one stage which is then used at a subsequent stage. If you cannot establish the result, then do not give up. Try and use it in the subsequent section. You will often find that this will give you the necessary insight to allow you to go back and establish the earlier result. For most students, frequent study sessions of not too long a duration are more productive than occasional long drawn out sessions. The best study environment varies greatly from one individual to another. Try experimenting with different environments to find out what is the most effective for you.

As far as initial conditions are concerned, that is assumptions about your background, it is difficult to be precise, because you can probably get by with much less than the book might seem to indicate (see §1.5). Added to which, there is a big difference between understanding a topic fully and only having some vague acquaintance with it. On the mathematical side, you certainly need to know calculus, up to and including partial differentiation, and solution of simple ordinary differential equations. Basic algebra is assumed and some matrix theory, although you can probably take eigenvalues and diagonalisation on trust. Familiarity with vectors and some exposure to vector fields is assumed. It would also be good to have met the ideas of a vector space and bases. We use Taylor's theorem a lot, but probably knowledge of Maclaurin's will be sufficient. On the Physics side, you obviously need to know Newton's laws and Newtonian gravitation. It would be helpful also to know a little about the potential formulation of gravitation (though, again, just the basics will do). The book assumes familiarity with electromagnetism (Maxwell's equations, in particular) and fluid dynamics (the Navier–Stokes equation, in particular), but neither of these are absolutely essential. It would be very helpful to have met some ideas about waves (such as the fundamental relationship $c = \lambda v$) and the wave equation in particular. In cosmology, it is assumed that you know something about basic astronomy.

Having listed all these topics, then, if you are still unsure about your background, my approach would be to say: try the book and see how you get on, if it gets beyond you (and it is not a level two section) press on for a bit and, if things do not get any better, then cut out. Hopefully, you may still have learnt a lot, and you can always come back to it when your background is stronger. In fact, it should not require much background to get started, for part A on special relativity assumes very little. After that you hit part B, and this is where your motivation will be seriously tested. I hope you make it through because the pickings on the other side are very rich indeed. So, finally, good luck!

1.2 Acknowledgements

Very little of this book is wholly original. When I drew up the notes, I decided from the outset that I would collect together the best approaches to the material which were known to me. Thus, to take an example right from the beginning of the book, I believe that the k-calculus provides the best introduction to special relativity, because it offers insight from the outset through the simple diagrams that can be drawn. Indeed one of the themes of this book is the provision of a large number of illustrative diagrams (over 200 in fact). The visual sense is the most immediate we possess and helps lead directly to a better comprehension. A good subtitle for the book would be, **An approach to relativity theory via space-time pictures**. The k-calculus is an approach developed by H. Bondi from the earlier ideas of A. Milne. My use of it is not surprising since I spent my years as a research student at King's College, London, in the era of Hermann Bondi and Felix Pirani, and many colleagues will detect their influences throughout the book. So the fact is that many of the approaches in the book have been borrowed from one author or another; there is little that I have written completely afresh. My intention has been to organize the material in such a way that it is the more readily accessible to the majority of students.

General relativity has the reputation of being intellectually very demanding. There is the apocryphal story, I think attributed to Sir Arthur Eddington, who, when asked whether he believed it true that only three people in the world understood general relativity, replied, 'Who is the third?'

Indeed, the intellectual leap required by Einstein to move from the special theory to the general theory is, there can be little doubt, one of the greatest in the history of human thought. So it is not surprising that the theory has the reputation it does. However, general relativity has been with us for some three-quarters of a century and our understanding is such that we can now build it up in a series of simple logical steps. This brings the theory within the grasp of most undergraduates equipped with the right background.

Quite clearly, I owe a huge debt to all the authors who have provided the source material for and inspiration of this book, However, I cannot make the proper detailed acknowledgements to all these authors, because some of them are not known even to me, and I would otherwise run the risk of leaving somebody out. Most of the sources can be found in the bibliography given at the end of the book, and some specific references can be found in the section on further reading. I sincerely hope I have not offended anyone (authors or publishers) in adopting this approach. I have written this book in the spirit that any explanation that aids understanding should ultimately reside in the pool of human knowledge and thence in the public domain. None the less, I would like to thank all those who, wittingly or unwittingly, have made this book possible. In particular, I would like to thank my old Oxford tutor, Alan Tayler, since it was largely his backing that led finally to the book being produced. In the process of converting the notes to a book, I have made a number of changes, and have added sections, further exercises, and answers. Consequently this new material, unlike the earlier, has not been vetted by the student body and it seems more than likely that it may contain errors of one sort or another. If this is the case, I hope that it does not detract too much from the book and, of course, I would be delighted to receive corrections from readers. However, I have sought some help and, in this respect, I would particularly like to thank my colleague James Vickers for a critical reading of much of the book.

Having said I do not wish to cite my sources, I now wish to make one important exception. I think it would generally be accepted in the relativity community that the most authoritative text in existence in the field is **The large scale structure of space-time** by Stephen Hawking and George Ellis (published by Cambridge University Press). Indeed, this has taken on something akin to the status of the Bible in the field. However, it is written at a level which is perhaps too sophisticated for most undergraduates (in parts too sophisticated for most specialists!). When I compiled the notes, I had in mind the aspiration that they might provide a small stepping stone to Hawking and Ellis. In particular, I hoped it might become the next port of call for anyone wishing to pursue their interest further. To that end, and because I cannot improve on it, I have in places included extracts from that source virtually verbatim. I felt that, if students were to consult this text, then the familiarity of some of the material might instil confidence and encourage them to delve deeper. I am hugely indebted to the authors for allowing me to borrow from their superb book.

1.3 A brief survey of relativity theory

It might be useful, before embarking on the course proper, to attempt to give some impression of the areas which come under the umbrella of relativity theory. I have attempted this schematically in Fig. 1.1. This is a rather partial

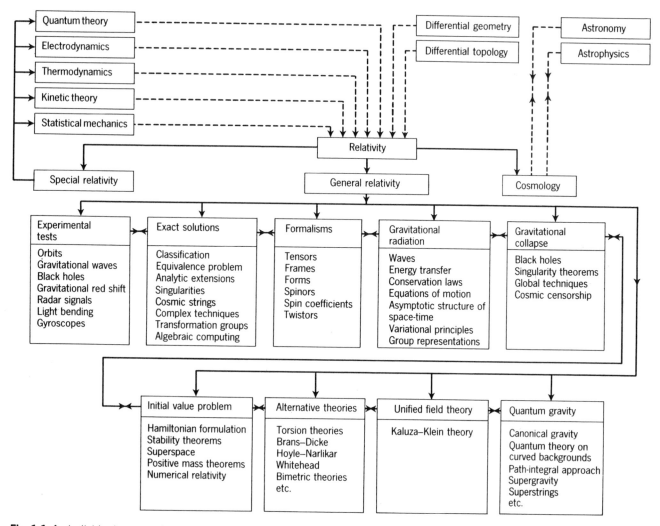

Fig. 1.1 An individual survey of relativity.

and incomplete view, but should help to convey some idea of our planned route. Most of the topics mentioned are being actively researched today. Of course, they are interrelated in a much more complex way than the diagram suggests.

Every few years since 1955 (in fact every three since 1959), the relativity community comes together in an international conference of general relativity and gravitation. The first such conference held in Berne in 1955 is now referred to as GR0, with the subsequent ones numbered accordingly. The list, to date, of the GR conferences is given in Table 1.1. At these conferences, there are specialist discussion groups which are held covering the whole area of interest. Prior to GR8, a list was published giving some detailed idea of what each discussion group would cover. This is presented below and may be used as an alternative to Fig. 1.1 to give an idea of the topics which comprise the subject.

Table 1.1

GR0	1955	Bern, Switzerland
GR1	1957	Chapel Hill, North Carolina, USA
GR2	1959	Royaumont, France
GR3	1962	Jablonna, Poland
GR4	1965	London, England
GR5	1968	Tblisi, USSR
GR6	1971	Copenhagen, Denmark
GR7	1974	Tel-Aviv, Israel
GR8	1977	Waterloo, Canada
GR9	1980	Jena, DDR
GR10	1983	Padua, Italy
GR11	1986	Stockholm, Sweden
GR12	1989	Boulder, Colorado, USA

I. Relativity and astrophysics

Relativistic stars and binaries; pulsars and quasars; gravitational waves and gravitational collapse; black holes; X-ray sources and accretion models.

II. Relativity and classical physics

Equations of motion; conservation laws; kinetic theory; asymptotic flatness and the positivity of energy; Hamiltonian theory, Lagrangians, and field theory; relativistic continuum mechanics, electrodynamics, and thermodynamics.

III. Mathematical relativity

Differential geometry and fibre bundles; the topology of manifolds; applications of complex manifolds; twistors; causal and conformal structures; partial differential equations and exact solutions; stability; geometric singularities and catastrophe theory; spin and torsion: Einstein–Cartan theory.

IV. Relativity and quantum physics

Quantum theory on curved backgrounds; quantum gravity; gravitation and elementary particles; black hole evaporation; quantum cosmology.

V. Cosmology

Galaxy formation; super-clustering; cosmological consequences of spontaneous symmetry breakdown: domain structures; current estimates of cosmological parameters; radio source counts; microwave background; the isotropy of the universe; singularities.

VI. Observational and experimental relativity

Theoretical frameworks and viable theories; tests of relativity; gravitational wave detection; solar oblateness.

VII. Computers in relativity

Numerical methods; solution of field equations; symbolic manipulation systems in general relativity.

1.4 Notes for the teacher

In my twenty years as a university lecturer, I have undergone two major conversions which have profoundly affected the way I teach. These have, in their way, contributed to the existence of this book. The first conversion was to the efficacy of the printed word. I began teaching, probably like most of my colleagues, by giving lectures using the medium of chalk and talk. I soon discovered that this led to something of a conflict in that the main thing that students want from a course (apart from success in the exam) is a good set of lecture notes, whereas what I really wanted was that they should understand the course. The process of trying to give students a good set of lecture notes meant that there was, to me, a lot of time wasted in the process of note taking. I am sure colleagues know the caricature of the conventional lecture: notes are copied from the lecturer's notebook to the student's notebook without their going through the heads of either — a definition which is perhaps too close for comfort. I was converted at an early stage to the desirability of providing students with printed notes. The main advantage is that it frees up the lecture period from the time-consuming process of note copying, and the time released can be used more effectively for developing and explaining the course at a rate which the students are able to cope with. I still find that there is something rather final and definitive about the printed word. This has the effect on me of making me think more carefully about what goes into a course and how best to organize it. Thus, printed notes have the added advantage of making me put more into the preparation of a course than I would have done otherwise. It must be admitted that there are some disadvantages with using printed notes, but this is not the place to elaborate on them.

My second conversion was to the efficacy of self-study. This is a rather elaborate title for the concept of students studying and learning on their own from books or prepared materials. It is still a surprise to me just how little of this actually goes on in certain disciplines. And yet you would think that one of the main objectives of a university education is to teach students how to use books. My experience is that, in mathematics particularly, students find this hard to do. This is not so surprising since it requires high-level skills which many do not come to university equipped with. So one needs a mechanism which encourages students to use books. My first experience was in designing a Keller-type (self-paced) self-study course, where the students study from specially prepared units and are required to pass periodic tests before they move on to new topics. This eventually led me in other courses to use a coursework component counting towards a final assessment as a mechanism for helping to get students to study on their own. I have been involved in a good deal of research into this approach and the most frequent remark students make about coursework is that 'it gets me to work'. The coursework approach was particularly important in the design of the general relativity course for reasons which I believe are worth exploring.

In the mid-1970s, there were very few undergraduate courses in general relativity in existence in the UK. Those that there were usually only got as far as the Schwarzschild solution and then stopped. This was because the bulk of the course was devoted to developing the necessary expertise in tensors and there did not seem to be any short cut. This meant, from the viewpoint of both the student and the teacher, that the course ended just as things were beginning to get really interesting. It was clear to me that what students really wanted to know about most were the topics of black holes, gravitational waves, and cosmology. So, from the outset, the object was to design a course which made this possible. It was achieved by separating out what is Part B of this book as a self-study unit on tensors. The notes were distributed at the beginning of the course and the students were instructed to begin immediately working through the self-study part and attempting all the exercises. The fact that most students put in the bulk of their efforts in their other courses towards the end of these courses helped in this respect, since they were less heavily loaded at the outset. The students were offered some optional tutorials in case they got stuck (as some undertaking individual study for the first time invariably did). The inducement for doing the exercises was that they counted towards the final assessment (by some 35 per cent currently). The deadline for completing the exercises was set for about a third of the way through the course. While the students were busy in their own time working on the tensors, the lecture course began by revising the key ideas in special relativity. The special theory was then formulated in a tensorial way, making use of the new language and so providing some initial motivation. This was followed by a detailed and deliberate development of the principles underlying general relativity. Tensors are then used in earnest for the first time in deriving the equation of geodesic deviation of Chapter 10. It is by about this time that the students have gained considerable expertise in manipulating tensors and the lectures help to provide further motivation and consolidation. This device means that the Schwarzschild solution can be reached by not much more than half-way through the course. Another important advantage of printed lecture notes is that one has much more control over the speed at which the course is delivered, and one can to some extent tune the speed to fit the capabilities of the class.

The Southampton course is some thirty-six lectures in length. In the early years, when the students had a good background in special relativity, I was able to cover all three end topics. Indeed, in the first year of operation, I ended up in the final week by organizing five seminars given by outside speakers which all the students attended and which attempted to show how the work we had covered related to some topics of current research interest. In more recent years, the preparation of the students in special relativity has been more patchy, and so I have taken this more on board and have been somewhat less ambitious. This has usually meant leaving out a topic such as rotating black holes or gravitational radiation. Of course, since these are contained in the notes, the students are able to fill in these gaps if they so choose.

I have been encouraged to write up the notes in book form for a number of reasons. The course has been running for some fifteen years and several hundred students have been through it, so that I have a good deal of consumer experience to draw upon. Not only has the course proved popular, but it seems to have coped surprisingly successfully with students of a wide ability range. This may in part be because I have included many of the more detailed steps in the text itself (and where these have been left out they have often been relegated to straightforward exercises). In fact, the notes are sold to the students to cover the cost of production. It has been gratifying that each year a number of students who are not on the course, sometimes not even in a related discipline, but who have by chance come across the notes, purchase a copy for themselves. Finally, a number of my relativity colleagues both in the UK and abroad have asked for copies and used them in varying degrees in their own courses. So, despite the fact that there are a number of fine texts around in the area, I have agreed to present the notes in book form. I hope you, the teacher, find them a valuable resource in your teaching.

1.5 A final note for the less able student

I was far from being a child prodigy, and yet I learnt relativity at the age of 15! Let me elaborate. As testimony to my intellectual ordinariness, I had left my junior school at the age of 11 having achieved the unremarkable feat of coming 22nd in the class in my final set of examinations. Yet I really did know some relativity four years on — and I don't just mean the special theory, but the general theory (up to and including the Schwarzschild solution and the classical tests). I remember detecting a hint of disbelief when I recounted this to the same Alan Tayler, who was later to become my tutor, in an Oxford entrance interview. He followed up by asking me to define a tensor, and when I rattled off a definition, he seemed somewhat surprised. Indeed, as it turned out, we did not cover very much more than I first knew in the Oxford third year specialist course on general relativity. So how was this possible?

I, too, had heard the story about how only a few people in the world really understood relativity, and it had aroused my curiosity. I went to the local library and, as luck would have it, I pulled out a book entitled **Einstein's Theory of Relativity** by Lillian Lieber (1949). This is a very bizarre book in appearance. The book is not set out in the usual way but rather as though it were concrete poetry. Moreover, it is interspersed by surrealist drawings by Hugh Lieber involving the symbols from the text. I must confess that at first sight the book looks rather cranky; but it is not. I worked through it, filling in all the details missing from the calculations as I went. What was amazing was that the book did not make too many assumptions about what mathematics the reader needed to know. For example, I had not then met partial differentiation in my school mathematics, and yet there was sufficient coverage in the book for me to cope. It felt almost as if the book had been written just for me. The combination of the intrinsic interest of the material and the success I had in doing the intervening calculations provided sufficient motivation for me to see the enterprise through to the end.

Perhaps, if you consider yourself a less able student, you are a bit daunted by the intellectual challenge that lies ahead. I will not deny that the book includes some very demanding ideas (indeed, I do not understand every facet of all of these ideas myself). But I hope the two facts that the arguments are broken down into small steps and that the calculations are doable, will help you on your way. Even if you decide to cut out after part C, you will have come a long way. Take heart from my little story—I am certain that if you persevere you will consider it worth the effort in the end.

Exercises

1.1 (§1.3) Go to the library and see if you can locate current copies of the following journals:

(i) **Journal of General Relativity and Gravitation**;
(ii) **Classical and Quantum Gravity**;
(iii) **Journal of Mathematical Physics**;
(iv) **Physical Review** D.

See if you can relate any of the articles in them to any of the topics contained in Fig. 1.1.

1.2 Look back through copies of **Scientific American** for future reference, to see what articles there have been in recent years on relativity theory, especially black holes, gravitational waves, and cosmology.

1.3 Read a biography of Einstein (see Part A of the Selected Bibliography at the end of this book).

A. Special Relativity

The *k*-calculus

2

2.1 Model building

Before we start, we should be clear what we are about. The essential activity of mathematical physics, or theoretical physics, is that of **modelling** or **model building**. The activity consists of constructing a mathematical model which we hope in some way captures the essentials of the phenomena we are investigating. I think we should never fail to be surprised that this turns out to be such a productive activity. After all, the first thing you notice about the world we inhabit is that it is an extremely complex place. The fact that so much of this rich structure can be captured by what are, in essence, a set of simple formulae is to me quite astonishing. Just think how simple Newton's universal law of gravitation is; and yet it encompasses a whole spectrum of phenomena from a falling apple to the shape of a globular cluster of stars. As Einstein said, 'The most incomprehensible thing about the world is that it is comprehensible.'

The very success of the activity of modelling has, throughout the history of science, turned out to be counterproductive. Time and again, the successful model has been confused with the ultimate reality, and this in turn has stultified progress. Newtonian theory provides an outstanding example of this. So successful had it been in explaining a wide range of phenomena, that, after more than two centuries of success, the laws had taken on an absolute character. Thus it was that, when at the end of the nineteenth century it was becoming increasingly clear that something was fundamentally wrong with the current theories, there was considerable reluctance to make any fundamental changes to them. Instead, a number of artificial assumptions were made in an attempt to explain the unexpected phenomena. It eventually required the genius of Einstein to overthrow the prejudices of centuries and demonstrate in a number of simple thought experiments that some of the most cherished assumptions of Newtonian theory were untenable. This he did in a number of brilliant papers in 1905 proposing a theory which has become known today as the **special theory of relativity**.

We should perhaps be discouraged from using words like right or wrong when discussing a physical theory. Remembering that the essential activity is model building, a model should then rather be described as good or bad, depending on how well it describes the phenomena it encompasses. Thus, Newtonian theory is an excellent theory for describing a whole range of phenomena. For example, if one is concerned with describing the motion of a car, then the Newtonian framework is likely to be the appropriate one.

However, it fails to be appropriate if we are interested in very high speeds (comparable with the speed of light) or very intense gravitational fields (such as in the nucleus of a galaxy). To put it another way: together with every theory, there should go its **range of validity**. Thus, to be more precise, we should say that Newtonian theory is an excellent theory within its range of validity. From this point of view, developing our models of the physical world does not involve us in constantly throwing theories out, since they are perceived to be wrong, or unlearning them, but rather it consists more of a process of refinement in order to increase their range of validity. So the moral of this section is that, for all their remarkable success, one must not confuse theoretical models with the ultimate reality they seek to describe.

2.2 Historical background

In 1865, James Clerk Maxwell put forward the theory of electromagnetism. One of the triumphs of the theory was the discovery that light waves are electromagnetic in character. Since all other known wave phenomena required a material medium in which the oscillations were carried, it was postulated that there existed an all-pervading medium, called the 'luminiferous ether', which carried the oscillations of electromagnetism. It was then anticipated that experiments with light would allow the absolute motion of a body through the ether to be detected. Such hopes were upset by the null result of the famous Michelson–Morley experiment (1881) which attempted to measure the velocity of the earth relative to the ether and found it to be undetectably small. In order to explain this null result, two **ad hoc** hypotheses were put forward by Lorentz, Fitzgerald, and Poincaré (1895), namely, the contraction of rigid bodies and the slowing down of clocks when moving through the ether. These effects were contained is some simple formulae called the 'Lorentz transformations'. This would affect every apparatus designed to measure the motion relative to the ether so as to neutralize exactly all expected results. Although this theory was consistent with the observations, it had the philosophical defect that its fundamental assumptions were unverifiable.

In fact, the essence of the special theory of relativity is contained in the Lorentz transformations. However, Einstein was able to derive them from two postulates, the first being called the 'principle of special relativity'—a principle which Poincaré had also suggested independently in 1904—and the second concerning the constancy of the velocity of light. In so doing, he was forced to re-evaluate our ideas of space and time and he demonstrated through a number of simple thought experiments that the source of the limitations of the classical theory lay in the concept of **simultaneity**. Thus, although in a sense Einstein found nothing new in that he rederived the Lorentz transformations, his derivation was physically meaningful and in the process revealed the inadequacy of some of the fundamental assumptions of classical thought. Herein lies his chief contribution.

2.3 Newtonian framework

We start by outlining the Newtonian framework. An **event** intuitively means something happening in a fairly limited region of space and for a short duration in time. Mathematically, we idealize this concept to become a point

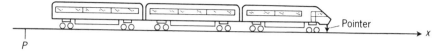
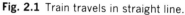

Fig. 2.1 Train travels in straight line.

in space and an instant in time. Everything that happens in the universe is an event or collection of events. Consider a train travelling from one station P to another R, leaving at 10 a.m. and arriving at 11 a.m. We can illustrate this in the following way: for simplicity, let us assume that the motion takes place in a straight line (say along the x-axis (Fig. 2.1); then we can represent the motion by a **space-time diagram** (Fig. 2.2) in which we plot the position of some fixed point on the train, which we represent by a pointer, against time. The curve in the diagram is called the **history** or **world-line** of the pointer. Notice that at Q the train was stationary for a period.

We shall call individuals equipped with a clock and a measuring rod or ruler **observers**. Had we looked out of the train window on our journey at a clock in a passing station then we would have expected it to agree with our watch. One of the central assumptions of the Newtonian framework is that two observers will, once they have synchronized their clocks, always agree about the time of an event, irrespective of their relative motion. This implies that for all observers time is an **absolute** concept. In particular, all observers can agree on an origin of time. In order to fix an event in space, an observer may choose a convenient origin in space together with a set of three Cartesian coordinate axes. We shall refer to an observer's clock, ruler, and coordinate axes as a **frame of reference** (Fig. 2.3). Then an observer is able to **coordinatize** events, that is, determine the time t an event occurs and its relative position (x, y, z).

We have set the stage with space and time; they provide the backcloth, but what is the story about? The stuff which provides the events of the universe is **matter**. For the moment, we shall idealize lumps of matter into objects called **bodies**. If the body has no physical extent, we refer to it as a **particle** or **point mass**. Thus, the role of observers in Newtonian theory is to chart the history of bodies.

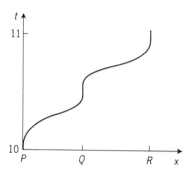

Fig. 2.2 Space-time diagram of pointer.

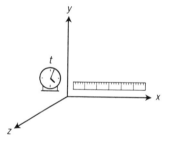

Fig. 2.3 Observer's frame of reference.

2.4 Galilean transformations

Now, relativity theory is concerned with the way different observers see the same phenomena. One can ask: are the laws of physics the same for all observers or are there preferred states of motion, preferred reference systems, and so on? Newtonian theory postulates the existence of preferred frames of reference. This is contained essentially in the first law, which we shall call N1 and state in the following form:

N1: Every body continues in its state of rest or of uniform motion in a straight line unless it is compelled to change that state by forces acting on it.

Thus, there exists a privileged set of bodies, namely those not acted on by forces. The frame of reference of a co-moving observer is called an **inertial** frame (Fig. 2.4). It follows that, once we have found one inertial frame, then all

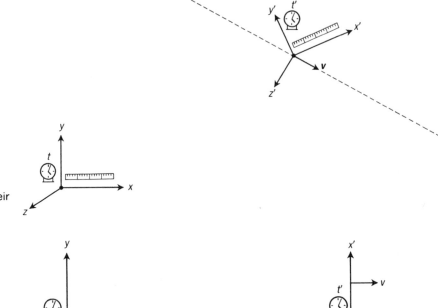

Fig. 2.4 Two observed bodies and their inertial frames.

Fig. 2.5 Two frames in standard configuration at time *t*.

others are at rest or travel with constant velocity relative to it (for otherwise Newton's first law would no longer be true). The transformation which connects one inertial frame with another is called a **Galilean transformation**. To fix ideas, let us consider two inertial frames called S and S' in **standard configuration**, that is, with axes parallel and S' moving along S's positive x-axis with constant velocity (Fig. 2.5). We also assume that the observers synchronize their clocks so that the origins of time are set when the origins of the frames coincide. It follows from Fig. 2.5 that the Galilean transformation connecting the two frames is given by

$$x = x' + vt, \qquad y = y', \qquad z = z', \qquad t = t'. \qquad (2.1)$$

The last equation provides a manifestation of the assumption of absolute time in Newtonian theory. Now, Newton's laws hold only in inertial frames. From a mathematical viewpoint, this means that Newton's laws must be **invariant** under a Galilean transformation.

2.5 The principle of special relativity

We begin by stating the relativity principle which underpins Newtonian theory

Restricted principle of special relativity:
All inertial observers are equivalent as far as dynamical experiments are concerned.

This means that, if one inertial observer carries out some dynamical experiments and discovers a physical law, then any other inertial observer performing the same experiments must discover the same law. Put another way, these laws must be invariant under a Galilean transformation. That is to say, if the law involves the coordinates x, y, z, t of an inertial observer S, then the law relative to another observer S' will be the same with x, y, z, t replaced by x', y', z', t', respectively. Many fundamental principles of physics are statements of **impossibility**, and the above statement of the relativity principle is equivalent to the statement of the impossibility of deciding, by performing dynamical experiments, whether a body is absolutely at rest or in uniform motion. In Newtonian theory, we cannot determine the **absolute** position in space of an event, but only its position **relative** to some other event. In exactly the same way, uniform velocity has only a relative significance; we can only talk about the velocity of a body relative to some other. Thus, both position and velocity are **relative** concepts.

Einstein realized that the principle as stated above is empty because there is no such thing as a purely **dynamical** experiment. Even on a very elementary level, any dynamical experiment we think of performing involves observation, i.e. **looking**, and looking is a part of optics, not dynamics. In fact, the more one analyses any one experiment, the more it becomes apparent that practically all the branches of physics are involved in the experiment. Thus, Einstein took the logical step of removing the restriction of dynamics in the principle and took the following as his first postulate.

Postulate I. Principle of special relativity:
  All inertial observers are equivalent.

Hence we see that this principle is in no way a contradiction of Newtonian thought, but rather constitutes its logical completion.

2.6 The constancy of the velocity of light

We previously defined an observer in Newtonian theory as someone equipped with a clock and ruler with which to map the events of the universe. However, the approach of the k-calculus is to dispense with the rigid ruler and use radar methods for measuring distances. (What is rigidity anyway? If a moving frame appears non-rigid in another frame, which, if either, is the rigid one?) Thus, an observer measures the distance of an object by sending out a light signal which is reflected off the object and received back by the observer. The **distance** is then simply defined as **half the time difference between emission and reception**. Note that by this method **distances** are measured in intervals of time, like the light year or the light second ($\sim 10^{10}$ cm).

Why use light? The reason is that we know that the velocity of light is independent of many things. Observations from double stars tell us that the velocity of light **in vacuo** is independent of the motion of the sources as well as independent of colour, intensity, etc. For, if we suppose that the velocity of light were dependent on the motion of the source relative to an observer (so that if the source was coming towards us the light would be travelling faster and vice versa) then we would no longer see double stars moving in Keplerian orbits (circles, ellipses) about each other: their orbits would appear distorted; yet no such distortion is observed. There are many experiments which confirm this assumption. However, these were not known to Einstein in 1905, who adopted the second postulate purely on heuristic grounds. We state the second postulate in the following form.

> **Postulate II. Constancy of velocity of light:**
> The velocity of light is the same in all inertial systems.

Or stated another way: there is no overtaking of light by light in empty space. The speed of light is conventionally denoted by c and has numerical value $2.998 \times 10^8 \, \text{m s}^{-1}$, but in this chapter we shall adopt **relativistic units** in which c is taken to be unity (i.e. $c = 1$). Note, in passing, that another reason for using radar methods is that other methods are totally impracticable for large distances. In fact, these days, distances from the Earth to the Moon and Venus can be measured very accurately by bouncing radar signals off them.

2.7 The *k*-factor

For simplicity, we shall begin by working in two dimensions, one spatial dimension and one time dimension. Thus, we consider a system of observers distributed along a straight line, each equipped with a clock and a flashlight. We plot the events they map in a two-dimensional space-time diagram. Let us assume we have two observers, A at rest and B moving away from A with uniform (constant) speed. Then, in a space-time diagram, the world-line of A will be represented by a vertical straight line and the world-line of B by a straight line at an angle to A's, as shown in Fig. 2.6.

A light signal in the diagram will be denoted by a straight line making an angle $\frac{1}{4}\pi$ with the axes, because we are taking the speed of light to be 1. Now, suppose A sends out a series of flashes of light to B, where the interval between the flashes is denoted by T according to A's clock. Then it is plausible to assume that the intervals of reception by B's clock are proportional to T, say kT. Moreover, the quantity k, which we call the **k-factor**, is

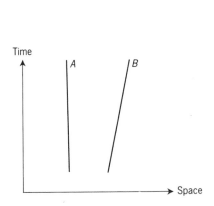

Fig. 2.6 The world-lines of observers A and B.

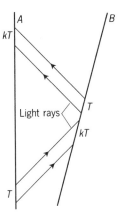

Fig. 2.7 The reciprocal nature of the *k*-factor.

clearly a characteristic of the motion of B relative to A. We now assume that if A and B are **inertial** observers, then k is a **constant** in time. (In fact, there is a hidden assumption here, since how do we know that B's world-line will be a **straight** line as indicated in the diagram? Strictly speaking, we are assuming that there is a **linear** relationship between the space and time coordinates of A and B.) Then the principle of special relativity requires that the relationship between A and B must be reciprocal, so that, if B emits two signals with a time lapse of T according to B's clock, then A receives them after a time lapse of kT according to A's clock (Fig. 2.7). Note that, from B's point of view, A is moving away from B with the same relative speed.

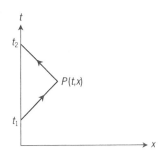

Fig. 2.8 Coordinatizing events.

Observer A assigns coordinates to an event P by bouncing a light signal off it. So that if a light signal is sent out at a time $t = t_1$, and received back at a time $t = t_2$ (Fig. 2.8), then, according to our radar definition of distances, the coordinates of P are given by

$$(t, x) = \left(\tfrac{1}{2}(t_1 + t_2), \tfrac{1}{2}(t_2 - t_1)\right), \tag{2.2}$$

remembering that the velocity of light is 1.

We now use the k-factor to develop the k-calculus.

2.8 Relative speed of two inertial observers

Consider the configuration shown in Fig. 2.9 and assume that A and B synchronize their clocks to zero when they cross at event O. After a time T, A sends a signal to B, which is reflected back at event P. From B's point of view, a light signal is sent to A after a time lapse of kT by B's clock. It follows from the definition of the k-factor that A receives this signal after a time lapse of $k(kT)$. Then, using (2.2) with $t_1 = T$ and $t_2 = k^2T$, we find the coordinates of P according to A's clock are given by

$$(t, x) = \left(\tfrac{1}{2}(k^2 + 1)T, \tfrac{1}{2}(k^2 - 1)T\right). \tag{2.3}$$

Thus, as T varies, this gives the coordinates of the events which constitute B's world-line. Hence, if v is the velocity of B relative to A, we find

$$v = \frac{x}{t} = \frac{k^2 - 1}{k^2 + 1}.$$

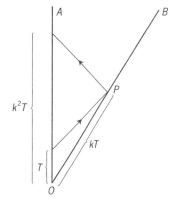

Fig. 2.9 Relating the k-factor to the relative speed of separation.

Solving for k in terms of v, and noting from the diagram that k must be greater than 1 if the observers are separating, we find

$$k = \left(\frac{1 + v}{1 - v}\right)^{\frac{1}{2}}. \tag{2.4}$$

We shall see in the next chapter that this is the usual relativistic formula for the radial Doppler shift. If B is moving away from A then $k > 1$ which represents a 'red' shift, whereas if B is approaching A then $k < 1$ which represents a 'blue' shift. Note that the transformation $v \rightarrow -v$ corresponds to $k \rightarrow 1/k$. Moreover,

$$v = 0 \quad \Leftrightarrow \quad k = 1,$$

as we should expect for observers relatively at rest: once they have synchronized their clocks, the synchronization remains (Fig. 2.10).

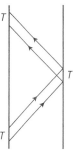

Fig. 2.10 Observers relatively at rest ($k = 1$).

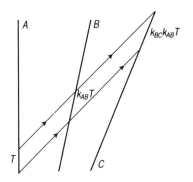

Fig. 2.11 Composition of *k*-factors.

2.9 Composition law for velocities

Consider the situation in Fig. 2.11, where k_{AB} denotes the *k*-factor between A and B, with k_{BC} and k_{AC} defined similarly. It follows immediately that

$$k_{AC} = k_{AB}k_{BC}. \tag{2.5}$$

Using (2.4), we find the corresponding **composition law** for velocities:

$$v_{AC} = \frac{v_{AB} + v_{BC}}{1 + v_{AB}v_{BC}}. \tag{2.6}$$

This formula has been confirmed by Fizeau's experiment in which the speed of light in a moving fluid is measured and turns out not to be simply the sum of the speed of light and the moving fluid but rather obeys the more complicated law (2.6) to higher order. Note that, if v_{AB} and v_{BC} are small compared with the speed of light, i.e.

$$v_{AB} \ll 1, \qquad v_{BC} \ll 1,$$

then we obtain the classical Newtonian formula

$$v_{AC} = v_{AB} + v_{BC}$$

to lowest order. Although the composition law for velocities is not simple, the one for *k*-factors is, and in special relativity it is the *k*-factors which are the directly measurable quantities. Note also that, formally, if we substitute $v_{BC} = 1$, representing the speed of a light signal **relative to B**, in (2.6), then the resulting speed of the light signal relative to A is

$$v_{AC} = \frac{v_{AB} + 1}{1 + v_{AB}} = 1,$$

in agreement with the constancy of the velocity of light postulate.

From the composition law, we can show that, if we add two speeds less than the speed of light, then we again obtain a speed less than the speed of light. This does not mean, as is sometimes stated, that nothing can move faster than the speed of light in special relativity, but rather that the speed of light is a border which can not be crossed or even reached. More precisely, special relativity allows for the existence of three classes of particles.

1. Particles that move slower than the speed of light are called **subluminal** particles. They include material particles and elementary particles such as electrons and neutrons.

2. Particles that move with the speed of light are called **luminal** particles. They include the carrier of the electromagnetic field interaction, the photon, and theoretically the carrier of the gravitational field interaction, called the graviton. These are both particles with zero rest mass (see §4.5). It was thought that neutrinos also had zero rest mass, but more recent evidence suggests they may have a tiny mass.

3. Particles that move faster than the speed of light are called **superluminal** particles or **tachyons**. There was some excitement in the 1970s surrounding the possible existence of tachyons, but all attempts to detect them to date have failed. This suggests two likely possibilities: either tachyons do

not exist or, if they do, they do not interact with ordinary matter. This would seem to be just as well, for otherwise they could be used to signal back into the past and so would appear to violate causality. For example, it would be possible theoretically to construct a device which sent out a tachyon at a given time and which would trigger a mechanism in the device to blow it up **before** the tachyon was sent out!

2.10 The relativity of simultaneity

Consider two events P and Q which take place at the same time, according to A, and also at points equal but opposite distances away. A could establish this by sending out and receiving the light rays as shown in Fig. 2.12 (continuous lines). Suppose now that another inertial observer B meets A at the time these events occur **according to** A. B also sends out light rays RQU and SPV to illuminate the events, as shown (dashed lines). By symmetry $RU = SV$ and so these events are equidistant according to B. However, the signal RQ was sent before the signal SP and so B concludes that the event Q took place **well before** P. Hence, events that A judges to be simultaneous, B judges not to be simultaneous. Similarly, A maintains that P, O, and Q occurred simultaneously, whereas B maintains that they occurred in the order Q, then O, and then P.

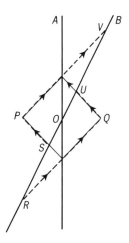

Fig. 2.12 Relativity of simultaneity.

This relativity of simultaneity lies at the very heart of special relativity and resolves many of the paradoxes that the classical theory gives rise to, such as the Michelson–Morley experiment. Einstein realized the crucial role that simultaneity plays in the theory and gave the following simple thought experiment to illustrate its dependence on the observer. Imagine a train travelling along a straight track with velocity v relative to an observer A on the bank of the track. In the train, B is an observer situated at the centre of one of the carriages. We assume that there are two electrical devices on the track which are the length of the carriage apart and equidistant from A. When the carriage containing B goes over these devices, they fire and activate two light sources situated at each end of the carriage (Fig. 2.13). From the configuration, it is clear that A will judge that the two events, when the light sources first switch on, occur **simultaneously**. However, B is travelling towards the light emanating from light source 2 and away from the light emanating from light source 1. Since the speed of light is a constant, B will see the light from source 2 before seeing the light from source 1, and so will conclude that one light source comes on **before** the other.

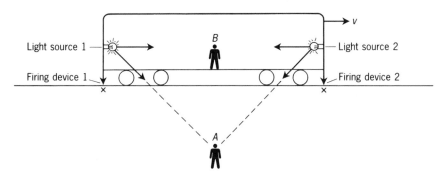

Fig. 2.13 Light signals emanating from the two sources.

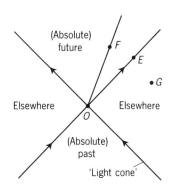

Fig. 2.14 Event relationships in special relativity.

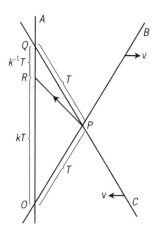

Fig. 2.15 The clock paradox.

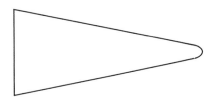

Fig. 2.16 Spatial analogue of clock paradox.

We can now classify event relationships in space and time in the following manner. Consider any event O on A's world-line and the four regions, as shown in Fig. 2.14, given by the light rays ending and commencing at O. Then the event E is on the light ray leaving O and so occurs **after** O. Any other inertial observer agrees on this; that is, no observer sees E illuminated before A sends out the signal from O. The fact that E is illuminated (because A originally sends out a signal at O) **subsequent** to O is a manifestation of **causality** — the event O ultimately causes the event E. Similarly, the event F can be reached by an inertial observer travelling from O with finite speed. Again, all inertial observers agree that F occurs after O. Hence all the events in this region are called the **absolute future of O**. In the same way, any event occurring in the region vertically below takes place in O's **absolute past**. However, the temporal relationship to O of events in the other two regions, called **elsewhere** (or sometimes the **relative past** and **relative future**) will not be something all observers will agree upon. For example, one class of observers will say that G took place after O, another class before, and a third class will say they took place simultaneously. The light rays entering and leaving O constitute what is called the **light cone** or **null cone** at O (the fact that it is a cone will become clearer later when we take all the spatial dimensions into account). Note that the world-line of any inertial observer or material particle passing through O must lie within the light cone at O.

2.11 The clock paradox

Consider three inertial observers as shown in Fig. 2.15, with the relative velocity $v_{AC} = -v_{AB}$. Assume that A and B synchronize their clocks at O and that C's clock is synchronized with B's at P. Let B and C meet after a time T according to B, whereupon they emit a light signal to A. According to the *k*-calculus, A receives the signal at R after a time kT since meeting B. Remembering that C is moving with the opposite velocity to B (so that $k \to k^{-1}$), then A will meet C at Q after a subsequent time lapse of $k^{-1}T$. The total time that A records between events O and Q is therefore $(k + k^{-1})T$. For $k \neq 1$, this is **greater** than the combined time intervals $2T$ recorded between events OP and PQ by B and C. But should not the time lapse between the two events agree? This is one form of the so-called **clock paradox**.

However, it is not really a paradox, but rather what it shows is that in relativity time, like distance, is a route-dependent quantity. The point is that the $2T$ measurement is made by **two** inertial observers, not one. Some people have tried to reverse the argument by setting B and C to rest, but this is not possible since they are in relative motion to each other. Another argument says that, when B and C meet, C should take B's clock and use it. But, in this case, the clock would have to be **accelerated** when being transferred to C and so it is no longer inertial. Again, some opponents of special relativity (e.g. H. Dingle) have argued that the short period of acceleration should not make such a difference, but this is analogous to saying that a journey between two points which is straight nearly all the time is about the same length as one which is wholly straight (as shown), which is absurd (Fig. 2.16). The moral is that in special relativity time is a more difficult concept to work with than the absolute time of Newton.

A more subtle point revolves around the implicit assumption that the clocks of A and B are 'good' clocks, i.e. that the seconds of A's clock are the same as those of B's clock. One suggestion is that A has two clocks and adjusts the tick rate until they are the same and then sends one of them to B at a very slow rate of acceleration. The assumption here is that the very slow rate of acceleration will not affect the tick rate of the clock. However, what is there to say that a clock may not be able to somehow add up the small bits of acceleration and so affect its performance. A more satisfactory approach would be for A and B to use identically constructed atomic clocks (which is after all what physicists use today to measure time). The objection then arises that their construction is based on ideas in quantum physics which is, a priori, outside the scope of special relativity. However, this is a manifestation of a point raised earlier, that virtually any real experiment which one can imagine carrying out involves more than one branch of physics. The whole structure is intertwined in a way which cannot easily be separated.

2.12 The Lorentz transformations

We have derived a number of important results in special relativity, which only involve one spatial dimension, by use of the k-calculus. Other results follow essentially from the transformations connecting inertial observers, the famous Lorentz transformations. We shall finally use the k-calculus to derive these transformations.

Let event P have coordinates (t, x) relative to A and (t', x') relative to B (Fig. 2.17). Observer A must send out a light ray at time $t - x$ to illuminate P at time t and also receive the reflected ray back at $t + x$ (check this from (2.2)). The world-line of A is given by $x = 0$, and the origin of A's time coordinate t is arbitrary. Similar remarks apply to B, where we use primed quantities for B's coordinates (t', x'). Assuming A and B synchronize their clocks when they meet, then the k-calculus immediately gives

$$t' - x' = k(t - x), \qquad t + x = k(t' + x'). \tag{2.7}$$

After some rearrangement, and using equation (2.4), we obtain the so-called **special Lorentz transformation**

$$t' = \frac{t - vx}{(1 - v^2)^{\frac{1}{2}}}, \qquad x' = \frac{x - vt}{(1 - v^2)^{\frac{1}{2}}}. \tag{2.8}$$

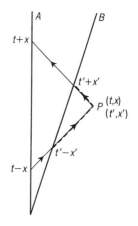

Fig. 2.17 Coordinatization of events by inertial observers.

This is also referred to as a **boost in the x-direction with speed v**, since it takes one from A's coordinates to B's coordinates and B is moving away from A with speed v. Some simple algebra reveals the result (exercise)

$$t'^2 - x'^2 = t^2 - x^2,$$

showing that the quantity $t^2 - x^2$ is an invariant under a special Lorentz transformation or boost.

To obtain the corresponding formulae in the case of three spatial dimensions we consider Fig. 2.5 with two inertial frames in standard configuration. Now, since by assumption the xz-plane ($y = 0$) of A must coincide with the $x'z'$-plane ($y' = 0$) of B, then the y and y' coordinates must be connected by a transformation of the form

$$y = ny', \tag{2.9}$$

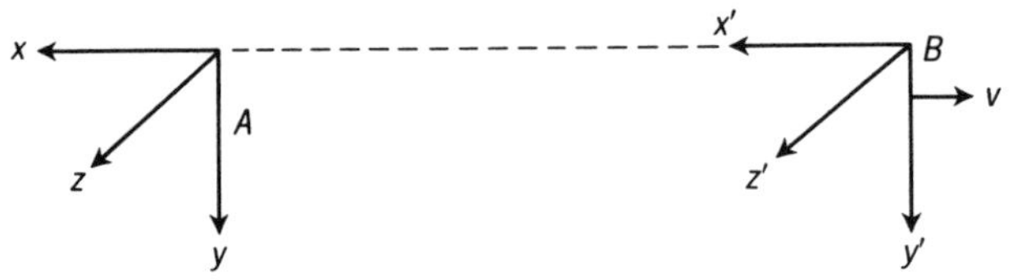

Fig. 2.18 The *x*- and *y*-axes reversed in Fig. 2.5.

Fig. 2.19 Figure 2.18 from *B*'s point of view.

because

$$y = 0 \quad \Leftrightarrow \quad y' = 0.$$

We now make the assumption that space is **isotropic**, that is, it is the same in any direction. We then reverse the direction of the *x*- and *y*-axes of *A* and *B* and consider the motion from *B*'s point of view (see Figs. 2.18 and 2.19). Clearly, from *B*'s point of view, the roles of *A* and *B* have interchanged. Hence, by symmetry, we must have

$$y' = ny. \tag{2.10}$$

Combining (2.9) and (2.10), we find

$$n^2 = 1 \quad \Rightarrow \quad n = \pm 1.$$

The negative sign can be dismissed since, as $v \to 0$, we must have $y' \to y$, in which case $n = 1$. Hence, we find $y' = y$, and a similar argument for z produces $z' = z$.

2.13 The four-dimensional world view

We now compare the special Lorentz transformation of the last section in relativistic units with the Galilean transformation connecting inertial observers in standard configuration (see Table 2.1). In a Galilean transformation, the absolute time coordinate remains invariant. However, in a

Table 2.1

Galilean transformation	Lorentz transformation
$t' = t$	$t' = \dfrac{t - vx}{(1 - v^2)^{\frac{1}{2}}}$
$x' = x - vt$	$x' = \dfrac{x - vt}{(1 - v^2)^{\frac{1}{2}}}$
$y' = y$	$y' = y$
$z' = z$	$z' = z$

Lorentz transformation, the time and space coordinates get mixed up (note the symmetry in x and t). In the words of Minkowski, 'Henceforth space by itself, and time by itself are doomed to fade away into mere shadows, and only a kind of union of the two will preserve an independent reality.'

In the old Newtonian picture, time is split off from three-dimensional Euclidean space. Moreover, since we have an absolute concept of simultaneity, we can consider two simultaneous events with coordinates (t, x_1, y_1, z_1) and (t, x_2, y_2, z_2), and then the square of the Euclidean **distance** between them,

$$\sigma^2 = (x_1 - x_2)^2 + (y_1 - y_2)^2 + (z_1 - z_2)^2, \tag{2.11}$$

is invariant under a Galilean transformation. In the new special relativity picture, time and space merge together into a four-dimensional continuum called **space-time**. In this picture, the square of the **interval** between any two events (t_1, x_1, y_1, z_1) and (t_2, x_2, y_2, z_2) is defined by

$$s^2 = (t_1 - t_2)^2 - (x_1 - x_2)^2 - (y_1 - y_2)^2 - (z_1 - z_2)^2, \tag{2.12}$$

and it is this quantity which is invariant under a Lorentz transformation. Note that we always denote the square of the interval by s^2, but the quantity s is only defined if the right-hand side of (2.12) is non-negative. If we consider two events separated infinitesimally, (t, x, y, z) and $(t + dt, x + dx, y + dy, z + dz)$, then this equation becomes

$$ds^2 = dt^2 - dx^2 - dy^2 - dz^2 \tag{2.13}$$

where all the infinitesimals are squared in (2.13). A four-dimensional space-time continuum in which the above form is invariant is called **Minkowski space-time** and it provides the background geometry for special relativity.

So far, we have only met a special Lorentz transformation which connects two inertial frames in standard configuration. A **full Lorentz transformation** connects two frames in general position (Fig. 2.20). It can be shown that a full Lorentz transformation can be decomposed into an ordinary spatial rotation, followed by a boost, followed by a further ordinary rotation. Physically, the first rotation lines up the x-axis of S with the velocity v of S'. Then a boost in

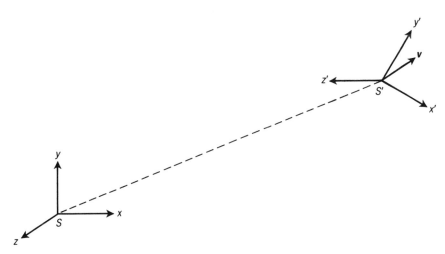

Fig. 2.20 Two frames in general position.

this direction with speed v transforms S to a frame which is at rest relative to S'. A final rotation lines up the coordinate frame with that of S'. The spatial rotations introduce no new physics. The only new physical information arises from the boost and that is why we can, without loss of generality, restrict our attention to a special Lorentz transformation.

Exercises

2.1 (§**2.4**) Write down the Galilean transformation from observer S to observer S', where S' has velocity v_1 relative to S. Find the transformation from S' to S and state in simple terms how the transformations are related. Write down the Galilean transformation from S' to S'', where S'' has velocity v_2 relative to S'. Find the transformation from S to S''. Prove that the Galilean transformations form an Abelian (commutative) group.

2.2 (§**2.7**) Draw the four fundamental k-factor diagrams (see Fig. 2.7) for the cases of two inertial observers A and B approaching and receding with uniform velocity v:

   (i) as seen by A;
   (ii) as seen by B.

2.3 (§**2.8**) Show that $v \to -v$ corresponds to $k \to k^{-1}$. If $k > 1$ corresponds physically to a red shift of recession, what does $k < 1$ correspond to?

2.4 (§**2.9**) Show that (2.6) follows from (2.5). Use the composition law for velocities to prove that if $0 < v_{AB} < 1$ and $0 < v_{BC} < 1$, then $0 < v_{AC} < 1$.

2.5 (§**2.9**) Establish the fact that if v_{AB} and v_{BC} are small compared with the velocity of light, then the composition law for velocities reduces to the standard additive law of Newtonian theory.

2.6 (§**2.10**) In the event diagram of Fig. 2.14, find a geometrical construction for the world-line of an inertial observer passing through O who considers event G as occurring

simultaneously with O. Hence describe the world-lines of inertial observers passing through O who consider G as occurring before or after O.

2.7 (§**2.11**) Draw Fig. 2.15 from B's point of view. Co-ordinatize the events O, R, and Q with respect to B and find the times between O and R, and R and Q, and compare them with A's timings.

2.8 (§**2.12**) Deduce (2.8) from (2.7). Use (2.7) to deduce directly that

$$t'^2 - x'^2 = t^2 - x^2.$$

Confirm the equality under the transformation formula (2.8).

2.9 (§**2.12**) In S, two events occur at the origin and a distance X along the x-axis simultaneously at $t = 0$. The time interval between the events in S' is T. Show that the spatial distance between the events in S' is $(X^2 + T^2)^{\frac{1}{2}}$ and determine the relative velocity v of the frames in terms of X and T.

2.10 (§**2.13**) Show that the interval between two events (t_1, x_1, y_1, z_1) and (t_2, x_2, y_2, z_2) defined by

$$s^2 = (t_1 - t_2)^2 - (x_1 - x_2)^2 - (y_1 - y_2)^2 - (z_1 - z_2)^2$$

is invariant under a special Lorentz transformation. Deduce the Minkowski line element (2.13) for infinitesimally separated events. What does s^2 become if $t_1 = t_2$, and how is it related to the Euclidean distance σ between the two events?

The key attributes of special relativity

3

3.1 Standard derivation of the Lorentz transformations

We start this chapter by deriving again the Lorentz transformations, but this time by using a more standard approach. We shall work in non-relativistic units in which the speed of light is denoted by c. We restrict attention to two inertial observers S and S' in standard configuration. As before, we shall show that the Lorentz transformations follow from the two postulates, namely, the principle of special relativity and the constancy of the velocity of light.

Now, by the first postulate, if the observer S sees a **free** particle, that is, a particle with no forces acting on it, travelling in a straight line with constant velocity, then so will S'. Thus, using vector notation, it follows that under a transformation connecting the two frames

$$\boldsymbol{r} = \boldsymbol{r}_0 + \boldsymbol{u}t \quad \Leftrightarrow \quad \boldsymbol{r}' = \boldsymbol{r}_0' + \boldsymbol{u}'t'.$$

Since straight lines get mapped into straight lines, it suggests that the transformation between the frames is **linear** and so we shall assume that the transformation from S to S' can be written in matrix form

$$\begin{bmatrix} t' \\ x' \\ y' \\ z' \end{bmatrix} = L \begin{bmatrix} t \\ x \\ y \\ z \end{bmatrix}, \tag{3.1}$$

where L is a 4×4 matrix of quantities which can only depend on the speed of separation v. Using exactly the same argument as we used at the end of §2.12, the assumption that space is isotropic leads to the transformations of y and z being

$$y' = y \quad \text{and} \quad z' = z. \tag{3.2}$$

We next use the second postulate. Let us assume that, when the origins of S and S' are coincident, they zero their clocks, i.e. $t = t' = 0$, and emit a flash of light. Then, according to S, the light flash moves out radially from the origin with speed c. The wave front of light will constitute a sphere. If we define the quantity I by

$$I(t, x, y, z) = x^2 + y^2 + z^2 - c^2 t^2,$$

then the events comprising this sphere must satisfy $I = 0$. By the second postulate, S' must also see the light move out in a spherical wave front with speed c and satisfy

$$I' = x'^2 + y'^2 + z'^2 - c^2 t'^2 = 0.$$

Thus it follows that, under a transformation connecting S and S',

$$I = 0 \quad \Leftrightarrow \quad I' = 0, \tag{3.3}$$

from which one may conclude

$$I = nI', \tag{3.4}$$

where n is a quantity which can only depend on v. Using the same argument as we did in §2.12, we can reverse the role of S and S' and so by the relativity principle we must also have

$$I' = nI. \tag{3.5}$$

Combining the last two equations we find

$$n^2 = 1 \quad \Rightarrow \quad n = \pm 1.$$

In the limit as $v \to 0$, the two frames coincide and $I' \to I$, from which we conclude that we must take $n = 1$.

Substituting $n = 1$ in (3.4), this becomes

$$x^2 + y^2 + z^2 - c^2 t^2 = x'^2 + y'^2 + z'^2 - c^2 t'^2,$$

and, using (3.2), this reduces to

$$x^2 - c^2 t^2 = x'^2 - c^2 t'^2. \tag{3.6}$$

We next introduce imaginary time coordinates T and T' defined by

$$T = \mathrm{i}ct, \tag{3.7}$$

$$T' = \mathrm{i}ct', \tag{3.8}$$

in which case equation (3.6) becomes

$$x^2 + T^2 = x'^2 + T'^2.$$

In a two-dimensional (x, T)-space, the quantity $x^2 + T^2$ represents the distance of a point P from the origin. This will only remain invariant under a **rotation** in (x, T)-space (Fig. 3.1). If we denote the angle of rotation by θ, then a rotation is given by

$$x' = x \cos \theta + T \sin \theta, \tag{3.9}$$

$$T' = -x \sin \theta + T \cos \theta. \tag{3.10}$$

Now, the origin of S' ($x' = 0$), as seen by S, moves along the positive x-axis of S with speed v and so must satisfy $x = vt$. Thus, we require

$$x' = 0 \quad \Leftrightarrow \quad x = vt \quad \Leftrightarrow \quad x = vT/\mathrm{i}c,$$

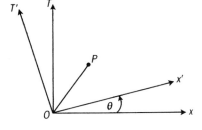

Fig. 3.1 A rotation in (x, T)-space.

using (3.7). Substituting this into (3.9) gives

$$\tan \theta = \mathrm{i}v/c, \tag{3.11}$$

from which we see that the angle θ is imaginary as well. We can obtain an expression for $\cos \theta$, using

$$\cos \theta = \frac{1}{\sec \theta} = \frac{1}{(1 + \tan^2 \theta)^{\frac{1}{2}}} = \frac{1}{(1 - v^2/c^2)^{\frac{1}{2}}}.$$

If we use the conventional symbol β for this last expression, i.e.

$$\beta \equiv \frac{1}{(1 - v^2/c^2)^{\frac{1}{2}}},$$

where the symbol $\equiv$ here means 'is defined to be', then (3.9) gives

$$x' = \cos\theta(x + T\tan\theta) = \beta[x + ict(iv/c)] = \beta(x - vt).$$

Similarly, (3.10) gives

$$T' = ict' = \cos\theta(-x\tan\theta + T) = \beta[-x(iv/c) + ict],$$

from which we find

$$t' = \beta(t - vx/c^2).$$

Thus, collecting the results together, we have rederived the special Lorentz transformation or boost (in non-relativistic units):

$$t' = \beta(t - vx/c^2), \qquad x' = \beta(x - vt), \qquad y' = y, \qquad z' = z. \qquad (3.12)$$

If we put $c = 1$, this takes the same form as we found in §2.13.

3.2 Mathematical properties of Lorentz transformations

From the results of the last section, we find the following properties of a special Lorentz transformation or boost.

1. Using the imaginary time coordinate T, a boost along the x-axis of speed v is equivalent to an imaginary rotation in (x, T)-space through an angle θ given by $\tan\theta = iv/c$.

2. If we consider v to be very small compared with c, for which we use the notation $v \ll c$, and neglect terms of order v^2/c^2, then we regain a Galilean transformation

$$t' = t, \qquad x' = x - vt, \qquad y' = y, \qquad z' = z.$$

We can obtain this result formally by taking the limit $c \to \infty$ in (3.12).

3. If we solve (3.12) for the unprimed coordinates, we get

$$t = \beta(t' + vx'/c^2), \qquad x = \beta(x' + vt'), \qquad y = y', \qquad z = z'.$$

This can be obtained formally from (3.12) by interchanging primed and unprimed coordinates and replacing v by $-v$. This we should expect from physical reasons, since, if S' moves along the positive x-axis of S with speed v, then S moves along the negative x'-axis of S' with speed v, or, equivalently, S moves along the positive x'-axis of S' with speed $-v$.

4. Special Lorentz transformations form a **group**:
(a) The identity element is given by $v = 0$.
(b) The inverse element is given by $-v$ (as in 3 above).

(c) The product of two boosts with velocities v and v' is another boost with velocity v''. Since v and v' correspond to rotations in (x, T)-space of θ and θ', where

$$\tan\theta = iv/c \quad \text{and} \quad \tan\theta' = iv'/c,$$

then their resultant is a rotation of $\theta'' = \theta + \theta'$, where

$$iv''/c = \tan\theta'' = \tan(\theta + \theta') = \frac{\tan\theta + \tan\theta'}{1 - \tan\theta\tan\theta'},$$

from which we find

$$v'' = \frac{v + v'}{1 + vv'/c^2}.$$

Compare this with equation (2.6) in relativistic units.
(d) Associativity is left as an exercise.

5. The square of the infinitesimal interval between infinitesimally separated events (see (2.13)),

$$ds^2 = c^2\,dt^2 - dx^2 - dy^2 - dz^2 , \qquad (3.13)$$

is invariant under a Lorentz transformation.

We now turn to the key physical attributes of Lorentz transformations. Throughout the remaining sections, we shall assume that S and S' are in standard configuration with non-zero relative velocity v.

3.3 Length contraction

Consider a rod fixed in S' with endpoints x'_A and x'_B, as shown in Fig. 3.2. In S, the ends have coordinates x_A and x_B (which, of course, vary in time) given by the Lorentz transformations

$$x'_A = \beta(x_A - vt_A), \qquad x'_B = \beta(x_B - vt_B). \qquad (3.14)$$

In order to measure the lengths of the rod according to S, we have to find the x-coordinates of the end points at the same time according to S. If we denote the **rest length**, namely, the length in S', by

$$l_0 = x'_A - x'_B$$

and the length in S at time $t = t_A = t_B$ by

$$l = x_B - x_A,$$

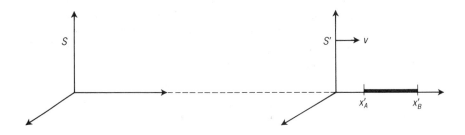

Fig. 3.2 A rod moving with velocity v relative to S.

then, subtracting the formulae in (3.14), we find the result

$$l = \beta^{-1} l_0. \tag{3.15}$$

Since

$$|v| < c \quad \Leftrightarrow \quad \beta > 1 \quad \Leftrightarrow \quad l < l_0,$$

the result shows that the length of a body in the direction of its motion with uniform velocity v is **reduced by a factor $(1 - v^2/c^2)^{\frac{1}{2}}$**. This phenomenon is called **length contraction**. Clearly, the body will have greatest length in its rest frame, in which case it is called the rest length or **proper length**. Note also that the length approaches zero as the velocity approaches the velocity of light.

In an attempt to explain the null result of the Michelson–Morley experiment, Fitzgerald had suggested the apparent shortening of a body in motion relative to the ether. This is rather different from the length contraction of special relativity, which is not to be regarded as illusory but is a very real effect. It is closely connected with the relativity of simultaneity and indeed can be deduced as a direct consequence of it. Unlike the Fitzgerald contraction, the effect is **relative**, i.e. a rod fixed in S appears contracted in S'. Note also that there are no contraction effects in directions transverse to the direction of motion.

3.4 Time dilation

Let a clock fixed at $x' = x'_A$ in S' record two successive events separated by an interval of time T_0 (Fig. 3.3). The successive events in S' are (x'_A, t'_1) and $(x'_A, t'_1 + T_0)$, say. Using the Lorentz transformation, we have in S

$$t_1 = \beta(t'_1 + vx'_A/c^2), \qquad t_2 = \beta(t'_1 + T_0 + vx'_A/c^2).$$

On subtracting, we find the time interval in S defined by

$$T = t_2 - t_1$$

is given by

$$T = \beta T_0. \tag{3.16}$$

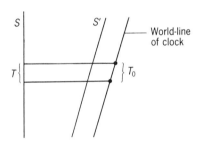

Fig. 3.3 Successive events recorded by a clock fixed in S'.

Thus, **moving clocks go slow by a factor $(1 - v^2/c^2)^{-\frac{1}{2}}$**. This phenomenon is called **time dilation**. The fastest rate of a clock is in its rest frame and is called its **proper rate**. Again, the effect has a reciprocal nature.

Let us now consider an accelerated clock. We define an **ideal clock** to be one unaffected by its acceleration; in other words, its instantaneous rate depends only on its instantaneous speed v, in accordance with the above phenomenon of time dilation. This is often referred to as the **clock hypothesis**. The time recorded by an ideal clock is called the **proper time** τ (Fig. 3.4). Thus, the proper time of an ideal clock between t_0 and t_1 is given by

$$\tau = \int_{t_0}^{t_1} \left(1 - \frac{v^2}{c^2}\right)^{\frac{1}{2}} dt. \tag{3.17}$$

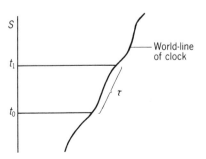

Fig. 3.4 Proper time recorded by an accelerated clock.

The general question of what constitutes a clock or an ideal clock is a non-trivial one. However, an experiment has been performed where an atomic clock was flown round the world and then compared with an identical clock left back on the ground. The travelling clock was found on return to be running slow by precisely the amount predicted by time dilation. Another instance occurs in the study of cosmic rays. Certain mesons reaching us from the top of the Earth's atmosphere are so short-lived that, even had they been travelling at the speed of light, their travel time in the absence of time dilation would exceed their known proper lifetimes by factors of the order of 10. However, these particles are in fact detected at the Earth's surface because their very high velocities keep them young, as it were. Of course, whether or not time dilation affects the human clock, that is, biological ageing, is still an open question. But the fact that we are ultimately made up of atoms, which do appear to suffer time dilation, would suggest that there is no reason by which we should be an exception.

3.5 Transformation of velocities

Consider a particle in motion (Fig. 3.5) with its Cartesian components of velocity being

$$(u_1, u_2, u_3) = \left(\frac{dx}{dt}, \frac{dy}{dt}, \frac{dz}{dt}\right) \quad \text{in } S$$

and

$$(u_1', u_2', u_3') = \left(\frac{dx'}{dt'}, \frac{dy'}{dt'}, \frac{dz'}{dt'}\right) \quad \text{in } S'.$$

Taking differentials of a Lorentz transformation

$$t' = \beta(t - vx/c^2), \qquad x' = \beta(x - vt), \qquad y' = y, \qquad z' = z,$$

we get

$$dt' = \beta(dt - v\,dx/c^2), \qquad dx' = \beta(dx - v\,dt), \qquad dy' = dy, \qquad dz' = dz,$$

and hence

$$u_1' = \frac{dx'}{dt'} = \frac{\beta(dx - v\,dt)}{\beta(dt - v\,dx/c^2)} = \frac{\dfrac{dx}{dt} - v}{1 - \dfrac{1}{c^2}\left(v\dfrac{dx}{dt}\right)} = \frac{u_1 - v}{1 - u_1 v/c^2}, \tag{3.18}$$

$$u_2' = \frac{dy'}{dt'} = \frac{dy}{\beta(dt - v\,dx/c^2)} = \frac{\dfrac{dy}{dt}}{\beta\left[1 - \dfrac{1}{c^2}\left(v\dfrac{dx}{dt}\right)\right]} = \frac{u_2}{\beta(1 - u_1 v/c^2)}, \tag{3.19}$$

Fig. 3.5 Particle in motion relative to S and S'.

$$u'_3 = \frac{dz'}{dt'} = \frac{dz}{\beta(dt - v\,dx/c^2)} = \frac{\dfrac{dz}{dt}}{\beta\left[1 - \dfrac{1}{c^2}\left(v\dfrac{dx}{dt}\right)\right]} = \frac{u_3}{\beta(1 - u_1 v/c^2)}. \quad (3.20)$$

Notice that the velocity components u_2 and u_3 transverse to the direction of motion of the frame S' are affected by the transformation. This is due to the time difference in the two frames. To obtain the inverse transformations, simply interchange primes and unprimes and replace v by $-v$.

3.6 Relationship between space-time diagrams of inertial observers

We now show how to relate the space-time diagrams of S and S' (see Fig. 3.6). We start by taking ct and x as the coordinate axes of S, so that a light ray has slope $\frac{1}{4}\pi$ (as in relativistic units). Then, to draw the ct'- and x'-axes of S', we note from the Lorentz transformation equations (3.12)

$$ct' = 0 \quad \Leftrightarrow \quad ct = (v/c)x,$$

that is, the x'-axis, $ct' = 0$, is the straight line $ct = (v/c)x$ with slope $v/c < 1$. Similarly,

$$x' = 0 \quad \Leftrightarrow \quad ct = (c/v)x,$$

that is, the ct'-axis, $x' = 0$, is the straight line $ct = (c/v)x$ with slope $c/v > 1$. The lines parallel to $O(ct')$ are the world-lines of fixed points in S'. The lines parallel to Ox' are the lines connecting points at a fixed time according to S' and are called **lines of simultaneity in S'**. The coordinates of a general event P are $(ct, x) = (OR, OQ)$ relative to S and $(ct', x') = (OV, OU)$ relative to S'. However, the diagram is somewhat misleading because the length scales along the axes are not the same. To relate them, we draw in the hyperbolae

$$x^2 - c^2 t^2 = x'^2 - c^2 t'^2 = \pm 1,$$

as shown in Fig. 3.7. Then, if we first consider the positive sign, setting $ct' = 0$, we get $x' = \pm 1$. It follows that OA is a unit distance on Ox'. Similarly, taking the negative sign and setting $x' = 0$ we get $ct' = \pm 1$ and so OB is the unit measure on Oct'. Then the coordinates of P in the frame S' are given by

$$(ct', x') = \left(\frac{OU}{OA}, \frac{OV}{OB}\right).$$

Note the following properties from Fig. 3.7.

1. A boost can be thought of as a rotation through an imaginary angle in the (x, T)-plane, where T is imaginary time. We have seen that this is equivalent, in the real (x, ct)-plane, to a skewing of the coordinate axes inwards through the same angle. (This was not appreciated by some past opponents of special relativity, who gave some erroneous counter-arguments based on the mistaken idea that a boost could be represented by a real rotation in the (x, ct)-plane.)
2. The hyperbolae are the same for all frames and so we can draw in any number of frames in the same diagram and use the hyperbolae to calibrate them.

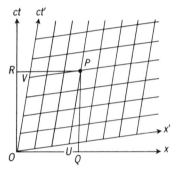

Fig. 3.6 The world-lines in S of the fixed points and simultaneity lines of S'.

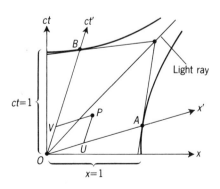

Fig. 3.7 Length scales in S and S'.

3. The length contraction and time dilation effects can be read off directly from the diagram. For example, the world-lines of the endpoints of a unit rod OA in S', namely $x' = 0$ and $x' = 1$, cut Ox in less than unit distance. Similarly world-lines $x = 0$ and $x = 1$ in S cut Ox' inside OE, from which the reciprocal nature of length contraction is evident.

4. Even A has coordinates $(ct', x') = (0, 1)$ relative to S', and hence by a Lorentz transformation coordinates $(ct, x) = (\beta v/c, \beta)$ relative to S. The quantity OA defined by

$$OA = (c^2 t^2 + x^2)^{\frac{1}{2}} = \beta(1 + v^2/c^2)^{\frac{1}{2}}$$

is a measure of the calibration factor

$$\left(\frac{1 + v^2/c^2}{1 - v^2/c^2}\right)^{\frac{1}{2}}.$$

3.7 Acceleration in special relativity

We start with the inverse transformation of (3.18), namely,

$$u_1 = \frac{u_1' + v}{1 + u_1' v/c^2},$$

from which we find the differential

$$\begin{aligned}
\mathrm{d}u_1 &= \frac{\mathrm{d}u_1'}{1 + u_1' v/c^2} - \left(\frac{u_1' + v}{(1 + u_1' v/c^2)^2}\right)\frac{v}{c^2}\,\mathrm{d}u_1' \\
&= \frac{1}{\beta^2}\frac{\mathrm{d}u_1'}{(1 + u_1' v/c^2)^2}.
\end{aligned}$$

Similarly, from the inverse Lorentz transformation

$$t = \beta(t' + x' v/c^2),$$

we find the differential

$$\mathrm{d}t = \beta(\mathrm{d}t' + \mathrm{d}x' v/c^2) = \beta(1 + u_1' v/c^2)\,\mathrm{d}t'.$$

Combining these results, we find that the x-component of the acceleration transforms according to

$$\frac{\mathrm{d}u_1}{\mathrm{d}t} = \frac{1}{\beta^3(1 + u_1' v/c^2)^3}\frac{\mathrm{d}u_1'}{\mathrm{d}t'}. \tag{3.21}$$

Similarly, we find

$$\frac{\mathrm{d}u_2}{\mathrm{d}t} = \frac{1}{\beta^2(1 + u_1' v/c^2)^2}\frac{\mathrm{d}u_2'}{\mathrm{d}t'} - \frac{v u_2'}{c^2 \beta^2(1 + u_1' v/c^2)^3}\frac{\mathrm{d}u_1'}{\mathrm{d}t'}, \tag{3.22}$$

$$\frac{\mathrm{d}u_3}{\mathrm{d}t} = \frac{1}{\beta^2(1 + u_1' v/c^2)^3}\frac{\mathrm{d}u_3'}{\mathrm{d}t'} - \frac{v u_3'}{c^2 \beta^2(1 + u_1' v/c^2)^3}\frac{\mathrm{d}u_1'}{\mathrm{d}t'}. \tag{3.23}$$

The inverse transformations can be found in the usual way.

It follows from the transformation formulae that acceleration is not an invariant in special relativity. However, it is clear from the formulae that acceleration is an **absolute** quantity, that is, all observers agree whether a body is accelerating or not. Put another way, if the acceleration is zero in one frame, then it is necessarily zero in any other frame. We shall see that this is

Table 3.1

Theory	Position	Velocity	Time	Acceleration
Newtonian	Relative	Relative	Absolute	Absolute
Special relativity	Relative	Relative	Relative	Absolute
General relativity	Relative	Relative	Relative	Relative

no longer the case in general relativity. We summarize the situation in Table 3.1, which indicates why the subject matter of the book is 'relativity' theory.

3.8 Uniform acceleration

The Newtonian definition of a particle moving under uniform acceleration is

$$\frac{\mathrm{d}u}{\mathrm{d}t} = \text{constant}.$$

This turns out to be inappropriate in special relativity since it would imply that $u \to \infty$ as $t \to \infty$, which we know is impossible. We therefore adopt a different definition. Acceleration is said to be **uniform** in special relativity if it has the same value in any **co-moving frame**, that is, at each instant, the acceleration in an inertial frame travelling with the same velocity as the particle has the same value. This is analogous to the idea in Newtonian theory of motion under a constant force. For example, a spaceship whose motor is set at a constant emission rate would be uniformly accelerated in this sense. Taking the velocity of the particle to be $u = u(t)$ relative to an inertial frame S, then at any instant in a co-moving frame S', it follows that $v = u$, the velocity relative to S' is zero, i.e. $u' = 0$, and the acceleration is a constant, a say, i.e. $\mathrm{d}u'/\mathrm{d}t' = a$. Using (3.21), we find

$$\frac{\mathrm{d}u}{\mathrm{d}t} = \frac{1}{\beta^3} a = \left(1 - \frac{u^2}{c^2}\right)^{\frac{3}{2}} a.$$

We can solve this differential equation by separating the variables

$$\frac{\mathrm{d}u}{(1 - u^2/c^2)^{\frac{3}{2}}} = a \, \mathrm{d}t$$

and integrating both sides. Assuming that the particle starts from rest at $t = t_0$, we find

$$\frac{u}{(1 - u^2/c^2)^{\frac{1}{2}}} = a(t - t_0).$$

Solving for u, we get

$$u = \frac{\mathrm{d}x}{\mathrm{d}t} = \frac{a(t - t_0)}{[1 + a^2(t - t_0)^2/c^2]^{\frac{1}{2}}}.$$

Next, integrating with respect to t, and setting $x = x_0$ at $t = t_0$, produces

$$(x - x_0) = \frac{c}{a}[c^2 + a^2(t - t_0)^2]^{\frac{1}{2}} - \frac{c^2}{a}.$$

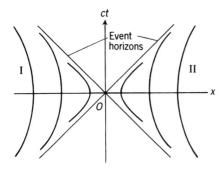

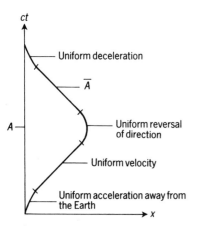

Fig. 3.8 Hyperbolic motions.

This can be rewritten in the form

$$\frac{(x - x_0 + c^2/a)^2}{(c^2/a)^2} - \frac{(ct - ct_0)^2}{(c^2/a)^2} = 1, \qquad (3.24)$$

which is a hyperbola in (x, ct)-space. If, in particular, we take $x_0 - c^2/a = t_0 = 0$, then we obtain a family of hyperbolae for different values of a (Fig. 3.8). These world-lines are known as **hyperbolic motions** and, as we shall see in Chapter 23, they have significance in cosmology. It can be shown that the radar distance between the world-lines is a constant. Moreover, consider the regions I and II bounded by the light rays passing through O, and a system of particles undergoing hyperbolic motions as shown in Fig. 3.8 (in some cosmological models, the particles would be galaxies). Then, remembering that light rays emanating from any point in the diagram do so at 45°, no particle in region I can communicate with another particle in region II, and vice versa. The light rays are called **event horizons** and act as barriers beyond which no knowledge can ever be gained. We shall see that event horizons will play an important role later in this book.

3.9 The twin paradox

This is a form of the clock paradox which has caused the most controversy — a controversy which raged on and off for over 50 years. The paradox concerns two twins whom we shall call A and $\bar{A}$. The twin $\bar{A}$ takes off in a spaceship for a return trip to some distant star. The assumption is that $\bar{A}$ is uniformly accelerated to some given velocity which is retained until the star is reached, whereupon the motion is uniformly reversed, as shown in Fig. 3.9. According to A, $\bar{A}$'s clock records slowly on the outward and return journeys and so, on return, $\bar{A}$ will be younger than A. If the periods of acceleration are negligible compared with the periods of uniform velocity, then could not $\bar{A}$ reverse the argument and conclude that it is A who should appear to be the younger? This is the basis of the paradox.

The resolution rests on the fact that the accelerations, however brief, have immediate and finite effects on $\bar{A}$ but not on A who remains inertial throughout. One striking way of seeing this effect is to draw in the simultaneity lines of $\bar{A}$ for the periods of uniform velocity, as in Fig. 3.10. Clearly, the period of uniform reversal has a marked effect on the simultaneity lines. Another way of looking at it is to see the effect that the periods of acceleration have on shortening the length of the journey as viewed by $\bar{A}$. Let us be specific: we assume that the periods of acceleration are T_1, T_2, and T_3, and that, after the period T_1, $\bar{A}$ has attained a speed $v = \sqrt{3}c/2$. Then, from A's viewpoint, during the period T_1, $\bar{A}$ finds that more than half the outward journey has been accomplished, in that $\bar{A}$ has transferred to a frame in which the distance between the Earth and the star is more than halved by length contraction. Thus, $\bar{A}$ accomplishes the outward trip in about half the time which A ascribes to it, and the same applies to the return trip. In fact, we could use the machinery of previous sections to calculate the time elapsed in both the periods of uniform acceleration and uniform velocity, and we would again reach the conclusion that on return $\bar{A}$ will be younger than A. As we have said before, this points out the fact that in special relativity time is a route-dependent quantity. The fact that in Fig. 3.9 $\bar{A}$'s world-line is longer

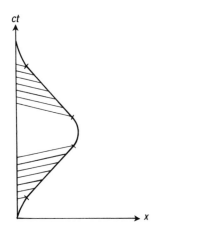

Fig. 3.9 The twin paradox.

Fig. 3.10 Simultaneity lines of $\bar{A}$ on the outward and return journeys.

than A's, and yet takes **less** time to travel, is connected with the Minkow-skian metric

$$ds^2 = c^2\,dt^2 - dx^2 - dy^2 - dz^2$$

and the **negative** signs which appear in it compared with the positive signs occurring in the usual three-dimensional Euclidean metric.

3.10 The Doppler effect

All kinds of waves appear lengthened when the source recedes from the observer: sounds are deepened, light is reddened. Exactly the opposite occurs when the source, instead, approaches the observer. We first of all calculate the **classical** Doppler effect.

Consider a source of light emitting radiation whose wavelength in its rest frame is λ_0. Consider an observer S relative to whose frame the source is in motion with radial velocity u_r. Then, if two successive pulses are emitted at time differing by dt' as measured by S', the distance these pulses have to travel will differ by an amount $u_r\,dt'$ (see Fig. 3.11). Since the pulses travel with speed c, it follows that they arrive at S with a time difference

$$\Delta t = dt' + u_r\,dt'/c,$$

giving

$$\Delta t/dt' = 1 + u_r/c.$$

Now, using the fundamental relationship between wavelength and velocity, set

$$\lambda = c\Delta t \quad \text{and} \quad \lambda_0 = c\,dt'.$$

We then obtain the **classical Doppler formula**

$$\lambda/\lambda_0 = 1 + u_r/c. \tag{3.25}$$

Let us now consider the special relativistic formula. Because of time dilation (see Fig. 3.3), the time inverval between successive pulses according to S is $\beta\,dt'$ (Fig. 3.12). Hence, by the same argument, the pulses arrive at S with a time difference

$$\Delta t = \beta\,dt' + u_r\beta\,dt'/c$$

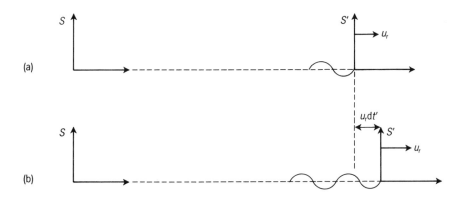

(a)

(b)

Fig. 3.11 The Doppler effect: (a) first pulse; (b) second pulse.

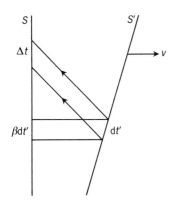

Fig. 3.12 The special relativistic Doppler shift.

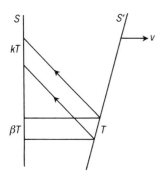

Fig. 3.13 The radial Doppler shift k.

and so this time we find that the **special relativistic Doppler formula** is

$$\frac{\lambda}{\lambda_0} = \frac{1 + u_r/c}{(1 - v^2/c^2)^{\frac{1}{2}}}. \qquad (3.26)$$

If the velocity of the source is purely radial, then $u_r = v$ and (3.26) reduces to

$$\frac{\lambda}{\lambda_0} = \left(\frac{1 + v/c}{1 - v/c}\right)^{\frac{1}{2}}. \qquad (3.27)$$

This is the **radial Doppler shift**, and, if we set $c = 1$, we obtain (2.4), which is the formula for the **k-factor**. Combining Figs. 2.7 and 3.12, the radial Doppler shift is illustrated in Fig. 3.13, where dt' is replaced by T. From equation (3.26), we see that there is also a change in wavelength, even when the radial velocity of the source is zero. For example, if the source is moving in a circle about the origin of S with speed v (as measured by an instantaneous co-moving frame), then the **transverse Doppler shift** is given by

$$\frac{\lambda}{\lambda_0} = \frac{1}{(1 - v^2/c^2)^{\frac{1}{2}}}. \qquad (3.28)$$

This is a purely relativistic effect due to the time dilation of the moving source. Experiments with revolving apparatus using the so-called 'Mössbauer effect' have directly confirmed the transverse Doppler shift in full agreement with the relativistic formula, thus providing another striking verification of the phenomenon of time dilation.

Exercises

3.1 (§3.1) S and S' are in standard configuration with $v = \alpha c$ ($0 < \alpha < 1$). If a rod at rest in S' makes an angle of $45°$ with Ox in S and $30°$ with $O'x$ in S', then find α.

3.2 (§3.1) Note from the previous question that perpendicular lines in one frame need not be perpendicular in another frame. This shows that there is no obvious meaning to the phrase 'two inertial frames are parallel', unless their relative velocity is along a common axis, because the axes of either frame need not appear rectangular in the other. Verify that the Lorentz transformation between frames in standard configuration with relative velocity $v = (v, 0, 0)$ may be written in vector form

$$r' = r + \left(\frac{v \cdot r}{v^2}(\beta - 1) - \beta t\right)v, \qquad t' = \beta\left(t - \frac{v \cdot r}{c^2}\right),$$

where $r = (x, y, z)$. The formulae are said to comprise the 'Lorentz transformation without relative rotation'. Justify

this name by showing that the formulae remain valid when the frames are not in standard configuration, but are parallel in the sense that the same rotation must be applied to each frame to bring the two into standard configuration (in which case v is the velocity of S' relative to S, but $v = (v, 0, 0)$ no longer applies).

3.3 (§3.1) Prove that the first two equations of the special Lorentz transformation can be written in the form

$$ct' = -x\sinh\phi + ct\cosh\phi, \qquad x' = x\cosh\phi - ct\sinh\phi,$$

where the **rapidity** ϕ is defined by $\phi = \tanh^{-1}(v/c)$. Establish also the following version of these equations:

$$ct' + x' = e^{-\phi}(ct + x),$$
$$ct' - x' = e^{\phi}(ct - x),$$
$$e^{2\phi} = (1 + v/c)/(1 - v/c).$$

What relation does ϕ have to θ in equation (3.11)?

3.4 (§**3.1**) **Aberration** refers to the fact that the direction of travel of a light ray depends on the motion of the observer. Hence, if a telescope observes a star at an inclination θ' to the horizontal, then show that **classically** the 'true' inclination θ of the star is related to θ' by

$$\tan \theta' = \frac{\sin \theta}{\cos \theta + v/c},$$

where v is the velocity of the telescope relative to the star. Show that the corresponding relativistic formula is

$$\tan \theta' = \frac{\sin \theta}{\beta(\cos \theta + v/c)}.$$

3.5 (§**3.2**) Show that special Lorentz transformations are associative, that is, if $O(v_1)$ represents the transformation from observer S to S', then show that

$$\big(O(v_1)O(v_2)\big)O(v_3) = O(v_1)\big(O(v_2)O(v_3)\big).$$

3.6 (§**3.3**) An athlete carrying a horizontal 20-ft-long pole runs at a speed v such that $(1 - v^2/c^2)^{-\frac{1}{2}} = 2$ into a 10-ft-long room and closes the door. Explain, in the athlete's frame, in which the room is only 5 ft long, how this is possible. [Hint: no effect travels faster than light.] Show that the minimum length of the room for the performance of this trick is $20/(\sqrt{3} + 2)$ ft. Draw a space-time diagram to indicate what is going on in the rest frame of the athlete.

3.7 (§**3.5**) A particle has velocity $u = (u_1, u_2, u_3)$ in S and $u' = (u'_1, u'_2, u'_3)$ in S'. Prove from the velocity transformation formulae that

$$c^2 - u^2 = \frac{c^2(c^2 - u'^2)(c^2 - v^2)}{(c^2 + u'_1 v)^2}.$$

Deduce that, if the speed of a particle is less than c in any one inertial frame, then it is less than c in every inertial frame.

3.8 (§**3.7**) Check the transformation formulae for the components of acceleration (3.21)–(3.23). Deduce that acceleration is an absolute quantity in special relativity.

3.9 (§**3.8**) A particle moves from rest at the origin of a frame S along the x-axis, with constant acceleration α (as measured in an instantaneous rest frame). Show that the equation of motion is

$$\alpha x^2 + 2c^2 x - \alpha c^2 t^2 = 0,$$

and prove that the light signals emitted after time $t = c/\alpha$ at the origin will never reach the receding particle. A standard clock carried along with the particle is set to read zero at the beginning of the motion and reads τ at time t in S. Using the clock hypothesis, prove the following relationships:

$$\frac{u}{c} = \tanh \frac{\alpha \tau}{c}, \qquad \left(1 - \frac{u^2}{c^2}\right)^{-\frac{1}{2}} = \cosh \frac{\alpha \tau}{c},$$

$$\frac{\alpha t}{c} = \sinh \frac{\alpha \tau}{c}, \qquad x = \frac{c^2}{\alpha}\left(\cosh \frac{\alpha \tau}{c} - 1\right).$$

Show that, if $T^2 \ll c^2/\alpha^2$, then, during an elapsed time T in the inertial system, the particle clock will record approximately the time $T(1 - \alpha^2 T^2/6c^2)$.

If $\alpha = 3g$, find the difference in recorded times by the spaceship clock and those of the inertial system

   (a) after 1 hour;
   (b) after 10 days.

3.10 (§**3.9**) A space traveller $\bar{A}$ travels through space with uniform acceleration g (to ensure maximum comfort). Find the distance covered in 22 years of $\bar{A}$'s time. [Hint: using years and light years as time and distance units, respectively, then $g = 1.03$]. If on the other hand, $\bar{A}$ describes a straight double path $XYZYX$, with acceleration g on XY and ZY, and deceleration on YZ and YX, for 6 years each, then draw a space-time diagram as seen from the Earth and find by how much the Earth would have aged in 24 years of $\bar{A}$'s time.

3.11 (§**3.10**) Let the relative velocity between a source of light and an observer be u, and establish the **classical** Doppler formulae for the frequency shift:

source moving, observer at rest: $\quad v = \dfrac{v_0}{1 + u/c}$,

observer moving, source at rest: $\quad v = (1 - u/c)v_0$,

where v_0 is the frequency in the rest frame of the source. What are the corresponding relativistic results?

3.12 (§**3.10**) How fast would you need to drive towards a red traffic light for the light to appear green? [Hint: $\lambda_{\text{red}} \approx 7 \times 10^{-5}$ cm, $\lambda_{\text{green}} \approx 5 \times 10^{-5}$ cm.]

4 The elements of relativistic mechanisms

4.1 Newtonian theory

Before discussing relativistic mechanics, we shall review some basic ideas of Newtonian theory. We have met Newton's first law in § 2.4, and it states that a body not acted upon by a force moves in a straight line with uniform velocity. The second law describes what happens if an object changes its velocity. In this case, something is causing it to change its velocity and this something is called a **force**. For the moment, let us think of a force as something tangible like a push or a pull. Now, we know from experience that it is more difficult to push a more massive body and get it moving than it is to push a less massive body. This resistance of a body to motion, or rather change in motion, is called its **inertia**. To every body, we can ascribe, at least at one particular time, a number measuring its inertia, which (again for the moment) we shall call its **mass** m. If a body is moving with velocity v, we define its **linear momentum** p to be the product of its mass and velocity. Then Newton's second law (N2) states that the force acting on a body is equal to the rate of change of linear momentum. The third law (N3) is less general and talks about a restricted class of forces called **internal** forces, namely, forces acting on a body due to the influence of other bodies in a system. The third law states that the force acting on a body due to the influence of the other bodies, the so-called **action**, is equal and opposite to the force acting on these other bodies due to the influence of the first body, the so-called **reaction**. We state the two laws below.

> **N2:** The rate of change of momentum of a body is equal to the force acting on it, and is in the direction of the force.

> **N3:** To every action there is an equal and opposite reaction.

Then, for a body of mass m with a force F acting on it, Newton's second law states

$$F = \frac{\mathrm{d}p}{\mathrm{d}t} = \frac{\mathrm{d}(mv)}{\mathrm{d}t}. \tag{4.1}$$

If, in particular, the mass is a constant, then

$$F = m\frac{\mathrm{d}v}{\mathrm{d}t} = ma \qquad (4.2)$$

where a is the acceleration.

Now, strictly speaking, in Newtonian theory, all observable quantities should be defined in terms of their measurement. We have seen how an observer equipped with a frame of reference, ruler, and clock can map the events of the universe, and hence measure such quantities as position, velocity, and acceleration. However, Newton's laws introduce the new concepts of force and mass, and so we should give a prescription for their measurement. Unfortunately, any experiment designed to measure these quantities involves Newton's laws themselves in its interpretation. Thus, Newtonian mechanics has the rather unexpected property that the operational definitions of force and mass which are required to make the laws physically significant are actually contained in the laws themselves.

To make this more precise, let us discuss how we might use the laws to measure the mass of a body. We consider two bodies isolated from all other influences other than the force acting on one due to the influence of the other and vice versa (Fig. 4.1). Since the masses are assumed to be constant, we have, by Newton's second law in the form (4.2),

$$F_1 = m_1 a_1 \quad \text{and} \quad F_2 = m_2 a_2.$$

In addition, by Newton's third law, $F_1 = -F_2$. Hence, we have

$$m_1 a_1 = - m_2 a_2. \qquad (4.3)$$

Therefore, if we take one standard body and define it to have **unit** mass, then we can find the mass of the other body, by using (4.3). We can keep doing this with any other body and in this way we can calibrate masses. In fact, this method is commonly used for comparing the masses of elementary particles. Of course, in practice, we cannot remove all other influences, but it may be possible to keep them almost constant and so neglect them.

We have described how to use Newton's laws to measure mass. How do we measure force? One approach is simply to use Newton's second law, work out ma for a body and then read off from the law the force acting on m. This is consistent, although rather circular, especially since a force has independent properties of its own. For example, Newton has provided us with a way for working out the force in the case of gravitation in his **universal law of gravitation** (UG).

UG: Two particles attract each other with a force directly proportional to their masses and inversely proportional to the distance between them.

If we denote the constant of proportionality by G (with value 6.67×10^{-11} in m.k.s. units), the so-called Newtonian constant, then the law is (see Fig. 4.2)

$$F = - G\frac{m_1 m_2}{r^2}\hat{r}, \qquad (4.4)$$

Fig. 4.1 Measuring mass by mutually induced accelerations.

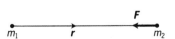

Fig. 4.2 Newton's universal law of gravitation.

where a hat denotes a unit vector. There are other force laws which can be stated separately. Again, another independent property which holds for certain forces is contained in Newton's third law. The standard approach to defining force is to consider it as being **fundamental**, in which case force laws can be stated separately or they can be worked out from other considerations. We postpone a more detailed critique of Newton's laws until Part C of the book.

Special relativity is concerned with the behaviour of material bodies and light rays **in the absence of gravitation**. So we shall also postpone a detailed consideration of gravitation until we discuss general relativity in Part C of the book. However, since we have stated Newton's universal laws of gravitation in (4.4), we should, for completeness, include a statement of Newtonian gravitation for a distribution of matter. A distribution of matter of mass density $\rho = \rho(x, y, z, t)$ gives rise to a gravitational potential ϕ which satisfies **Poisson's equation**

$$\nabla^2 \phi = 4\pi G \rho \qquad (4.5)$$

at points inside the distribution, where the Laplacian operator ∇^2 is given in Cartesian coordinates by

$$\nabla^2 = \frac{\partial^2}{\partial x^2} + \frac{\partial^2}{\partial y^2} + \frac{\partial^2}{\partial z^2}.$$

At points external to the distribution, this reduces to **Laplace's equation**

$$\nabla^2 \phi = 0. \qquad (4.6)$$

We assume that the reader is familiar with this background to Newtonian theory.

4.2 Isolated systems of particles in Newtonian mechanics

In this section, we shall, for completeness, derive the conservation of linear momentum in Newtonian mechanics for a system of n particles. Let the ith particle have constant mass m_i and position vector r_i relative to some arbitrary origin. Then the ith particle possesses linear momentum p_i defined by $p_i = m_i \dot{r}_i$, where the dot denotes differentiation with respect to time t. If F_i is the total force on m_i, then, by Newton's second law, we have

$$F_i = \dot{p}_i = m_i \ddot{r}_i. \qquad (4.7)$$

The total force F_i on the ith particle can be divided into an external force F_i^{ext} due to any external fields present and to the resultant of the internal forces. We write

$$F_i = F_i^{\text{ext}} + \sum_{j=1}^{n} F_{ij},$$

where F_{ij} is the force or the ith particle due to the jth particle and where, for convenience, we define $F_{ii} = 0$. If we sum over i in (4.7), we find

$$\frac{d}{dt} \sum_{i=1}^{n} p_i = \sum_{i=1}^{n} \frac{dp_i}{dt} = \sum_{i=1}^{n} F_i^{ext} + \sum_{i,j=1}^{n} F_{ij}.$$

Using Newton's third law, namely, $F_{ij} = -F_{ji}$, then the last term is zero and we obtain $\dot{P} = F^{ext}$, where $P = \sum_{i=1}^{n} p_i$ is termed the **total linear momentum** of the system and $F^{ext} = \sum_{i=1}^{n} F_i^{ext}$ is the **total external force** on the system. If, in particular, the system of particles is **isolated**, then

$$F^{ext} = 0 \quad \Rightarrow \quad P = c,$$

where c is a constant vector. This leads to the law of the **conservation of linear momentum** of the system, namely,

$$P_{initial} = P_{final}. \tag{4.8}$$

4.3 Relativistic mass

The transition from Newtonian to relativistic mechanics is not, in fact, completely straightforward, because it involves at some point or another the introduction of *ad hoc* assumptions about the behaviour of particles in relativistic situations. We shall adopt the approach of trying to keep as close to the non-relativistic definition of energy and momentum as we can. This leads to results which in the end must be confronted with experiment. The ultimate justification of the formulae we shall derive resides in the fact that they have been repeatedly confirmed in numerous laboratory experiments in particle physics. We shall only derive them in a simple case and state that the arguments can be extended to a more general situation.

It would seem plausible that, since length and time measurements are dependent on the observer, then mass should also be an observer-dependent quantity. We thus assume that a particle which is moving with a velocity u relative to an inertial observer has a mass, which we shall term its **relativistic mass**, which is some function of u, that is,

$$m = m(u), \tag{4.9}$$

where the problem is to find the explicit dependence of m on u. We restrict attention to motion along a straight line and consider the special case of two equal particles colliding **inelastically** (in which case they stick together), and look at the collision from the point of view of two inertial observers S and S' (see Fig. 4.3). Let one of the particles be at rest in the frame S and the other possess a velocity u before they collide. We then assume that they coalesce and that the combined object moves with velocity U. The masses of the two particles are respectively $m(0)$ and $m(u)$ by (4.9). We denote $m(0)$ by m_0 and term it the **rest mass** of the particle. In addition, we denote the mass of the combined object by $M(U)$. If we take S' to be the **centre-of-mass frame**, then it should be clear that, relative to S', the two equal particles collide with equal and opposite speeds, leaving the combined object with mass M_0 at rest. It follows that S' must have velocity U relative to S.

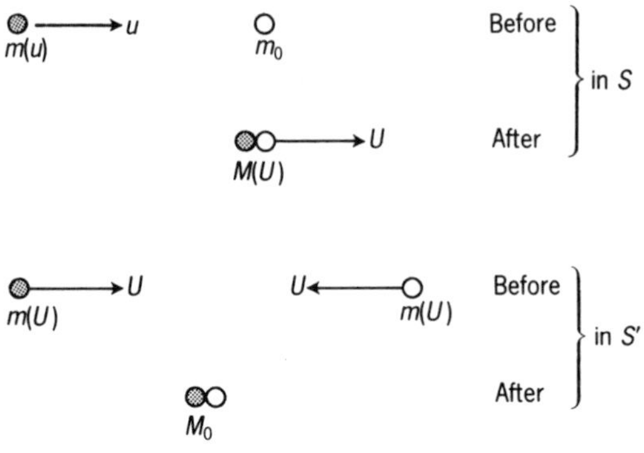

Fig. 4.3 The inelastic collision in the frames S and S'.

We shall assume both conservation of mass and conservation of linear momentum and see what this leads to. In the frame S, we obtain

$$m(u) + m_0 = M(U), \qquad m(u)u + 0 = M(U)U,$$

from which we get, eliminating $M(U)$,

$$m(u) = m_0 \left(\frac{U}{u - U} \right). \tag{4.10}$$

The left-hand particle has a velocity U relative to S', which in turn has a velocity U relative to S. Hence, using the composition of velocities law, we can compose these two velocities and the resultant velocity must be identical with the velocity u of the left-hand particle in S. Thus, by (2.6) in non-relativistic units,

$$u = \frac{2U}{(1 + U^2/c^2)}.$$

Solving for U in terms of u, we obtain the quadratic

$$U^2 - \left(\frac{2c^2}{u} \right) U + c^2 = 0,$$

which has roots

$$U = \frac{c^2}{u} \pm \left[\left(\frac{c^2}{u} \right)^2 - c^2 \right]^{\frac{1}{2}} = \frac{c^2}{u} \left[1 \pm \left(1 - \frac{u^2}{c^2} \right)^{\frac{1}{2}} \right].$$

In the limit $u \to 0$, this must produce a finite result, so we must take the negative sign (check), and, substituting in (4.10), we find finally

$$m(u) = \gamma m_0, \tag{4.11}$$

where

$$\gamma \equiv (1 - u^2/c^2)^{-\frac{1}{2}}. \tag{4.12}$$

This is the basic result which relates the relativistic mass of a moving particle to its rest mass. Note that this is the same in structure as the time dilation formula (3.16), i.e. $T = \beta T_0$, where $\beta = (1 - v^2/c^2)^{-\frac{1}{2}}$, except that time dilation involves the factor β which depends on the velocity v of the frame S' relative to S, whereas γ depends on the velocity u of the particle relative to S. If we plot m against u, we see that relativistic mass increases without bound as u approaches c (Fig. 4.4).

It is possible to extend the above argument to establish (4.11) in more general situations. However, we emphasize that it is not possible to derive the result a priori, but only with the help of extra assumptions. However it is produced, the only real test of the validity of the result is in the experimental arena and here it has been extensively confirmed.

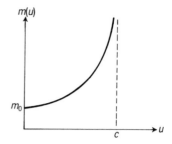

Fig. 4.4 Relativistic mass as a function of velocity.

4.4 Relativistic energy

Let us expand the expression for the relativistic mass, namely,

$$m(u) = \gamma m_0 = m_0(1 - u^2/c^2)^{-\frac{1}{2}},$$

in the case when the velocity u is small compared with the speed of light c. Then we get

$$m(u) = m_0 + \frac{1}{c^2}(\tfrac{1}{2}m_0 u^2) + O\left(\frac{u^4}{c^4}\right), \tag{4.13}$$

where the final term stands for all terms of order $(u/c)^4$ and higher. If we multiply both sides by c^2, then, apart from the constant $m_0 c^2$, the right-hand side is to first approximation the classical kinetic energy (k.e.), that is,

$$mc^2 = m_0 c^2 + \tfrac{1}{2}m_0 u^2 + \cdots \simeq \text{constant} + \text{k.e.} \tag{4.14}$$

We have seen that relativistic mass contains within it the expression for classical kinetic energy. In fact, it can be shown that the conservation of relativistic mass leads to the conservation of kinetic energy in the Newtonian approximation. As a simple example, consider the collision of two particles with rest mass m_0 and $\bar{m}_0$, initial velocities v_1 and $\bar{v}_1$, and final velocities v_2 and $\bar{v}_2$, respectively (Fig. 4.5). Conservation of relativistic mass gives

$$m_0(1 - v_1^2/c^2)^{-\frac{1}{2}} + \bar{m}_0(1 - \bar{v}_1^2/c^2)^{-\frac{1}{2}} = m_0(1 - v_2^2/c^2)^{-\frac{1}{2}}$$
$$+ \bar{m}_0(1 - \bar{v}_2^2/c^2)^{-\frac{1}{2}}. \tag{4.15}$$

If we now assume that $v_1, v_2, \bar{v}_1$, and $\bar{v}_2$ are all small compared with c, then we find (exercise) that the leading terms in the expansion of (4.15) give

$$\tfrac{1}{2}m_0 v_1^2 + \tfrac{1}{2}\bar{m}_0 \bar{v}_1^2 = \tfrac{1}{2}m_0 v_2^2 + \tfrac{1}{2}\bar{m}_0 \bar{v}_2^2, \tag{4.16}$$

which is the usual conservation of energy equation. Thus, in this sense, conservation of relativistic mass includes within it conservation of energy. Now, since energy is only defined up to the addition of a constant, the result

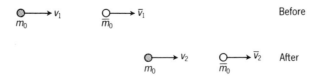

Before

After

Fig. 4.5 Two colliding particles.

(4.14) suggest that we regard the **energy** E of a particle as given by

$$E = mc^2. \tag{4.17}$$

This is one of the most famous equations in physics. However, it is not just a mathematical relationship between two different quantities, namely energy and mass, but rather states that energy and mass are **equivalent** concepts. Because of the arbitrariness in the actual value of E, a better way of stating the relationship is to say that a change in energy is equal to a change in relativistic mass, namely,

$$\Delta E = \Delta mc^2$$

Using conventional units, c^2 is a large number and indicates that a small change in mass is equivalent to an enormous change in energy. As is well known, this relationship and the deep implications it carries with it for peace and war, have been amply verified. For obvious reasons, the term $m_0 c^2$ is termed the **rest energy** of the particle. Finally, we point out that conservation of linear momentum, using relativistic mass, leads to the usual conservation law in the Newtonian approximation. For example (exercise), the collision problem considered above leads to the usual conservation of linear momentum equation for slow-moving particles:

$$m_0 v_1 + \bar{m}_0 \bar{v}_1 = m_0 v_2 + \bar{m}_0 \bar{v}_2. \tag{4.18}$$

Extending these ideas to three spatial dimensions, then a particle moving with velocity u relative to an inertial frame S has relativistic mass m, energy E, and linear momentum p given by

$$m = \gamma m_0, \qquad E = mc^2, \qquad p = mu. \tag{4.19}$$

Some straightforward algebra (exercise) reveals that

$$(E/c)^2 - p_x^2 - p_y^2 - p_z^2 = (m_0 c)^2, \tag{4.20}$$

where $m_0 c$ is an invariant, since it is the same for all inertial observers. If we compare this with the invariant (3.13), i.e.

$$(ct)^2 - x^2 - y^2 - z^2 = s^2,$$

then it suggests that the quantities $(E/c, p_x, p_y, p_z)$ transform under a Lorentz transformation in the same way as the quantities (ct, x, y, z). We shall see in Part C that the language of tensors provides a better framework for discussing transformation laws. For the moment, we shall assume that energy and momentum transform in an identical manner and quote the results. Thus, in a frame S' moving in standard configuration with velocity v relative to S, the transformation equations are (see (3.12))

$$E' = \beta(E - vp_x), \qquad p_x' = \beta(p_x - vE/c^2), \qquad p_y' = p_y, \qquad p_z' = p_z. \tag{4.21}$$

The inverse transformations are obtained in the usual way, namely, by interchanging primes and unprimes and replacing v by $-v$, which gives

$$E = \beta(E' + vp'_x), \quad p_x = \beta(p'_x + vE'/c^2), \quad p_y = p'_y, \quad p_z = p'_z. \quad (4.22)$$

If, in particular, we take S' to be the instantaneous rest frame of the particle, then $p' = 0$ and $E' = E_0 = m_0 c^2$. Substituting in (4.22), we find

$$E = \beta E' = \frac{m_0 c^2}{(1 - v^2/c^2)^{\frac{1}{2}}} = mc^2,$$

where $m = m_0(1 - v^2/c^2)^{-\frac{1}{2}}$ and $p = (\beta v E'/c^2, 0, 0) = (mv, 0, 0) = mv$, which are precisely the values of the energy, mass, and momentum arrived at in (4.19) with u replaced by v.

4.5 Photons

At the end of the last century, there was considerable conflict between theory and experiment in the investigation of radiation in enclosed volumes. In an attempt to resolve the difficulties, Max Planck proposed that light and other electromagnetic radiation consisted of individual 'packets' of energy, which he called **quanta**. He suggested that the energy E of each quantum was to depend on its frequency v, and proposed the simple law, called **Planck's hypothesis**,

$$E = hv, \quad (4.23)$$

where h is a universal constant known now as **Planck's constant**. The idea of the quantum was developed further by Einstein, especially in attempting to explain the photoelectric effect. The effect is to do with the ejection of electrons from a metal surface by incident light (especially ultraviolet) and is strongly in support of Planck's quantum hypothesis. Nowadays, the quantum theory is well established and applications of it to explain properties of molecules, atoms, and fundamental particles are at the heart of modern physics. Theories of light now give it a dual wave–particle nature. Some properties, such as diffraction and interference, are wavelike in nature, while the photoelectric effect and other cases of the interaction of light and atoms are best described on a particle basis.

The particle description of light consists in treating it as a stream of quanta called **photons**. Using equation (4.19) and substituting in the speed of light, $u = c$, we find

$$m_0 = \gamma^{-1} m = (1 - u^2/c^2)^{\frac{1}{2}} m = 0, \quad (4.24)$$

that is, the rest mass of a photon must be zero! This is not so bizarre as it first seems, since no inertial observer ever sees a photon at rest — its speed is always c — and so the rest mass of a photon is merely a notional quantity. If we let $\hat{n}$ be a unit vector denoting the direction of travel of the photon, then

$$p = (p_x, p_y, p_z) = p\hat{n},$$

and equation (4.20) becomes

$$(E/c)^2 - p^2 = 0.$$

Taking square roots (and remembering c and p are positive), we find that the energy E of a photon is related to the magnitude p of its momentum by

$$E = pc. \tag{4.25}$$

Finally, using the energy–mass relationship $E = mc^2$, we find that the relativistic mass of a photon is non-zero and is given by

$$m = p/c. \tag{4.26}$$

Combining these results with Planck's hypothesis, we obtain the following formulae for the energy E, relativistic mass m, and linear momentum $\boldsymbol{p}$ of the photon:

$$E = h\nu, \qquad m = h\nu/c^2, \qquad \boldsymbol{p} = (h\nu/c)\hat{\boldsymbol{n}}. \tag{4.27}$$

It is gratifying to discover that special relativity, which was born to reconcile conflicts in the kinematical properties of light and matter, also includes their mechanical properties in a single all-inclusive system.

We finish this section with an argument which shows that Planck's hypothesis can be derived directly within the framework of special relativity. We have already seen in the last chapter that the radial Doppler effect for a moving source is given by (3.27), namely

$$\frac{\lambda}{\lambda_0} = \left(\frac{1 + v/c}{1 - v/c}\right)^{\frac{1}{2}},$$

where λ_0 is the wavelength in the frame of the source and λ is the wavelength in the frame of the observer. We write this result, instead, in terms of frequency, using the fundamental relationships $c = \lambda\nu$ and $c = \lambda_0\nu_0$, to obtain

$$\frac{\nu_0}{\nu} = \left(\frac{1 + v/c}{1 - v/c}\right)^{\frac{1}{2}}. \tag{4.28}$$

Now, suppose that the source emits a light flash of total energy E_0. Let us use the equations (4.22) to find the energy received in the frame of the observer S. Since, recalling Fig. 3.11, the light flash is travelling along the negative x-direction of both frames, the relationship (4.25) leads to the result $p'_x = -E_0/c$, with the other primed components of momentum zero. Substituting in the first equation of (4.22), namely,

$$E = \beta(E' + vp'_x),$$

we get

$$E = \beta(E_0 - vE_0/c) = \frac{E_0(1 - v/c)}{(1 - v^2/c^2)^{\frac{1}{2}}} = E_0\left(\frac{1 - v/c}{1 + v/c}\right)^{\frac{1}{2}},$$

or

$$\frac{E_0}{E} = \left(\frac{1 + v/c}{1 - v/c}\right)^{\frac{1}{2}}. \tag{4.29}$$

Combining this with equation (4.28), we obtain

$$\frac{E_0}{\nu_0} = \frac{E}{\nu}.$$

Since this relationship holds for **any** pair of inertial observers, it follows that the ratio must be a universal constant, which we call h. Thus, we have derived Planck's hypothesis $E = hv$.

We leave our considerations of special relativity at this point and turn our attention to the formalism of tensors. This will enable us to reformulate special relativity in a way which will aid our transition to general relativity, that is, to a theory of gravitation consistent with special relativity.

Exercises

4.1 (§4.1) Discuss the possibility of using force rather than mass as the basic quantity, taking, for example, a standard weight at a given latitude as the unit of force. How should one then define and measure the mass of a body?

4.2 (§4.3) Show that, in the inelastic collision considered in §4.3, the rest mass of the combined object is greater than the sum of the original rest masses. Where does this increase derive from?

4.3 (§4.3) A particle of rest mass $\bar{m}_0$ and speed u strikes a stationary particle of rest mass m_0. If the collision is perfectly inelastic, then find the rest mass of the composite particle.

4.4 (§4.4) (i) Establish the transition from equation (4.15) to (4.16).
(ii) Establish the Newtonian approximation equation (4.18).

4.5 (§4.4) Show that (4.19) leads to (4.20). Deduce (4.21).

4.6 (§4.4) Newton's second law for a particle of relativistic mass m is

$$F = \frac{d}{dt}(mu).$$

Define the work done dE in moving the particle from r to $r + dr$. Show that the rate of doing work is given by

$$\frac{dE}{dt} = \frac{d(mu)}{dt} \cdot u.$$

Use the definition of relativistic mass to obtain the result

$$\frac{dE}{dt} = \frac{m_0}{(1 - u^2/c^2)^{3/2}} u \frac{du}{dt} \quad \left[\text{Hint: } u \cdot \frac{du}{dt} = u \frac{du}{dt}\right].$$

Express this last result in terms of dm/dt and integrate to obtain

$$E = mc^2 + \text{constant}.$$

4.7 (§4.4) Two particles whose rest masses are m_1 and m_2 move along a straight line with velocities u_1 and u_2, measured in the same direction. They collide inelastically to form a new particle. Show that the rest mass and velocity of the

new particle are m_3 and u_3, respectively, where

$$m_3^2 = m_1^2 + m_2^2 + 2m_1 m_2 \gamma_1 \gamma_2 (1 - u_1 u_2/c^2),$$

$$u_3 = \frac{m_1 \gamma_1 u_1 + m_2 \gamma_2 u_2}{m_1 \gamma_1 + m_2 \gamma_2},$$

with

$$\gamma_1 = (1 - u_1^2/c^2)^{-\frac{1}{2}}, \qquad \gamma_2 = (1 - u_2^2/c^2)^{-\frac{1}{2}}.$$

4.8 (§4.4) A particle of rest mass m_0, energy e_0, and momentum p_0 suffers a head on elastic collision (i.e. masses of particles unaltered) with a stationary mass M. In the collision, M is knocked straight forward, with energy E and momentum P, leaving the first particle with energy e and p. Prove that

$$P = \frac{2p_0 M(e_0 + Mc^2)}{2Me_0 + M^2c^2 + m_0^2 c^2}$$

and

$$p = \frac{P_0(m^2 c^2 - M^2 c^2)}{2Me_0 + M^2 c^2 + m_0^2 c^2}.$$

What do these formulae become in the classical limit?

4.9 (§4.4) Assume that the formulae (4.19) hold for a tachyon, which travels with speed $v > c$. Taking the energy to be a measurable quantity, then deduce that the rest mass of a tachyon is imaginary and define the real quantity μ_0 by $m_0 = i\mu_0$.

If the tachyon moves along the x-axis and if we assume that the x-component of the momentum is a real positive quantity, then deduce

$$m = \frac{v}{|v|} \alpha \mu_0, \qquad p = \mu_0 |v| \alpha, \qquad E = mc^2,$$

where $\alpha = (v^2/c^2 - 1)^{-\frac{1}{2}}$.

Plot $E/m_0 c^2$ against v/c for both tachyons and subluminal particles.

4.10 (§4.5) Two light rays in the (x, y)-plane of an inertial observer, making angles θ and $-\theta$, respectively, with the positive x axis, collide at the origin. What is the velocity v of the inertial observer (travelling in standard configuration) who sees the light rays collide head on?

4.11 (§4.5) An atom of rest mass m_0 is at rest in a laboratory and absorbs a photon of frequency v. Find the velocity and mass of the recoiling particle.

4.12 (§4.5) An atom at rest in a laboratory emits a photon and recoils. If its initial mass is m_0 and it loses the rest energy e in the emission, prove that the frequency of the emitted photon is given by

$$v = \frac{e}{h}(1 - e/2m_0c^2).$$

B. The Formalism of Tensors

Tensor algebra 5

5.1 Introduction

To work effectively in Newtonian theory, one really needs the language of
vectors. This language, first of all, is more succinct, since it summarizes a set
of three equations in one. Moreover, the formalism of vectors helps to solve
certain problems more readily, and, most important of all, the language
reveals structure and thereby offers insight. In exactly the same way, in
relativity theory, one needs the language of tensors. Again, the language helps
to summarize sets of equations succinctly and to solve problems more readily,
and it reveals structure in the equations. This part of the book is devoted to
learning the formalism of tensors which is a pre-condition for the rest of the
book.

The approach we adopt is to concentrate on the technique of tensors
without taking into account the deeper geometrical significance behind the
theory. We shall be concerned more with what you do with tensors rather
than what tensors actually are. There are two distinct approaches to the
teaching of tensors: the abstract or index-free (coordinate-free) approach and
the conventional approach based on indices. There has been a move in recent
years in some quarters to introduce tensors from the start using the more
modern abstract approach (although some have subsequently changed their
mind and reverted to the conventional approach). The main advantage of this
approach is that it offers deeper geometrical insight. However, it has two
disadvantages. First of all, it requires much more of a mathematical background, which in turn takes time to develop. The other disadvantage is that,
for all its elegance, when one wants to do a real calculation with tensors, as
one frequently needs to, then recourse has to be made to indices. We shall
adopt the more conventional index approach, because it will prove faster and
more practical. However, we advise those who wish to take their study of the
subject further to look at the index-free approach at the first opportunity.

We repeat that the exercises are seen as integral to this part of the book and
should not be omitted.

5.2 Manifolds and coordinates

We shall start by working with tensors defined in n dimensions since, and it is
part of the power of the formalism, there is little extra effort involved. A
tensor is an object defined on a geometric entry called a (differential)
manifold. We shall not define a manifold precisely because it would involve

us in too much of a digression. But, in simple terms, a manifold is something which 'locally' looks like a bit of n-dimensional Euclidean space $\mathbb{R}^n$. For example, compare a 2-sphere S^2 with the Euclidean plane $\mathbb{R}^2$. They are clearly different. But a small bit of S^2 looks very much like a small bit of $\mathbb{R}^2$ (if we neglect metrical properties). The fact that S^2 is 'compact', i.e. in some sense finite, whereas $\mathbb{R}^2$ 'goes off to infinity' is a **global** property rather than a local property. We shall not say anything precise about global properties — the **topology** of the manifold —, although the issue will surface when we start to look carefully at solutions of Einstein's equations in general relativity.

We shall simply take an n-dimensional manifold M to be a set of points such that each point possesses a set of n **coordinates** $(x^1, x^2, \ldots, x^n)$, where each coordinate ranges over a subset of the reals, which may, in particular, range from $-\infty$ to $+\infty$. To start off with, we can think of these coordinates as corresponding to distances or angles in Euclidean space. The reason why the coordinates are written as superscripts rather than subscripts will become clear later. Now the key thing about a manifold is that it may not be possible to cover the whole manifold by one **non-degenerate** coordinate system, namely, one which ascribes a **unique** set of n coordinate numbers to each point. Sometimes it is simply convenient to use coordinate systems with **degenerate** points. For example, plane polar coordinates (R, ϕ) in the plane have a degeneracy at the origin because ϕ is indeterminate there (Fig. 5.1). However, here we could avoid the degeneracy at the origin by using Cartesian coordinates. But in other circumstances we have no choice in the matter. For example, it can be shown that there is no coordinate system which covers the whole of a 2-sphere S^2 without degeneracy. The smallest number needed is two, which is shown schematically in Fig. 5.2. We therefore

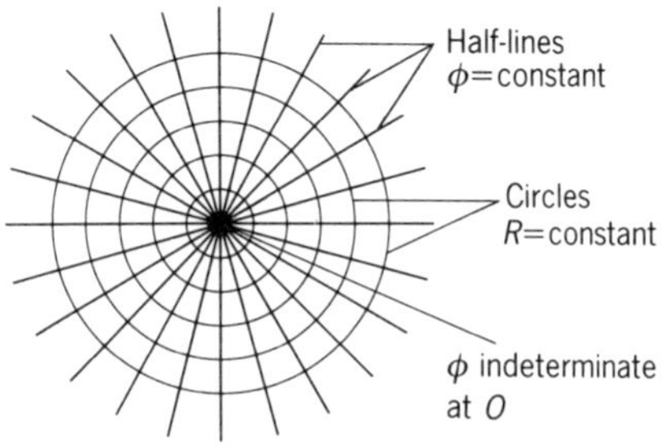

Fig. 5.1 Plane polar coordinate curves.

Half-lines
ϕ=constant

Circles
R=constant

ϕ indeterminate
at O

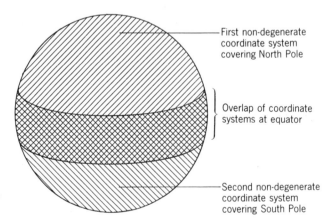

First non-degenerate
coordinate system
covering North Pole

Overlap of coordinate
systems at equator

Second non-degenerate
coordinate system
covering South Pole

Fig. 5.2 Two non-degenerate coordinate systems covering an S^2.

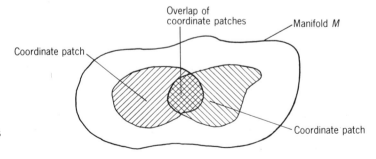

Overlap of
coordinate patches

Manifold M

Coordinate patch

Coordinate patch

Fig. 5.3 Overlapping coordinate patches in a manifold.

work with coordinate systems which cover only a portion of the manifold and which are called **coordinate patches**. Figure 5.3 indicates this schematically. A set of coordinate patches which covers the whole manifold is called an **atlas**. The theory of manifolds tells us how to get from one coordinate patch to another by a coordinate transformation in the overlap region. The behaviour of geometric quantities under coordinate transformations lies at the heart of tensor calculus.

5.3 Curves and surfaces

Given a manifold, we shall be concerned with points in it and subsets of points which define curves and surfaces of different dimensions. We shall frequently define these curves and surfaces **parametrically**. Thus (in exactly the same way as is done in Euclidean 2- and 3-space), since a curve has one degree of freedom it depends on one parameter and so we define a **curve** by the parametric equations

$$x^a = x^a(u) \quad (a = 1, 2, \cdots, n), \tag{5.1}$$

where u is the parameter and $x^1(u), x^2(u), \ldots, x^n(u)$ denote n functions of u. Similarly, since a **subspace** or **surface** of m dimensions ($m < n$) has m degrees of freedom, it depends on m parameters and it is given by the parametric equations

$$x^a = x^a(u^1, u^2, \ldots, u^m) \quad (a = 1, 2, \ldots, n). \tag{5.2}$$

If, in particular, $m = n - 1$, the subspace is called a **hypersurface**. In this case,

$$x^a = x^a(u^1, u^2, \ldots, u^{n-1}) \quad (a = 1, 2, \ldots, n)$$

and the $n - 1$ parameters can be eliminated from these n equations to give one equation connecting the coordinates, i.e.

$$f(x^1, x^2, \ldots, x^n) = 0. \tag{5.3}$$

From a different but equivalent point of view, a point in a general position in a manifold has n degrees of freedom. If it is restricted to lie in a hypersurface, an $(n - 1)$-subspace, then its coordinates must satisfy **one constraint**, namely,

$$f(x^1, x^2, \ldots, x^n) = 0,$$

which is the same as equation (5.3). Similarly, points in an m-dimensional subspace ($m < n$) must satisfy $n - m$ **constraints**

$$\left.\begin{aligned} f^1(x^1, x^2, \ldots, x^n) &= 0, \\ f^2(x^1, x^2, \ldots, x^n) &= 0, \\ &\vdots \\ f^{n-m}(x^1, x^2, \ldots, x^n) &= 0, \end{aligned}\right\} \tag{5.4}$$

which is an alternative to the parametric representation (5.2).

5.4 Transformation of coordinates

As we have seen, a point in a manifold can be covered by many different coordinate patches. The essential point about tensor calculus is that when we make a statement about tensors we do not wish it simply to hold just for one coordinate system but rather for **all** coordinate systems. Consequently, we need to find out how quantities behave when we go from one coordinate system to another one. We therefore consider the **change of coordinates** $x^a \rightarrow x'^a$ given by the n equations

$$x'^a = f^a(x^1, x^2, \dots, x^n) \quad (a = 1, 2, \dots, n), \tag{5.5}$$

where the f's are single-valued continuous differentiable functions, at least for certain ranges of their arguments. Hence, at this stage, we view a coordinate transformation **passively** as assigning to a point of the manifold whose old coordinates are $(x^1, x^2, \dots, x^n)$ the **new** primed coordinates $(x'^1, x'^2, \dots, x'^n)$. We can write (5.5) more succinctly as $x'^a = f^a(x)$, where, from now on, lower case Latin indices are assumed to run from 1 to n, the dimension of the manifold, and the f^a are all functions of the old unprimed coordinates. Furthermore, we can write the equation more simply still as

$$x'^a = x'^a(x), \tag{5.6}$$

where $x'^a(x)$ denote the n functions $f^a(x)$. Notation plays an important role in tensor calculus, and equation (5.6) is clearly easier to write than equation (5.5).

We next contemplate differentiating (5.6) with respect to each of the coordinates x^b. This produces the $n \times n$ **transformation matrix** of coefficients:

$$\left[\frac{\partial x'^a}{\partial x^b} \right] = \begin{bmatrix} \dfrac{\partial x'^1}{\partial x^1} & \dfrac{\partial x'^1}{\partial x^2} & \cdots & \dfrac{\partial x'^1}{\partial x^n} \\ \dfrac{\partial x'^2}{\partial x^1} & \dfrac{\partial x'^2}{\partial x^2} & \cdots & \dfrac{\partial x'^2}{\partial x^n} \\ \vdots & & & \\ \dfrac{\partial x'^n}{\partial x^1} & \dfrac{\partial x'^n}{\partial x^2} & \cdots & \dfrac{\partial x'^n}{\partial x^n} \end{bmatrix}. \tag{5.7}$$

The determinant J' of this matrix is called the **Jacobian** of the transformation:

$$J' = \left| \frac{\partial x'^a}{\partial x^b} \right|. \tag{5.8}$$

We shall assume that this in non-zero for some range of the coordinates x^b. Then it follows from the implicit function theorem that we can (in principle) solve equation (5.6) for the old coordinates x^a and obtain the **inverse** transformation equations

$$x^a = x^a(x'). \tag{5.9}$$

It follows from the product rule for determinants that, if we define the Jacobian of the inverse transformation by

$$J' = \left| \frac{\partial x^a}{\partial x'^b} \right|,$$

then $J = 1/J'$.

In three dimensions, the equation of a surface is given by $z = f(x, y)$, then its total differential is defined to be

$$dz = \frac{\partial f}{\partial x} dx + \frac{\partial f}{\partial y} dy.$$

Then, in an exactly analogous manner, starting from (5.6), we define the total differential

$$dx'^a = \frac{\partial x'^a}{\partial x^1} dx^1 + \frac{\partial x'^a}{\partial x^2} dx^2 + \cdots + \frac{\partial x'^a}{\partial x^n} dx^n$$

for each a running from 1 to n. We can write this more economically by introducing an explicit summation sign:

$$dx'^a = \sum_{b=1}^{n} \frac{\partial x'^a}{\partial x^b} dx^b. \tag{5.10}$$

This can be written more economically still by introducing the **Einstein summation convention**: whenever a literal index is repeated, it is understood to imply a summation over the index from 1 to n, the dimension of the manifold. Hence, we can write (5.10) simply as

$$dx'^a = \frac{\partial x'^a}{\partial x^b} dx^b. \tag{5.11}$$

The index a occurring on each side of this equation is said to be **free** and may take on separately any value from 1 to n. The index b on the right-hand side is **repeated** and hence there is an implied summation from 1 to n. A repeated index is called **bound** or **dummy** because it can be replaced by any other index not already in use. For example,

$$\frac{\partial x'^a}{\partial x^b} dx^b = \frac{\partial x'^a}{\partial x^c} dx^c$$

because c was not already in use in the expression. We define the **Kronecker delta** δ_b^a to be a quantity which is either 0 or 1 according to

$$\delta_b^a = \begin{cases} 1 & \text{if } a = b, \\ 0 & \text{if } a \neq b. \end{cases} \tag{5.12}$$

It therefore follows directly from the definition of partial differentiation (check) that

$$\frac{\partial x'^a}{\partial x'^b} = \frac{\partial x^a}{\partial x^b} = \delta_b^a. \tag{5.13}$$

5.5 Contravariant tensors

The approach we are going to adopt is to define a geometrical quantity in terms of its transformation properties under a coordinate transformation (5.6). We shall start with a prototype and then give the general definition. Consider two neighbouring points in the manifold P and Q with coordinates x^a and $x^a + dx^a$, respectively. The two points define an **infinitesimal displacement** or **infinitesimal vector** $\overrightarrow{PQ}$. The vector is not to be regarded as free, but as being attached to the point P (Fig. 5.4). The components of this vector in the x^a-coordinate system are dx^a. The components in another coordinate system, say the x'^a-coordinate system, are dx'^a which are connected to dx^a by (5.11), namely,

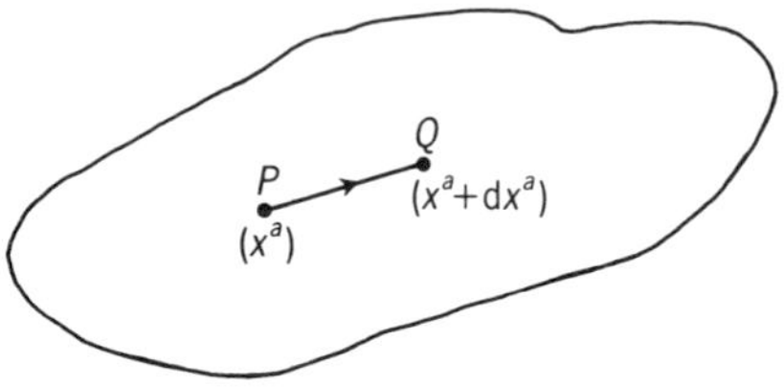

$$dx'^a = \frac{\partial x'^a}{\partial x^b}\, dx^b. \tag{5.14}$$

The transformation matrix appearing in this equation is to be regarded as being evaluated at the point P. i.e. strictly speaking we should write

$$dx'^a = \left[\frac{\partial x'^a}{\partial x^b}\right]_P dx^b, \tag{5.15}$$

but with this understood we shall stick to the notation of (5.14). Thus, $[\partial x'^a/\partial x^b]_P$ consists of an $n \times n$ matrix of real numbers. The transformation is therefore a linear homogeneous transformation. This is our prototype.

A **contravariant vector** or **contravariant tensor of rank (order) 1** is a set of quantities, written X^a in the x^a-coordinate system, associated with a point P, which transforms under a change of coordinates according to

$$X'^a = \frac{\partial x'^a}{\partial x^b} X^b, \tag{5.16}$$

where the transformation matrix is evaluated at P. The infinitesimal vector dx^a is a special case of (5.16) where the components X^a are infinitesimal. An example of a vector with finite components is provided by the **tangent vector** dx^a/du to the curve $x^a = x^a(u)$. It is important to distinguish between the actual geometric object like the tangent vector in Fig. 5.5 (depicted by an arrow) and its representation in a particular coordinate system, like the n numbers $[dx^a/du]_P$ in the x^a-coordinate system and the (in general) different numbers $[dx'^a/du]_P$ in the x'^a-coordinate system.

We now generalize the definition (5.16) to obtain contravariant tensors of higher rank or order. Thus, a **contravariant tensor of rank 2** is a set of n^2 quantities associated with a point P, denoted by X^{ab} in the x^a-coordinate system, which transform according to

$$X'^{ab} = \frac{\partial x'^a}{\partial x^c} \frac{\partial x'^b}{\partial x^d} X^{cd}. \tag{5.17}$$

The quantities X'^{ab} are the components in the x'^a-coordinate system, the transformation matrices are evaluated at P, and the law involves two dummy indices c and d. An example of such a quantity is provided by the product $Y^a Z^b$, say, of two contravariant vectors Y^a and Z^a. The definition of third- and higher-order contravariant tensors proceeds in an analogous manner. An

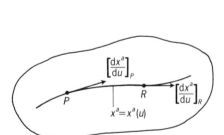

Fig. 5.4 Infinitesimal vector $\overrightarrow{PQ}$ attached to P.

Fig. 5.5 The tangent vector at two points of a curve $x^a = x^a(u)$.

important case is a tensor of zero rank, called a **scalar** or **scalar invariant** ϕ, which transforms according to

$$\phi' = \phi. \tag{5.18}$$

at P.

5.6 Covariant and mixed tensors

As in the last section, we begin by considering the transformation of a prototype quantity. Let

$$\phi = \phi(x^a) \tag{5.19}$$

be a real-valued function on the manifold, i.e. at every point P in the manifold, $\phi(P)$ produces a real number. We also assume that ϕ is continuous and differentiable, so that we can obtain the differential coefficients $\partial\phi/\partial x^a$.

Now, remembering from equation (5.9) that x^a can be thought of as a function of x'^b, equation (5.19) can be written equivalently as

$$\phi = \phi(x^a(x')).$$

Differentiating this with respect to x'^b, using the function of a function rule, we obtain

$$\frac{\partial\phi}{\partial x'^b} = \frac{\partial\phi}{\partial x^a}\frac{\partial x^a}{\partial x'^b}.$$

Then changing the order of the terms, the dummy index, and the free index (from b to a) gives

$$\frac{\partial\phi}{\partial x'^a} = \frac{\partial x^b}{\partial x'^a}\frac{\partial\phi}{\partial x^b}. \tag{5.20}$$

This is the prototype equation we are looking for. Notice that it involves the inverse transformation matrix $\partial x^b/\partial x'^a$. Thus, a **covariant vector** or **covariant tensor of rank (order) 1** is a set of quantities, written X_a in the x^a-coordinate system, associated with a point P, which transforms according to

$$X'_a = \frac{\partial x^b}{\partial x'^a}X_b. \tag{5.21}$$

Again, the transformation matrix occurring is assumed to be evaluated at P. Similarly, we define a covariant tensor of rank 2 by the transformation law

$$X'_{ab} = \frac{\partial x^c}{\partial x'^a}\frac{\partial x^d}{\partial x'^b}X_{cd}, \tag{5.22}$$

and so on for higher-rank tensors. Note the convention that contravariant tensors have raised indices whereas covariant tensors have lowered indices. (The way to remember this is that **co** goes be**low**.) The fact that the differentials dx^a transform as a contravariant vector explains the convention that the coordinates themselves are written as x^a rather than x_a, although note that it is only the differentials and not the coordinates which have tensorial character.

We can go on to define **mixed** tensors in the obvious way. For example, a mixed tensor of rank 3 — one contravariant rank and two covariant rank — satisfies

$$X'^a{}_{bc} = \frac{\partial x'^a}{\partial x^d} \frac{\partial x^e}{\partial x'^b} \frac{\partial x^f}{\partial x'^c} X^d{}_{ef} \tag{5.23}$$

If a mixed tensor has contravariant rank p and covariant rank q, then it is said to have **type** or **valence** (p, q).

We now come to the reason why tensors are important in mathematical physics. Let us illustrate the reason by way of an example. Suppose we find in one coordinate system that two tensors, X_{ab} and Y_{ab} say, are equal, i.e.

$$X_{ab} = Y_{ab}. \tag{5.24}$$

Let us multiply both sides by the matrices $\partial x^a/\partial x'^c$ and $\partial x^b/\partial x'^d$ and take the implied summations to get

$$\frac{\partial x^a}{\partial x'^c} \frac{\partial x^b}{\partial x'^d} X_{ab} = \frac{\partial x^a}{\partial x'^c} \frac{\partial x^b}{\partial x'^d} Y_{ab}.$$

Since X_{ab} and Y_{ab} are both covariant tensors of rank 2 it follows that $X'_{ab} = Y'_{ab}$. In other words, the equation (5.24) holds in **any** other coordinate system. In short, a tensor equation which holds in one coordinate system necessarily holds in **all** coordinate systems. Thus, although we introduce coordinate systems for convenience in tackling particular problems, if we work with tensorial equations then they hold in all coordinate systems. Put another way, tensorial equations are coordinate-independent. This is something that the index-free or coordinate-free approach makes clear from the outset.

5.7 Tensor fields

In vector analysis, a fixed vector is a vector associated with a point, whereas a vector **field** defined over a region is an association of a vector to every point in that region. In exactly the same way, a tensor is a set of quantities defined at one point in the manifold. A **tensor field** defined over some region of the manifold is an association of a tensor of the same valence to every point of the region, i.e.

$$P \to T^{a\cdots}_{b\cdots}(P),$$

where $T^{a\cdots}_{b\cdots}(P)$ is the value of the tensor at P. The tensor field is called continuous or differentiable if its components in all coordinate systems are continuous or differentiable functions of the coordinates. The tensor field is called smooth if its components are differentiable to all orders, which is denoted mathematically by saying that all the components are C^∞. Thus, for example, a contravariant vector field defined over a region is a set of n **functions** defined over that region, and the vector field is **smooth** if the functions are all C^∞. The transformation law for a contravariant vector field then becomes

$$X'^a(x') = \left[\frac{\partial x'^a}{\partial x^b} \right]_P X^b(x) \tag{5.25}$$

at each point P in the region, since the old components X^a are functions of the old x^a-coordinates and the new components X'^a are functions of the new x'^a-coordinates.

As in the case of vectors and vector fields in vector analysis, the distinction between a tensor and a tensor field is not always made completely clear. We shall for the most part be dealing with tensor fields from now on, but to conform with general usage we shall often refer to tensor fields simply as tensors. We will again shorten the transformation law such as (5.25) to the form (5.21) with everything else being implied. If we wish to emphasize that a tensor is a tensor field, we shall write it in functional form, namely, as $T^{a\cdots}_{b\cdots}(x)$.

5.8 Elementary operations with tensors

Tensor calculus is concerned with **tensorial operations**, that is, operations on tensors which result in quantities which are still tensors. A simple way of establishing whether or not a quantity is a tensor is to see how it transforms under a coordinate transformation. For example, we can deduce directly from the transformation law that two tensors of the same type can be added together to give a tensor of the same type, e.g.

$$X^a{}_{bc} = Y^a{}_{bc} + Z^a{}_{bc}. \tag{5.26}$$

The same holds true for subtraction and scalar multiplication.

A covariant tensor of rank 2 is said to be **symmetric** if $X_{ab} = X_{ba}$, in which case it has only $\frac{1}{2}n(n + 1)$ independent components (check this by establishing how many independent components there are of a symmetric matrix of order n). Symmetry is a tensorial property. A similar definition holds for a contravariant tensor X^{ab}. The tensor X_{ab} is said to be **anti-symmetric** or **skew symmetric** if $X_{ab} = -X_{ba}$, which has only $\frac{1}{2}n(n - 1)$ independent components; this is again a tensorial property. A notation frequently used to denote the symmetric part of a tensor is

$$X_{(ab)} = \tfrac{1}{2}(X_{ab} + X_{ba}) \tag{5.27}$$

and the anti-symmetric part is

$$X_{[ab]} = \tfrac{1}{2}(X_{ab} - X_{ba}). \tag{5.28}$$

In general,

$$X_{(a_1 a_2 \cdots a_r)} = \frac{1}{r!} \quad \text{(sum over all permutations of the indices } a_1 \text{ to } a_r)$$

and

$$X_{[a_1 a_1 \cdots a_r]} = \frac{1}{r!} \quad \text{(alternating sum over all permutations of the indices } a_1 \text{ to } a_r)$$

For example, we shall need to make use of the result

$$X_{[abc]} = \tfrac{1}{6}(X_{abc} - X_{acb} + X_{cab} - X_{cba} + X_{bca} - X_{bac}). \tag{5.29}$$

(A way to remember the above expression is to note that the positive terms are obtained by cycling the indices to the right and the corresponding negative terms by flipping the last two indices.) A **totally symmetric tensor** is defined to be one equal to its symmetric part, and a **totally anti-symmetric tensor** is one equal to its anti-symmetric part.

We can multiply two tensors of type (p_1, q_1) and (p_2, q_2) together and obtain a tensor of type $(p_1 + p_2, q_1 + q_2)$, e.g.

$$X^a{}_{bcd} = Y^a{}_b Z_{cd}. \tag{5.30}$$

In particular, a tensor of type (p, q) when multiplied by a scalar field ϕ is again a tensor of type (p, q). Given a tensor of mixed type (p, q), we can form a tensor of type $(p - 1, q - 1)$ by the process of **contraction**, which simply involves setting a raised and lowered index equal. For example,

$$X^a{}_{bcd} \xrightarrow{\text{contraction on } a \text{ and } b} X^a{}_{acd} = Y_{cd},$$

i.e. a tensor of type $(1, 3)$ has become a tensor of type $(0, 2)$. Notice that we can contract a tensor by multiplying by the Kronecker tensor δ^a_b, e.g.

$$X^a{}_{acd} = \delta^b_a X^a{}_{bcd}. \tag{5.31}$$

In effect, multiplying by δ^a_b turns the index b into a (or equivalently the index a into b).

5.9 Index-free interpretation of contravariant vector fields

As we pointed out in §5.5, we must distinguish between the actual geometric object itself and its components in a particular coordinate system. The important point about tensors is that we want to make statements which are independent of any particular coordinate system being used. This is abundantly clear in the index-free approach to tensors. We shall get a feel for this approach in this section by considering the special case of a contravariant vector field, although similar index-free interpretations can be given for any tensor field. The key idea is to interpret the vector field as an **operator** which maps real-valued functions into real-valued functions. Thus, if X represents a contravariant vector field, then X operates on any real-valued function f to produce another function g, i.e. $Xf = g$. We shall show how actually to compute Xf by introducing a coordinate system. However, as we shall see, we could equally well introduce any other coordinate system, and the computation would lead to the same result.

In the x^a-coordinate system, we introduce the notation

$$\partial_a \equiv \frac{\partial}{\partial x^a}$$

and then X is defined as the operator

$$X = X^a \partial_a, \tag{5.32}$$

so that

$$Xf = (X^a \partial_a) f = X^a (\partial_a f) \tag{5.33}$$

for any real-valued function f. Let us compute X in some other x'^a-coordinate system. We need to use the result (5.13) expressed in the following form: we may take x^a to be a function of x'^b by (5.9) and x'^b to be a function of x^c by (5.6), and so, using the function of a function rule, we find

$$\delta^a_b = \frac{\partial x^a}{\partial x^b} = \frac{\partial}{\partial x^b} x^a\big(x'^c(x^d)\big) = \frac{\partial x^a}{\partial x'^c} \frac{\partial x'^c}{\partial x^b}. \qquad (5.34)$$

Then, using the transformation law (5.16) and (5.20) together with the above trick, we get

$$X'^a \partial'_a = X'^a \frac{\partial}{\partial x'^a}$$

$$= \frac{\partial x'^a}{\partial x^b} X^b \frac{\partial x^c}{\partial x'^a} \frac{\partial}{\partial x^c}$$

$$= \frac{\partial x^c}{\partial x'^a} \frac{\partial x'^a}{\partial x^b} X^b \frac{\partial}{\partial x^c}$$

$$= \delta^c_b X^b \frac{\partial}{\partial x^b}$$

$$= X^b \frac{\partial}{\partial x^b}$$

$$= X^a \frac{\partial}{\partial x^a}$$

$$= X^a \partial_a.$$

Thus the result of operating on f by X will be the same **irrespective** of the coordinate system employed in (5.32).

In any coordinate system, we may think of the quantities $[\partial/\partial x_a]_P$ as forming a basis for all the vectors at P, since any vector at P is, by (5.32), given by

$$X_P = [X^a]_P \left[\frac{\partial}{\partial x^a}\right]_P,$$

that is, a linear combination of the $[\partial/\partial x^a]_P$. The vector space of all the contravariant vectors at P is known as the **tangent space** at P and is written $T_P(M)$ (Fig. 5.6). In general, the tangent space at any point in a manifold is

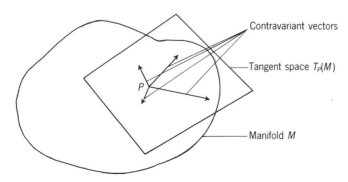

Fig. 5.6 The tangent space at P.

different from the underlying manifold. For this reason, we need to be careful in representing a finite contravariant vector by an arrow in our figures since, strictly speaking, the arrow lies in the tangent space not the manifold. Two exceptions to this are Euclidean space and Minkowski space-time, where the tangent space at each point coincides with the manifold.

Given two vector fields X and Y we can define a new vector field called the **commutator** or **Lie bracket** of X and Y by

$$[X, Y] = XY - YX. \tag{5.35}$$

Letting $[X, Y] = Z$ and operating with it on some arbitrary function f

$$
\begin{aligned}
Zf &= [X, Y]f \\
&= (XY - YX)f \\
&= X(Yf) - Y(Xf) \\
&= X(Y^a \partial_a f) - Y(X^a \partial_a f) \\
&= X^b \partial_b (Y^a \partial_a f) - Y^b \partial_b (X^a \partial_a f) \\
&= (X^b \partial_b Y^a - Y^b \partial_b X^a) \partial_a f - X^a Y^b (\partial_b \partial_a f - \partial_a \partial_b f).
\end{aligned}
$$

The least term vanishes since we assume commutativity of second mixed partial derivatives, i.e.

$$\partial_a \partial_b = \frac{\partial^2}{\partial x^a \partial x^b} = \frac{\partial^2}{\partial x^b \partial x^a} = \partial_b \partial_a.$$

Since f is arbitrary, we obtain the result

$$[X, Y]^a = Z^a = X^b \partial_b Y^a - Y^b \partial_b X^a \tag{5.36}$$

from which it clearly follows that the commutator of two vector fields is itself a vector field. It also follows, directly from the definition (5.35), that

$$[X, X] \equiv 0 \tag{5.37}$$

$$[X, Y] \equiv -[Y, X] \tag{5.38}$$

$$[X, [Y, Z]] + [Z, [X, Y]] + [Y, [Z, X]] \equiv 0. \tag{5.39}$$

Equation (5.38) shows that the Lie bracket is anti-commutative. The result (5.39) is known as **Jacobi's identity**. Notice it states that the left-hand side is not just equal to zero but is **identically** zero. What does this mean? The equation $x^2 - 4 = 0$ is only satisfied by particular values of x, namely, $+2$ and -2. The identity $x^2 - x^2 \equiv 0$ is satisfied for all values of x. But, you may argue, the x^2 terms cancel out, and this is precisely the point. An expression is identically zero if, when all the terms are written out fully, they all cancel in pairs.

Exercises

5.1 (§5.3) In Euclidean 3-space $\mathbb{R}^3$:

   (i) Write down the equation of a circle of radius a lying in the (x, y)-plane centred at the origin in (a) parametric form (b) constraint form.

   (ii) Write down the equation of a hypersurface consisting of a sphere of radius a centred at the origin in (a) parametric form (b) constraint form. Eliminate the parameters in form (a) to obtain form (b).

5.2 (§5.4) Write down the change of coordinates from Cartesian coordinates $(x^a) = (x, y, z)$ to spherical polar coordinates $(x'^a) = (r, \theta, \phi)$ in $\mathbb{R}^3$. Obtain the transformation matrices $[\partial x^a/\partial x'^b]$ and $[\partial x'^a/\partial x^b]$ expressing them both in terms of the primed coordinates. Obtain the Jacobians J and J'. Where is J' zero or infinite?

5.3 (§5.4) Show by manipulating the dummy indices that

$$(Z_{abc} + Z_{cab} + Z_{bca}) X^a X^b X^c = 3 Z_{abc} X^a X^b X^c.$$

5.4 (§5.4) Show that

   (i) $\delta_a^b X^a = X^b$,

   (ii) $\delta_a^b X_b = X_a$,

   (iii) $\delta_a^b \delta_b^c \delta_c^d = \delta_a^d$.

5.5 (§5.5) If Y^a and Z^a are contravariant vectors, then show that $Y^a Z^b$ is a contravariant tensor of rank 2.

5.6 (§5.5) Write down the change of coordinates from Cartesian coordinates $(x^a) = (x, y)$ to plane polar coordinates $(x'^a) = (R, \phi)$ in $\mathbb{R}^2$ and obtain the transformation matrix $[\partial x'^a/\partial x^b]$ expressed as a function of the primed coordinates. Find the components of the tangent vector to the curve consisting of a circle of radius a centred at the origin with the standard parametrization (see Exercise 5.1 (i)) and use (5.16) to find its components in the primed coordinate system.

5.7 (§5.6) Write down the definition of a tensor of type $(2, 1)$.

5.8 (§5.6) Prove that δ_a^b has the tensor character indicated. Prove also that δ_b^a is a **constant** or **numerical** tensor, that is, it has the same components in all coordinate systems.

5.9 (§5.6) Show, by differentiating (5.20) with respect to x'^c, that $\partial^2 \phi/\partial x^a \partial x^b$ is not a tensor.

5.10 (§5.8) Show that if $Y^a{}_{bc}$ and $Z^a{}_{bc}$ are tensors of the type indicated then so is their sum and difference.

5.11 (§5.8) (i) Show that the fact that a covariant second rank tensor is symmetric in one coordinate system is a tensorial property.

   (ii) If X^{ab} is anti-symmetric and Y_{ab} is symmetric then prove that $X^{ab} Y_{ab} = 0$.

5.12 (§5.8) Prove that any covariant (or contravariant) tensor of rank 2 can be written as the sum of a symmetric and an anti-symmetric tensor. [Hint: consider the identity $X_{ab} = \frac{1}{2}(X_{ab} + X_{ba}) + \frac{1}{2}(X_{ab} - X_{ba})$.]

5.13 (§5.8) If $X^a{}_{bc}$ is a tensor of the type indicated, then prove that the contracted quantity $Y_c = X^a{}_{ac}$ is a covariant vector.

5.14 (§5.8) Evaluate δ_a^a and $\delta_b^a \delta_a^b$ in n dimensions.

5.15 (§5.9) Check that the definition of the Lie bracket leads to the results (5.37), (5.38), and (5.39).

5.16 (§5.9) In $\mathbb{R}^2$, let $(x^a) = (x, y)$ denote Cartesian and $(x'^a) = (R, \phi)$ plane polar coordinates (see Exercise 5.6).

   (i) If the vector field X has components $X^a = (1, 0)$, then find X'^a.

   (ii) The operator grad can be written in each coordinate system as

$$\text{grad } f = \frac{\partial f}{\partial x} i + \frac{\partial f}{\partial y} j = \frac{\partial f}{\partial R} \hat{R} + \frac{\partial f}{\partial \phi} \frac{\hat{\phi}}{R},$$

   where f is an arbitrary function and

$$\hat{R} = \cos \phi\, i + \sin \phi\, j, \quad \hat{\phi} = -\sin \phi\, i + \cos \phi\, j.$$

   Take the scalar product of grad f with $i, j, \hat{R}$, and $\hat{\phi}$ in turn to find relationships between the operators $\partial/\partial x$, $\partial/\partial y$, $\partial/\partial R$, and $\partial/\partial \phi$.

   (iii) Express the vector field X as an operator in each coordinate system. Use part (ii) to show that these expressions are the same.

   (iv) If $Y^a = (0, 1)$ and $Z^a = (-y, x)$, then find Y'^a, Z'^a, Y, and Z.

   (v) Evaluate all the Lie brackets of X, Y, and Z.

6

Tensor calculus

6.1 Partial derivative of a tensor

In the last chapter, we met algebraic operations which are tensorial, that is, which convert tensors into tensors. The operations are addition, subtraction, multiplication, and contraction. The next question which arises is, What differential operations are there that are tensorial? The answer to this turns out to be very much more involved. The first thing we shall see is that partial differentiation of tensors is **not** tensorial. Different authors denote the partial derivative of a contravariant vector X^a by

$$\partial_b X^a \quad \text{or} \quad \frac{\partial X^a}{\partial x^b} \quad \text{or} \quad X^a{}_{,b} \quad \text{or} \quad X^a{}_{|b}$$

and similarly for higher-rank tensors. We shall use a mixture of all the first three notations. (Note that in the literature, the partial derivative of a tensor is often referred to as the **ordinary** derivative of a tensor, to distinguish it from the tensorial differentiation we shall shortly meet). Now differentiating (5.16) with respect to x'^c, we find

$$
\begin{aligned}
\partial'_c X'^a &= \frac{\partial}{\partial x'^c}\left(\frac{\partial x'^a}{\partial x^b} X^b\right) \\
&= \frac{\partial x^d}{\partial x'^c}\frac{\partial}{\partial x^d}\left(\frac{\partial x'^a}{\partial x^b} X^b\right) \\
&= \frac{\partial x'^a}{\partial x^b}\frac{\partial x^d}{\partial x'^c}\partial_d X^b + \frac{\partial^2 x'^a}{\partial x^b \partial x^d}\frac{\partial x^d}{\partial x'^c} X^b.
\end{aligned}
\tag{6.1}
$$

If the first term on the right-hand side alone were present, then this would be the usual tensor transformation law for a tensor of type $(1, 1)$. However, the presence of the second term prevents $\partial_b X^a$ from behaving like a tensor.

There is a fundamental reason why this is the case. By definition, the process of differentiation involves comparing a quantity evaluated at two neighbouring points, P and Q say, dividing by some parameter representing the separation of P and Q and then taking the limit as this parameter goes to zero. In the case of a contravariant vector field X^a, this would involve computing

$$\lim_{\delta u \to 0} \frac{[X^a]_P - [X^a]_Q}{\delta u}$$

for some appropriate parameter δu. However, from the transformation law in the form (5.25), we see that

$$X'^a_P = \left[\frac{\partial x'^a}{\partial x^b}\right]_P X^b_P \quad \text{and} \quad X'^a_Q = \left[\frac{\partial x'^a}{\partial x^b}\right]_Q X^b_Q.$$

This involves the transformation matrix evaluated at **different** points, from which it should be clear that $X^a_P - X^a_Q$ is not a tensor. Similar remarks hold for differentiating tensors in general.

It turns out that if we wish to differentiate a tensor in a tensorial manner then we need to introduce some auxiliary field onto the manifold. We shall meet three different types of differentiation. First of all, in the next section, we shall introduce a **contravariant vector field** onto the manifold and use it to define the **Lie derivative**. Then we shall introduce a quantity called an **affine connection** and use it to define **covariant differentiation**. Finally, we shall introduce a tensor called a **metric** and from it build a special affine connection, called the **metric connection**, and again define **covariant differentiation** but relative to this specific connection.

6.2 The Lie derivative

The argument we present in this section is rather intricate. It rests on the idea of interpreting a coordinate transformation **actively** as a point transformation, rather than **passively** as we have done up to now. The important results occur at the end of the section and consist of the formula for the Lie derivative of a general tensor field and the basic properties of Lie differentiation.

We start by considering a **congruence of curves** defined such that only one curve goes through each point in the manifold. Then, given any one curve of the congruence,

$$x^a = x^a(u),$$

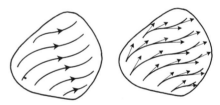

Fig. 6.1 The tangent vector field resulting from a congruence of curves.

we can use it to define the tangent vector field $\mathrm{d}x^a/\mathrm{d}u$ along the curve. If we do this for every curve in the congruence, then we end up with a vector field X^a (given by $\mathrm{d}x^a/\mathrm{d}u$ at each point) defined over the whole manifold (Fig. 6.1).

Conversely, given a non-zero vector field $X^a(x)$ defined over the manifold, then this can be used to define a congruence of curves in the manifold called the **orbits** or **trajectories** of X^a. The procedure is exactly the same as the way in which a vector field gives rise to field lines or streamlines in vector analysis. These curves are obtained by solving the ordinary differential equations

$$\frac{\mathrm{d}x^a}{\mathrm{d}u} = X^a\big(x(u)\big). \tag{6.2}$$

The existence and uniqueness theorem for ordinary differential equations guarantees a solution, at least for some subset of the reals. In what follows, we are really only interested in what happens locally (Fig. 6.2).

We therefore assume that X^a has been given and we have constructed the local congruence of curves. Suppose we have some tensor field $T^a_b\!\cdots\!(x)$ which we wish to differentiate using X^a. Then the essential idea is to use the congruence of curves to **drag** the tensor at some point P (i.e. $T^a_b\!\cdots\!(P)$) along the curve passing through P to some neighbouring point Q, and then compare this 'dragged-along tensor' with the tensor already there (i.e. $T^a_b\!\cdots\!(Q)$) (Fig. 6.3). Since the dragged-along tensor will be of the same type as

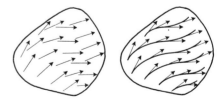

Fig. 6.2 The local congruence of curves resulting from a vector field.

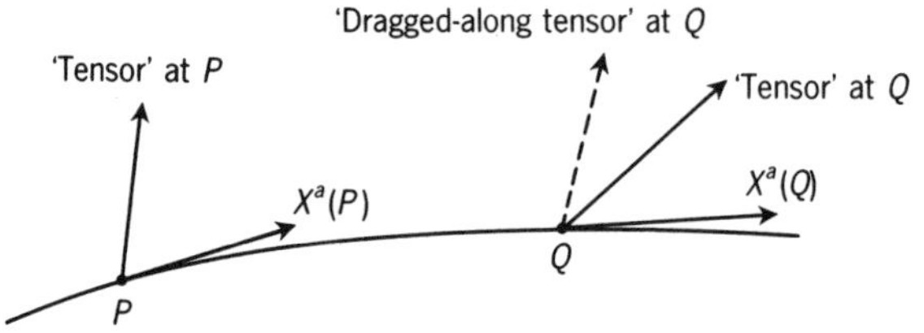

'Dragged-along tensor' at Q

'Tensor' at P

'Tensor' at Q

$X^a(P)$

$X^a(Q)$

Q

P

Fig. 6.3 Using the congruence to compare tensors at neighbouring points.

the tensor already at Q, we can **subtract the two tensors at Q** and so define a derivative by some limiting process as Q tends to P. The technique for dragging involves viewing the coordinate transformation from P to Q **actively**, and applying it to the usual transformation law for tensors. We shall consider the detailed calculation in the case of a contravariant tensor field of rank 2, $T^{ab}(x)$ say.

Consider the transformation

$$x'^a = x^a + \delta u \, X^a(x), \tag{6.3}$$

where δu is small. This is called a **point transformation** and is to be regarded actively as sending the point P, with coordinates x^a, to the point Q, with coordinates $x^a + \delta u \, X^a(x)$, where the coordinates of each point are given in the **same** x^a-coordinate system, i.e.

$$P \to Q$$
$$x^a \to x^a + \delta u \, X^a(x).$$

The point Q clearly lies on the curve of the congruence through P which X^a generates (Fig. 6.4). Differentiating (6.3), we get

$$\frac{\partial x'^a}{\partial x^b} = \delta^a_b + \delta u \, \partial_b X^a. \tag{6.4}$$

Next, consider the tensor field T^{ab} at the point P. Then its components at P are $T^{ab}(x)$ and, under the point transformation (6.3), we have the mapping

$$T^{ab}(x) \to T'^{ab}(x'),$$

i.e. the transformation 'drags' the tensor T^{ab} along from P to Q. The components of the dragged-along tensor are given by the usual transformation law for tensors (see (5.25)), and so, using (6.4),

$$T'^{ab}(x') = \frac{\partial x'^a}{\partial x^c} \frac{\partial x'^b}{\partial x^d} T^{cd}(x)$$

$$= (\delta^a_c + \delta u \, \partial_c X^a)(\delta^b_d + \delta u \, \partial_d X^b) T^{cd}(x)$$

$$= T^{ab}(x) + [\partial_c X^a T^{cb}(x) + \partial_d X^b T^{ad}(x)]\delta u + O(\delta u^2). \tag{6.5}$$

Applying Taylor's theorem to first order, we get

$$T^{ab}(x') = T^{ab}(x^c + \delta u \, X^c(x)) = T^{ab}(x) + \delta u \, X^c \partial_c T^{ab}. \tag{6.6}$$

We are now in a position to define the Lie derivative of T^{ab} with respect to

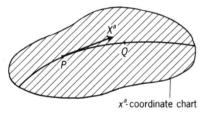

x^a

Q

P

x^a-coordinate chart

Fig. 6.4 The point P transformed to Q in the same x^a-coordinate system.

X^a, which is denoted by $L_X T^{ab}$, as

$$L_X T^{ab} = \lim_{\delta u \to 0} \frac{T^{ab}(x') - T'^{ab}(x')}{\delta u}. \tag{6.7}$$

This involves comparing the tensor $T^{ab}(x')$ already at Q with $T'^{ab}(x')$, the dragged-along tensor at Q. Using (6.5) and (6.6), we find

$$L_X T^{ab} = X^c \partial_c T^{ab} - T^{ac} \partial_c X^b - T^{cb} \partial_c X^a. \tag{6.8}$$

It can be shown that it is always possible to introduce a coordinate system such that the curve passing through P is given by x^1 varying, with $x^2, x^3 , \ldots, x^n$ all constant along the curve, and such that

$$X^a \overset{*}{=} \delta_1^a = (1, 0, 0, \ldots, 0) \tag{6.9}$$

along this curve. The notation $\overset{*}{=}$ used in (6.9) means that the equation holds only in a particular coordinate system. Then it follows that

$$X = X^a \partial_a \overset{*}{=} \partial_1,$$

and equation (6.8) reduces to

$$L_X T^{ab} \overset{*}{=} \partial_1 T^{ab}. \tag{6.10}$$

Thus, in this special coordinate system, Lie differentiation reduces to ordinary differentiation. In fact, one can define Lie differentiation starting from this viewpoint.

We end the section by collecting together some important properties of Lie differentiation with respect to X which follow from its definition.

1. It is **linear**; for example

$$L_X(\lambda Y^a + \mu Z^a) = \lambda L_X Y^a + \mu L_X Z^a, \tag{6.11}$$

where λ and μ are constants. Thus, in particular, the Lie derivative of the sum and difference of two tensors is the sum and difference, respectively, of the Lie derivatives of the two tensors.

2. It is **Leibniz**; that is, it satisfies the usual product rule for differentiation, for example

$$L_X(Y^a Z_{bc}) = Y^a(L_X Z_{bc}) + (L_X Y^a)Z_{bc}. \tag{6.12}$$

3. It is **type-preserving**; that is, the Lie derivative of a tensor of type (p, q) is again a tensor of type (p, q).
4. It commutes with contraction; for example

$$\delta_b^a L_X T^a{}_b = L_X T^a{}_a. \tag{6.13}$$

5. The Lie derivative of a **scalar field** ϕ is given by

$$\mathrm{L}_X\phi = X\phi = X^a\partial_a\phi \qquad (6.14)$$

6. The Lie derivative of a **contravariant vector field** Y^a is given by the Lie bracket of X and Y, that is,

$$\mathrm{L}_X Y^a = [X, Y]^a = X^b\partial_b Y^a - Y^b\partial_b X^a. \qquad (6.15)$$

7. The Lie derivative of a **covariant vector field** Y_a is given by

$$\mathrm{L}_X Y_a = X^b\partial_b Y_a + Y_b\partial_a X^b. \qquad (6.16)$$

8. The Lie derivative of a **general tensor field** $T^a_{b\cdots}$ is obtained as follows: we first partially differentiate the tensor and contract it with X. We then get an additional term for each index of the form of the last two terms in (6.15) and (6.16), where the corresponding sign is negative for a contravariant index and positive for a covariant index, that is,

$$\mathrm{L}_X T^a_{b\cdots} = X^c\partial_c T^a_{b\cdots} - T^c_{b\cdots}\partial_c X^a - \cdots + T^a_{c\cdots}\partial_b X^c + \cdots. \qquad (6.17)$$

6.3 The affine connection and covariant differentiation

Consider a contravariant vector field $X^a(x)$ evaluated at a point Q, with coordinates $x^a + \delta x^a$, near to a point P, with coordinates x^a. Then, by Taylor's theorem,

$$X^a(x + \delta x) = X^a(x) + \delta x^b \partial_b X^a \qquad (6.18)$$

to first order. If we denote the second term by $\delta X^a(x)$, i.e.

$$\delta X^a(x) = \delta x^b \partial_b X^a = X^a(x + \delta x) - X^a(x), \qquad (6.19)$$

then it is not tensorial since it involves subtracting tensors evaluated at two different points. We are going to define a tensorial derivative by introducing a vector at Q which in some general sense is 'parallel' to X^a at P. Since $x^a + \delta x^a$ is close to x^a, we can assume that the parallel vector only differs from $X^a(x)$ by a small amount, which we denote $\bar{\delta}X^a(x)$ (Fig. 6.5). By the same argument as in §6.1 above, $\bar{\delta}X^a(x)$ is not tensorial, but we shall construct it in such a way as to make the **difference** vector

$$X^a(x) + \delta X^a(x) - [X^a(x) + \bar{\delta}X^a(x)] = \delta X^a(x) - \bar{\delta}X^a(x) \qquad (6.20)$$

tensorial. It is natural to require that $\bar{\delta}X^a(x)$ should vanish whenever $X^a(x)$ or δx^a does. Then the simplest definition is to assume that $\bar{\delta}X^a$ is linear in both X^a and δx^a, which means that there exist multiplicative factors Γ^a_{bc}

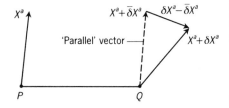

Fig. 6.5 The parallel vector $X^a + \delta X^a$ at Q.

where

$$\bar{\delta}X^a(x) = -\Gamma^a_{bc}(x)X^b(x)\delta x^c \tag{6.21}$$

and the minus sign is introduced to agree with convention.

We have therefore introduced a set of n^3 functions $\Gamma^a_{bc}(x)$ on the manifold, whose transformation properties have yet to be determined. This we do by defining the **covariant derivative** of X^a, written in one of the notations (where we shall use a mixture of the first two)

$$\nabla_c X^a \quad \text{or} \quad X^a{}_{;c} \quad \text{or} \quad X^a{}_{\|c},$$

by the limiting process

$$\nabla_c X^a = \lim_{\delta x^c \to 0} \frac{1}{\delta x^c} \{ X^a(x + \delta x) - [X^a(x) + \bar{\delta}X^a(x)] \}.$$

In other words, it is the difference between the vector $X^a(Q)$ and the vector at Q parallel to $X^a(P)$, divided by the coordinate differences, in the limit as these differences tend to zero. Using (6.18) and (6.21), we find

$$\nabla_c X^a = \partial_c X^a + \Gamma^a_{bc} X^b. \tag{6.22}$$

Note that in the formula the differentiation index c comes second in the downstairs indices of Γ. If we now demand that $\nabla_c X^a$ is a **tensor** of type $(1, 1)$, then a straightforward calculation (exercise) reveals that Γ^a_{bc} must transform according to

$$\Gamma'^a_{bc} = \frac{\partial x'^a}{\partial x^d} \frac{\partial x^e}{\partial x'^b} \frac{\partial x^f}{\partial x'^c} \Gamma^d_{ef} - \frac{\partial x^d}{\partial x'^b} \frac{\partial x^e}{\partial x'^c} \frac{\partial^2 x'^a}{\partial x^d \partial x^e}, \tag{6.23}$$

or equivalently (exercise)

$$\Gamma'^a_{bc} = \frac{\partial x'^a}{\partial x^d} \frac{\partial x^e}{\partial x'^b} \frac{\partial x^f}{\partial x'^c} \Gamma^d_{ef} + \frac{\partial x'^a}{\partial x^d} \frac{\partial^2 x^d}{\partial x'^b \partial x'^c}. \tag{6.24}$$

If the second term on the right-hand side were absent, then this would be the usual transformation law for a tensor of type $(1, 2)$. However, the presence of the second term reveals that the transformation law is linear **inhomogeneous**, and so Γ^a_{bc} is not a tensor. Any quantity Γ^a_{bc} which transforms according to (6.23) or (6.24) is called an **affine connection** or sometimes simply a **connection** or **affinity**. A manifold with a continuous connection prescribed on it is called an **affine manifold**. From another point of view, the existence of the inhomogeneous term in the transformation law is not surprising if we are to define a tensorial derivative, since its role is to compensate for the second term which occurs in (6.1).

We next define the covariant derivative of a scalar field to be the same as its ordinary derivative, i.e.

$$\nabla_a \phi = \partial_a \phi. \tag{6.25}$$

If we now demand that covariant differentiation satisfies the Leibniz rule, then we find

$$\nabla_c X_a = \partial_c X_a - \Gamma^b_{ac} X_b. \tag{6.26}$$

Notice again that the differentiation index comes last in the Γ-term and that this term enters with a minus sign. The name covariant derivative stems from the fact that the derivative of a tensor of type (p, q) is of type $(p, q + 1)$, i.e. it has one extra covariant rank. The expression in the case of a general tensor is (compare and contrast with (6.17))

$$\nabla_c T^{a\cdots}_{b\cdots} = \partial_c T^{a\cdots}_{b\cdots} + \Gamma^a_{dc} T^{d\cdots}_{b\cdots} + \cdots - \Gamma^d_{bc} T^{a\cdots}_{d\cdots} - \cdots. \tag{6.27}$$

It follows directly from the transformation laws that the sum of two connections is not a connection or a tensor. However, the **difference** of two connections is a tensor of valence $(1, 2)$, because the inhomogeneous term cancels out in the transformation. For the same reason, the anti-symmetric part of a Γ^a_{bc}, namely,

$$T^a_{bc} = \Gamma^a_{bc} - \Gamma^a_{cb}$$

is a tensor called the **torsion tensor**. If the torsion tensor vanishes, then the connection is **symmetric**, i.e.

$$\Gamma^a_{bc} = \Gamma^a_{cb}. \tag{6.28}$$

From now on, unless we state otherwise, we shall **restrict ourselves to symmetric connections**, in which case the torsion vanishes. The assumption that the connection is symmetric leads to the following useful result. In the expression for a Lie derivative of a tensor, **all** occurrences of the partial derivatives may be replaced by covariant derivatives. For example, in the case of a vector (exercise)

$$L_X Y^a = X^b \partial_b Y^a - Y^b \partial_b X^a = X^b \nabla_b Y^a - Y^b \nabla_b X^a. \tag{6.29}$$

6.4 Affine geodesics

If $T^{a\cdots}_{b\cdots}$ is any tensor, then we introduce the notation

$$\nabla_X T^{a\cdots}_{b\cdots} = X^c \nabla_c T^{a\cdots}_{b\cdots}, \tag{6.30}$$

that is, ∇_X of a tensor is its covariant derivative contracted with X. Now in §6.2 we saw that a contravariant vector field X determines a local congruence of curves,

$$x^a = x^a(u),$$

where the tangent vector field to the congruence is

$$\frac{dx^a}{du} = X^a.$$

We next define the **absolute derivative** of a tensor $T^{a\cdots}_{b\cdots}$ along a curve C of the congruence, written $DT^{a\cdots}_{b\cdots}/Du$, by

$$\frac{D}{Du}(T^{a\cdots}_{b\cdots}) = \nabla_X T^{a\cdots}_{b\cdots}. \qquad (6.31)$$

The tensor $T^{a\cdots}_{b\cdots}$ is said to be **parallely propagated** or **transported** along the curve C if

$$\frac{D}{Du}(T^{a\cdots}_{b\cdots}) = 0. \qquad (6.32)$$

This is a first-order ordinary differential equation for $T^{a\cdots}_{b\cdots}$, and so given an initial value for $T^{a\cdots}_{b\cdots}$, say $T^{a\cdots}_{b\cdots}(P)$, equation (6.32) determines a tensor along C which is everywhere parallel to $T^{a\cdots}_{b\cdots}(P)$.

Using this notation, an **affine geodesic** is defined as a privileged curve along which the tangent vector is propagated parallel to itself. In other words, the parallely propagated vector at any point of the curve is parallel, that is, proportional, to the tangent vector at that point:

$$\frac{D}{Du}\left(\frac{dx^a}{du}\right) = \lambda(u)\frac{dx^a}{du}.$$

Using (6.31), the equation for an affine geodesic can be written in the form

$$\nabla_X X^a = \lambda X^a, \qquad (6.33)$$

or equivalently (exercise)

$$\frac{d^2 x^a}{du^2} + \Gamma^a_{bc}\frac{dx^b}{du}\frac{dx^c}{du} = \lambda\frac{dx^a}{du}. \qquad (6.34)$$

The last result is very important and so we shall establish it afresh from first principles using the notation of the last section. Let the neighbouring points P and Q on C be given by $x^a(u)$ and

$$x^a(u + \delta u) = x^a(u) + \frac{dx^a}{du}\delta u$$

to first order in δu, respectively. Then in the notation of the last section

$$\delta x^a = \frac{dx^a}{du}\delta u. \qquad (6.35)$$

The vector $X^a(x)$ at P is now the tangent vector $(dx^a/du)(u)$. The vector at Q parallel to dx^a/du is, by (6.21) and (6.35),

$$\frac{dx^a}{du} - \Gamma^a_{bc} \frac{dx^b}{du} \frac{dx^c}{du} \delta u.$$

The vector already at Q is

$$\frac{dx^a}{du}(u + \delta u) = \frac{dx^a}{du} + \frac{d^2 x^a}{du^2} \delta u$$

to first order in δu. These last two vectors must be parallel, so we require

$$\frac{dx^a}{du} + \frac{d^2 x^a}{du^2} \delta u = [1 + \lambda(u)\delta u]\left(\frac{dx^a}{du} - \Gamma^a_{bc} \frac{dx^b}{du} \frac{dx^c}{du} \delta u\right),$$

where we have written the proportionality factor as $1 + \lambda(u)\delta u$ without loss of generality, since the equation must hold in the limit $\delta u \rightarrow 0$. Subtracting dx^a/du from each side, dividing by δu and taking the limit as δu tends to zero produces the result (6.34). Note that Γ^a_{bc} appears in the equation multiplied by the symmetric quantity $(dx^b/du)(dx^c/du)$, and so even if we had not assumed that Γ^a_{bc} was symmetric the equation picks out its symmetric part only.

If the curve is parametrized in such a way that λ vanishes (that is, by the above, so that the tangent vector is transported into itself), then the parameter is a privileged parameter called an **affine parameter**, often conventionally denoted by s, and the affine geodesic equation reduces to

$$\nabla_X X^a = 0, \tag{6.36}$$

or equivalently

$$\frac{d^2 x^a}{ds^2} + \Gamma^a_{bc} \frac{dx^b}{ds} \frac{dx^c}{ds} = 0. \tag{6.37}$$

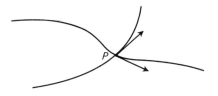

Fig. 6.6 Two affine geodesics passing through P with given directions.

An affine parameter s is only defined up to an **affine transformation** (exercise)

$$s \rightarrow \alpha s + \beta,$$

Fig. 6.7 Two affine geodesics from P refocusing at Q.

where α and β are constants. We can use the affine parameter s to define the **affine length** of the geodesic between two points P_1 and P_2 by $\int_{P_1}^{P_2} ds$, and so we can compare lengths on the **same geodesic**. However, we cannot compare lengths on different geodesics (without a metric) because of the arbitrariness in the parameter s. From the existence and uniqueness theorem for ordinary differential equations, it follows that corresponding to every direction at a point there is a unique geodesic passing through the point (Fig. 6.6). Similarly, any point can be joined to any other point, as long as the points are sufficiently 'close', by a unique geodesic. However, in the large, geodesics may **focus**, that is, meet again (Fig. 6.7).

6.5 The Riemann tensor

Covariant differentiation, unlike partial differentiation, is not in general commutative. For any tensor $T^{a\cdots}_{b\cdots}$, we define its **commutator** to be

$$\nabla_c \nabla_d T^{a\cdots}_{b\cdots} - \nabla_d \nabla_c T^{a\cdots}_{b\cdots}.$$

Let us work out the commutator in the case of a vector X^a. From (6.22), we see that

$$\nabla_c X^a = \partial_c X^a + \Gamma^a_{bc} X^b.$$

Remembering that this is a tensor of type $(1, 1)$ and using (6.27), we find

$$\nabla_d \nabla_c X^a = \partial_d(\partial_c X^a + \Gamma^a_{bc} X^b) + \Gamma^a_{ed}(\partial_c X^e + \Gamma^e_{bc} X^b) - \Gamma^e_{cd}(\partial_e X^a + \Gamma^a_{be} X^b),$$

with a similar expression for $\nabla_c \nabla_d X^a$, namely,

$$\nabla_c \nabla_d X^a = \partial_c(\partial_d X^a + \Gamma^a_{bd} X^b) + \Gamma^a_{ec}(\partial_d X^e + \Gamma^e_{bd} X^b) - \Gamma^e_{dc}(\partial_e X^a + \Gamma^a_{be} X^b).$$

Subtracting these last two equations and assuming that

$$\partial_d \partial_c X^a = \partial_c \partial_d X^a,$$

we obtain the result

$$\nabla_c \nabla_d X^a - \nabla_d \nabla_c X^a = R^a{}_{bcd} X^b + (\Gamma^e_{cd} - \Gamma^e_{dc}) \nabla_e X^a, \tag{6.38}$$

where $R^a{}_{bcd}$ is defined by

$$R^a{}_{bcd} = \partial_c \Gamma^a_{bd} - \partial_d \Gamma^a_{bc} + \Gamma^e_{bd} \Gamma^a_{ec} - \Gamma^e_{bc} \Gamma^a_{ed}. \tag{6.39}$$

Moreover, since we are only interested in torsion-free connections, the last term in (6.38) vanishes, namely, using (5.28),

$$\nabla_{[c} \nabla_{d]} X^a = \tfrac{1}{2} R^a{}_{bcd} X^b. \tag{6.40}$$

Since the left-hand side of (6.40) is a tensor, it follows that $R^a{}_{bcd}$ is a tensor of type $(1, 3)$. It is called the **Riemann tensor**. It can be shown that, for a symmetric connection, the commutator of any tensor can be expressed in terms of the tensor itself and the Riemann tensor. Thus, the vanishing of the Riemann tensor is a necessary and sufficient condition for the vanishing of the commutator of any tensor. In the section after next, we shall search for a geometrical characterization of the vanishing of the Riemann tensor.

6.6 Geodesic coordinates

We first prove a very useful result. At any point P in a manifold, we can introduce a special coordinate system, called a **geodesic coordinate system**, in which

$$[\Gamma^a_{bc}]_P \overset{*}{=} 0.$$

We can, without loss of generality, choose P to be at the origin of coordinates $x^a \overset{*}{=} 0$ and consider a transformation to a new coordinate system

$$x^a \to x'^a = x^a + \tfrac{1}{2} Q^a_{bc} x^b x^c, \tag{6.41}$$

where $Q^a_{bc} = Q^a_{cb}$ are constants to be determined. Differentiating (6.41), we get

$$\frac{\partial x'^a}{\partial x^d} = \delta^a_d + Q^a_{bd}x^d \quad \text{and} \quad \frac{\partial^2 x'^a}{\partial x^d \partial x^e} = Q^a_{de}.$$

Then, since x^a vanishes at P, we have

$$\left[\frac{\partial x'^a}{\partial x^b}\right]_P = \delta^a_b,$$

from which it follows immediately that the inverse matrix

$$\left[\frac{\partial x^a}{\partial x'^b}\right]_P = \delta^a_b.$$

Substituting these results in (6.23), we find

$$[\Gamma'^a_{bc}]_P = [\Gamma^a_{bc}]_P - Q^a_{bc}.$$

Since the connection is symmetric, we can choose the constants so that

$$Q^a_{bc} = [\Gamma^a_{bc}]_P,$$

and hence we obtain the promised result

$$[\Gamma'^a_{bc}]_P \overset{*}{=} 0. \tag{6.42}$$

Many tensorial equations can be established most easily in geodesic co-ordinates. Note that, although the connection vanishes at P,

$$[\Gamma'^a_{bc,d}]_P \overset{*}{\neq} 0$$

in general. It can be shown that the result can be extended to obtain a coordinate system in which the connection vanishes along a curve, but not in general over the whole manifold. If, however, there exists a special coordinate system in which the connection vanishes everywhere, then the manifold is called **affine flat** or simply **flat**. We shall next see that this is intimately connected with the vanishing of the Riemann tensor.

6.7 Affine flatness

In a general affine manifold, the intuitive concept of parallelism breaks down. For if we parallely transport a vector from one point to another along two different curves we will obtain two different vectors (Fig. 6.8). If, however, we can transport a vector from one point to any other and the resulting vector is independent of the path taken, then the connection is called **integrable**. Thus, for the usual concept of parallelism to hold, the manifold must possess an integrable connection. We now consider two lemmas which connect together the concepts of affine flatness, integrability, and vanishing Riemann tensor.

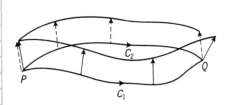

Fig. 6.8 Parallel transport round two curves in a general affine manifold.

Lemma: A necessary and sufficient condition for a connection to be integrable is that the Riemann tensor vanishes.

We consider, first, necessity. Since Γ^a_{bc} is integrable, we can start with a vector X^a at any point and from it construct a unique vector field $X^a(x)$ over the manifold by parallely propagating X^a. The equation for parallely propagating X^a is

$$\frac{DX^a}{Du} = \frac{dx^c}{du} \nabla_c X^a = 0,$$

and, since dx^c/du is arbitrary, it follows that the covariant derivative of X^a vanishes, i.e.

$$\nabla_c X^a = \partial_c X^a + \Gamma^a_{bc} X^b = 0. \tag{6.43}$$

Hence, this equation must possess solutions. A necessary condition for a solution of this first-order partial differential equation is

$$\partial_d \partial_c X^a = \partial_c \partial_d X^a. \tag{6.44}$$

In the previous section, we met the identity for the commutator of a vector field (6.38), namely

$$\nabla_c \nabla_d X^a - \nabla_d \nabla_c X^a = \partial_c \partial_d X^a - \partial_d \partial_c X^a + R^a{}_{bcd} X^b.$$

The left-hand side of this equation vanishes by construction, that is, by (6.43); hence it follows that (6.44) will hold if and only if

$$R^a{}_{bcd} X^b = 0.$$

Finally, since X^b is arbitrary at every point, a necessary condition for integrability is $R^a{}_{bcd} = 0$ everywhere.

We next prove sufficiency. We start by considering the difference in parallely propagating a vector X^a around an infinitesimal loop connecting x^a to $x^a + \delta x^a + dx^a$, first via $x^a + \delta x^a$ and then via $x^a + dx^a$ (Fig. 6.9). From §6.3, if we parallely transport X^a from x^a to $x^a + \delta x^a$, we obtain the vector

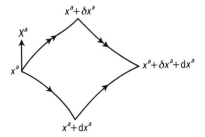

Fig. 6.9 Transporting X^a around an infinitesimal loop.

$$X^a(x + \delta x) = X^a(x) + \bar{\delta} X^a(x),$$

where, by (6.21),

$$\bar{\delta} X^a(x) = -\Gamma^a_{bc}(x) X^b(x) \delta x^c.$$

Similarly, if we transport this vector subsequently to $x^a + \delta x^a + dx^a$, we obtain the vector

$$X^a(x + \delta x + dx) = X^a(x + \delta x) + \bar{\delta} X^a(x + \delta x),$$

where, in this case,

$$\bar{\delta} X^a(x + \delta x) = -\Gamma^a_{bc}(x + \delta x) X^b(x + \delta x) dx^c.$$

Expanding by Taylor's theorem and using the previous results, we obtain (where everything is assumed evaluated at x^a)

$$\bar{\delta} X^a(x + \delta x) = -(\Gamma^a_{bc} + \partial_d \Gamma^a_{bc} \delta x^d)(X^b - \Gamma^b_{ef} X^e \delta x^f) dx^c$$

$$= -\Gamma^a_{bc} X^b dx^c - \partial_d \Gamma^a_{bc} X^b \delta x^d dx^c$$

$$+ \Gamma^a_{bc} \Gamma^b_{ef} X^e \delta x^f dx^c + \partial_d \Gamma^a_{bc} \Gamma^b_{ef} X^e \delta x^d \delta x^f dx^c.$$

Neglecting the last term, which is third order, we have

$$X^a(x + \delta x + dx)$$

$$= X^a - \Gamma^a_{bc} X^b \delta x^c - \Gamma^a_{bc} X^b dx^c - \partial_d \Gamma^a_{bc} X^b \delta x^d dx^c + \Gamma^a_{bc} \Gamma^b_{ef} X^e \delta x^f dx^c.$$

To obtain the equivalent result for the path connecting x^a to $x^a + \delta x^a + dx^a$

via $x^a + dx^a$, we simply interchange δx^a and dx^a to give

$$X^a(x + dx + \delta x)$$
$$= X^a - \Gamma^a_{bc} X^b dx^c - \Gamma^a_{bc} X^b \delta x^c - \partial_d \Gamma^a_{bc} X^b dx^d \delta x^c + \Gamma^a_{bc} \Gamma^b_{ef} X^e dx^f \delta x^c.$$

Hence, the difference between these two vectors is

$$\Delta X^a = X^a(x + \delta x + dx) - X^a(x + dx + \delta x)$$
$$= (\partial_d \Gamma^a_{bc} - \partial_c \Gamma^a_{bd} + \Gamma^a_{ed} \Gamma^e_{bc} - \Gamma^a_{ec} \Gamma^e_{bd}) X^b \delta x^c dx^d$$
$$= R^a{}_{bdc} X^b \delta x^c dx^d$$
$$= - R^a{}_{bcd} X^b \delta x^c dx^d$$

by (6.39) and the fact that the Riemann tensor is anti-symmetric on its last pair of indices (see (6.77)). Thus, the vector X^a will be the same at $x^a + \delta x^a + dx^a$, irrespective of which path is taken, if and only if $R^a{}_{bcd} = 0$. It follows that if the Riemann tensor vanishes then the vector X^a will not change if parallely transported around **any** infinitesimal closed loop. Using this result and assuming the manifold has no holes (that is, the manifold is **simply connected**), then we can continuously deform one curve into another by deforming the curves infinitesimally at each stage (Fig. 6.10), which establishes that the connection is integrable (check).

The second lemma is as follows.

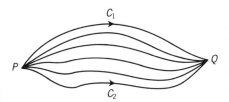

Fig. 6.10 Deforming C_1 into C_2 (infinitesimally at each stage).

> **Lemma:** A necessary and sufficient condition for a manifold to be affine flat is that the connection is symmetric and integrable.

Sufficiency is established by first choosing n linearly independent vectors

$$X_i{}^a \quad (i = 1, 2, \ldots, n)$$

at P, where the bold index i runs from 1 to n and labels the vectors. Using the integrability assumption we can construct the parallel vector fields $X_i{}^a(x)$ and these will also be linearly independent everywhere. Therefore, at each point P, $X_i{}^a(P)$ is a non-singular matrix of numbers and so we can construct its inverse, denoted by $X^i{}_b$, which must satisfy

$$X^i{}_b X_i{}^a = \delta^a_b, \tag{6.45}$$

where there is a summation over i. Multiplying the propagation equation

$$\partial_b X_i{}^a + \Gamma^a_{eb} X_i{}^e = 0$$

by $X^i{}_c$ produces

$$\Gamma^a_{cb} = - X^i{}_c \partial_b X_i{}^a. \tag{6.46}$$

Differentiating (6.45), we obtain

$$X_i{}^a \partial_c X^i{}_b = - X^i{}_b \partial_c X_i{}^a = \Gamma^a_{bc} \tag{6.47}$$

by (6.46). Using (6.47), we find that

$$X_i{}^a(\partial_c X^i{}_b - \partial_b X^i{}_c) = \Gamma^a_{bc} - \Gamma^a_{cb} = 0,$$

because the connection is symmetric by assumption. Since the determinant of $X_i{}^a$ is non-zero, it follows that the quantity in brackets must vanish, from which we get

$$\partial_c X^i{}_b = \partial_b X^i{}_c.$$

This in turn implies that $X^i{}_b$ must be the gradient of n scalar fields, $f^i(x)$ say, that is,

$$X^i{}_b = \partial_b f^i(x).$$

If we consider the transformation

$$x^a \to x'^a = f^a(x)$$

then

$$\frac{\partial x'^a}{\partial x^b} = \partial_b f^a(x) = X^a{}_b, \tag{6.48}$$

and so, taking inverses,

$$\frac{\partial x^a}{\partial x'^b} = X_b{}^a. \tag{6.49}$$

Multiplying (6.23) by $X_a{}^h$ and using (6.48) and (6.49) and then (6.45) and (6.47), we find

$$X_a{}^h \Gamma'^a{}_{bc} = X_a{}^h (X^a{}_d X_b{}^e X_c{}^f \Gamma^d{}_{ef} - X_b{}^e X_c{}^f \partial_e X^a{}_f)$$
$$= \delta^h_d X_b{}^e X_c{}^f \Gamma^d{}_{ef} - X_b{}^e X_c{}^f \Gamma^h{}_{ef} \equiv 0.$$

Again, since the determinant of $X_a{}^h$ is non-zero, $\Gamma'^a{}_{bc}$ vanishes everywhere in this coordinate system and hence the manifold is affine flat. The necessity is straightforward and is left as an exercise.

If we put these two lemmas together, we get the result we have been looking for.

Theorem: A necessary and sufficient condition for a manifold to be affine flat is that the Riemann tensor vanishes.

6.8 The metric

Any symmetric covariant tensor field of rank 2, say $g_{ab}(x)$, defines a **metric**. A manifold endowed with a metric is called a **Riemannian manifold**. A metric can be used to define distances and lengths of vectors. The infinitesimal **distance** (or **interval** in relativity), which we call ds, between two neighbouring points x^a and $x^a + \mathrm{d}x^a$ is defined by

$$\mathrm{d}s^2 = g_{ab}(x)\mathrm{d}x^a \mathrm{d}x^b. \tag{6.50}$$

Note that this gives the square of the infinitesimal distance, $(\mathrm{d}s)^2$, which is conventionally written as $\mathrm{d}s^2$. The equation (6.50) is also known as the **line element** and g_{ab} is also called the **metric form** or **first fundamental form**. The square of the **length** or **norm** of a contravariant vector X^a is defined by

$$X^2 = g_{ab}(x)X^a X^b. \tag{6.51}$$

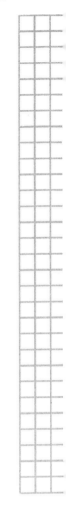

The metric is said to be **positive definite** or **negative definite** if, for all vectors X, $X^2 > 0$ or $X^2 < 0$, respectively. Otherwise, the metric is called **indefinite**. The **angle** between two vectors X^a and Y^a with $X^2 \neq 0$ and $Y^2 \neq 0$ is given by

$$\cos(X, Y) = \frac{g_{ab} X^a Y^b}{(|g_{cd} X^c X^d|)^{\frac{1}{2}} (|g_{ef} Y^e Y^f|)^{\frac{1}{2}}}. \tag{6.52}$$

In particular, the vectors X^a and Y^a are said to be **orthogonal** if

$$g_{ab} X^a Y^b = 0. \tag{6.53}$$

If the metric is indefinite (as in relativity theory), then there exist vectors which are orthogonal to themselves called **null vectors**, i.e.

$$g_{ab} X^a X^b = 0. \tag{6.54}$$

The **determinant** of the metric is denoted by

$$g = \det(g_{ab}) \tag{6.55}$$

The metric is **non-singular** if $g \neq 0$, in which case the **inverse** of g_{ab}, g^{ab}, is given by

$$g_{ab} g^{bc} = \delta_a^c. \tag{6.56}$$

It follows from this definition that g^{ab} is a contravariant tensor of rank 2 and it is called the **contravariant metric**. We may now use g_{ab} and g^{ab} to lower and raise tensorial indices by defining

$$T^{\cdots}_{\cdots a \cdots} = g_{ab} T^{\cdots b}_{\cdots \cdots} \tag{6.57}$$

and

$$T^{\cdots a \cdots}_{\cdots \cdots} = g^{ab} T^{\cdots}_{\cdots b \cdots}, \tag{6.58}$$

where we use the same kernel letter for the tensor. Since from now on we shall be working with a manifold endowed with a metric, we shall regard such associated contravariant and covariant tensors as representations of the **same** geometric object. Thus, in particular, g_{ab}, δ_a^b, and g^{ab} may all be thought of as different representations of the same geometric object, the metric g. Since we can raise and lower indices freely with the metric, we must be careful about the order in which we write contravariant and covariant indices. For example, in general, $X_a{}^b$ will be different from $X^b{}_a$.

6.9 Metric geodesics

Consider the timelike curve C with parametric equation $x^a = x^a(u)$. Dividing equation (6.50) by the square of du we get

$$\left(\frac{ds}{du}\right)^2 = g_{ab} \frac{dx^a}{du} \frac{dx^b}{du}. \tag{6.59}$$

Then the interval s between two points P_1 and P_2 on C is given by

$$s = \int_{P_1}^{P_2} ds = \int_{P_1}^{P_2} \frac{ds}{du} du = \int_{P_1}^{P_2} \left(g_{ab} \frac{dx^a}{du} \frac{dx^b}{du}\right)^{\frac{1}{2}} du. \tag{6.60}$$

We define a **timelike metric geodesic** between any two points P_1 and P_2 as the privileged curve joining them whose interval is **stationary** under small variations that vanish at the end points. Hence, the interval may be a maximum, a minimum, or a saddle point. Deriving the geodesic equations involves the calculus of variations and we postpone this to the next chapter. In that chapter, we shall see that the Euler–Lagrange equations result in the second-order differential equations

$$g_{ab} \frac{d^2 x^b}{du^2} + \{bc, a\} \frac{dx^b}{du} \frac{dx^c}{du} = \left(\frac{d^2 s}{du^2} \middle/ \frac{ds}{du} \right) g_{ab} \frac{dx^b}{du}, \qquad (6.61)$$

where the quantities in curly brackets are called the **Christoffel symbols of the first kind** and are defined in terms of derivatives of the metric by

$$\{ab, c\} = \tfrac{1}{2}(\partial_b g_{ac} + \partial_a g_{bc} - \partial_c g_{ab}). \qquad (6.62)$$

Multiplying through by g^{ad} and using (6.56), we get the equations

$$\frac{d^2 x^a}{du^2} + \left\{ \begin{matrix} a \\ bc \end{matrix} \right\} \frac{dx^b}{du} \frac{dx^c}{du} = \left(\frac{d^2 s}{du^2} \middle/ \frac{ds}{du} \right) \frac{dx^a}{du}, \qquad (6.63)$$

where $\left\{ \begin{smallmatrix} a \\ bc \end{smallmatrix} \right\}$ are the **Christoffel symbols of the second kind** defined by

$$\left\{ \begin{matrix} a \\ bc \end{matrix} \right\} = g^{ad}\{bc, d\}. \qquad (6.64)$$

In addition, the norm of the tangent vector dx^a/du is given by (6.59). If, in particular, we choose a parameter u which is linearly related to the interval s, that is,

$$u = \alpha s + \beta, \qquad (6.65)$$

where α and β are constants, then the right-hand side of (6.63) vanishes. In the special case when $u = s$, the **equations for a metric geodesic** become

$$\frac{d^2 x^a}{ds^2} + \left\{ \begin{matrix} a \\ bc \end{matrix} \right\} \frac{dx^b}{ds} \frac{dx^c}{ds} = 0 \qquad (6.66)$$

and

$$g_{ab} \frac{dx^a}{ds} \frac{dx^b}{ds} = 1, \qquad (6.67)$$

where we assume $ds \neq 0$.

Apart from trivial sign changes, similar results apply for spacelike geodesics, except that we replace s by σ, say, where

$$d\sigma^2 = -g_{ab}dx^a dx^b$$

However, in the case of an indefinite metric, there exist geodesics for which the distance between any two points is zero called **null geodesics**. It can also be shown that these curves can be parametrized by a special parameter u, called an **affine parameter**, such that their equation does not possess a right-hand side, that is,

$$\frac{d^2 x^a}{du^2} + \left\{\begin{matrix} a \\ bc \end{matrix}\right\} \frac{dx^b}{du} \frac{dx^c}{du} = 0, \tag{6.68}$$

where

$$g_{ab} \frac{dx^a}{du} \frac{dx^b}{du} = 0. \tag{6.69}$$

The last equation follows since the distance between any two points is zero, or equivalently the tangent vector is **null**. Again, any other affine parameter is related to u by the transformation

$$u \to \alpha u + \beta,$$

where α and β are constants.

6.10 The metric connection

In general, if we have a manifold endowed with both an affine connection and metric, then it possesses two classes of curves, affine geodesics and metric geodesics, which will be different (Fig. 6.11). However, comparing (6.37) with (6.66), the two classes will coincide if we take

$$\Gamma^a_{bc} = \left\{\begin{matrix} a \\ bc \end{matrix}\right\} \tag{6.70}$$

or, using (6.64) and (6.62), if

$$\Gamma^a_{bc} = \tfrac{1}{2} g^{ad}(\partial_b g_{dc} + \partial_c g_{db} - \partial_d g_{bc}). \tag{6.71}$$

It follows from the last equation that the connection is necessarily symmetric, i.e.

$$\Gamma^a_{bc} = \Gamma^a_{cb}. \tag{6.72}$$

In fact, if one checks the transformation properties of $\left\{\begin{smallmatrix} a \\ bc \end{smallmatrix}\right\}$ from first principles, it does indeed transform like a connection (exercise). This special connection built out of the metric and its derivatives is called the **metric connection**. From now on, we shall **always** work with the metric connection and we shall denote it by Γ^a_{bc} rather than $\left\{\begin{smallmatrix} a \\ bc \end{smallmatrix}\right\}$, where Γ^a_{bc} is defined by (6.71). This definition leads immediately to the identity (exercise)

$$\nabla_c g_{ab} \equiv 0. \tag{6.73}$$

Conversely, if we require that (6.73) holds for an arbitrary symmetric

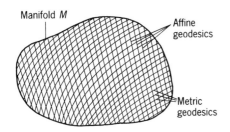

Manifold M — Affine geodesics — Metric geodesics

Fig. 6.11 Affine and metric geodesics on a manifold.

connection, then it can be deduced (exercise) that the connection is necessarily the metric connection. Thus, we have the following important result.

> **Theorem:** If ∇_a denotes covariant derivative with respect to the affine connection Γ^a_{bc}, then the necessary and sufficient condition for the covariant derivative of the metric to vanish is that the connection is the metric connection.

In addition, we can show that

$$\nabla_c \delta^a_b \equiv 0 \qquad (6.74)$$

and

$$\nabla_c g^{ab} \equiv 0. \qquad (6.75)$$

6.11 Metric flatness

Now at any point P of a manifold, g_{ab} is a symmetric matrix of real numbers. Therefore, by standard matrix theory, there exists a transformation which reduces the matrix to diagonal form with every diagonal term either $+1$ or -1. The excess of plus signs over minus signs in this form is called the **signature** of the metric. Assuming that the metric is continuous over the manifold and non-singular, then it follows that the signature is an invariant. In general, it will not be possible to find a coordinate system in which the metric reduces to this diagonal form everywhere. If, however, there does exist a coordinate system in which the metric reduces to diagonal form with ± 1 diagonal elements everywhere, then the metric is called **flat**.

How does metric flatness relate to affine flatness in the case we are interested in, that is, when the connection is the metric connection? The answer is contained in the following result.

> **Theorem:** A necessary and sufficient condition for a metric to be flat is that its Riemann tensor vanishes.

Necessity follows from the fact that there exists a coordinate system in which the metric is diagonal with ± 1 diagonal elements. Since the metric is constant everywhere, its partial derivatives vanish and therefore the metric connection Γ^a_{bc} vanishes as a consequence of the definition (6.71). Since Γ^a_{bc} vanishes **everywhere** then so must its derivatives. (One way to see this is to recall the definition of partial differentiation which involves subtracting quantities at neighbouring points. If the quantities are always zero, then their difference vanishes, and so does the resulting limit.) The Riemann tensor therefore vanishes by the definition (6.39).

Conversely, if the Riemann tensor vanishes, then by the theorem of §6.7, there exists a special coordinate system in which the connection vanishes everywhere. Since this is the metric connection, by (6.73),

$$\nabla_c g_{ab} = \partial_c g_{ab} - \Gamma^d_{ac} g_{db} - \Gamma^d_{bc} g_{ad} = 0,$$

from which we get

$$\partial_c g_{ab} = \Gamma^d_{ac} g_{db} + \Gamma^d_{bc} g_{ad}, \tag{6.76}$$

and it follows that $\partial_c g_{ab} = 0$. The metric is therefore constant everywhere and hence can be transformed into diagonal form with diagonal elements ± 1. Note the result (6.76) which expresses the ordinary derivative of the metric in terms of the connection. This equation will prove useful later.

Combining this theorem with the theorem of §6.7, we see that if we use the metric connection then metric flatness coincides with affine flatness.

6.12 The curvature tensor

The **curvature tensor** or **Riemann–Christoffel tensor** (Riemann tensor for short) is defined by (6.39), namely,

$$R^a_{bcd} = \partial_c \Gamma^a_{bd} - \partial_d \Gamma^a_{bc} + \Gamma^e_{bd} \Gamma^a_{ec} - \Gamma^e_{bc} \Gamma^a_{ed},$$

where Γ^a_{bc} is the metric connection, which by (6.71) is given as

$$\Gamma^a_{bc} = \tfrac{1}{2} g^{ad} (\partial_b g_{dc} + \partial_c g_{db} - \partial_d g_{bc}).$$

Thus, R^a_{bcd} depends on the metric and its first and second derivatives. It follows immediately from the definition that it is anti-symmetric on its last pair of indices

$$R^a_{bcd} = - R^a_{bdc}. \tag{6.77}$$

The fact that the connection is symmetric leads to the identity

$$R^a_{bcd} + R^a_{dbc} + R^a_{cdb} \equiv 0. \tag{6.78}$$

Lowering the first index with the metric, then it is easy to establish, for example by using geodesic coordinates, that the lowered tensor is symmetric under interchange of the first and last pair of indices, that is,

$$R_{abcd} = R_{cdab}. \tag{6.79}$$

Combining this with equation (6.77), we see that the lowered tensor is anti-symmetric on its first pair of indices as well:

$$R_{abcd} = - R_{bacd}. \tag{6.80}$$

Collecting these symmetries together, we see that the lowered curvature tensor satisfies

$$R_{abcd} = - R_{abdc} = - R_{bacd} = R_{cdab},$$
$$R_{abcd} + R_{adbc} + R_{acdb} \equiv 0. \tag{6.81}$$

These symmetries considerably reduce the number of independent components; in fact, in n dimensions, the number is reduced from n^4 to $\tfrac{1}{12} n^2 (n^2 - 1)$. In addition to the algebraic identities, it can be shown, again most easily by using geodesic coordinates, that the curvature tensor satisfies a set of

differential identities called the **Bianchi identities**:

$$\nabla_a R_{debc} + \nabla_c R_{deab} + \nabla_b R_{deca} \equiv 0. \qquad (6.82)$$

We can use the curvature tensor to define several other important tensors. The **Ricci tensor** is defined by the contraction

$$R_{ab} = R^c{}_{acb} = g^{cd} R_{dacb}, \qquad (6.83)$$

which by (6.79) is symmetric. A final contraction defines the **curvature scalar** or **Ricci scalar** R by

$$R = g^{ab} R_{ab}. \qquad (6.84)$$

These two tensors can be used to define the **Einstein tensor**

$$G_{ab} = R_{ab} - \tfrac{1}{2} g_{ab} R, \qquad (6.85)$$

which is also symmetric, and, by (6.82), the Einstein tensor can be shown to satisfy the **contracted Bianchi identities**

$$\nabla_b G_a{}^b \equiv 0. \qquad (6.86)$$

Note that some authors adopt a different sign convention, which leads to the Riemann tensor or the Ricci tensor having the opposite sign to ours.

6.13 The Weyl tensor

We shall mostly be concerned with tensors in four dimensions or less. The algebraic identities (6.81) lead to the following special cases for the curvature tensor:

(1) if $n = 1$, $R_{abcd} = 0$;

(2) if $n = 2$, R_{abcd} has one independent component — essentially R;

(3) if $n = 3$, R_{abcd} has six independent components — essentially R_{ab};

(4) if $n = 4$, R_{abcd} has twenty independent components — ten of which are given by R_{ab} and the remaining ten by the Weyl tensor.

The **Weyl tensor** or **conformal tensor** C_{abcd} is defined in n dimensions, $(n \geqslant 3)$ by

$$C_{abcd} = R_{abcd} + \frac{1}{n-2} (g_{ad} R_{cb} + g_{bc} R_{da} - g_{ac} R_{db} - g_{bd} R_{ca})$$

$$+ \frac{1}{(n-1)(n-2)} (g_{ac} g_{db} - g_{ad} g_{cb}) R.$$

Thus, in four dimensions, this becomes

$$C_{abcd} = R_{abcd} + \tfrac{1}{2}(g_{ad}R_{cb} + g_{bc}R_{da} - g_{ac}R_{db} - g_{bd}R_{ca})$$
$$+ \tfrac{1}{6}(g_{ac}g_{db} - g_{ad}g_{cb})R. \tag{6.87}$$

It is straightforward to show that the Weyl tensor possesses the same symmetries as the Riemann tensor, namely,

$$C_{abcd} = -C_{abdc} = -C_{bacd} = C_{cdab}, \tag{6.88}$$
$$C_{abcd} + C_{adbc} + C_{acdb} \equiv 0$$

However, it possesses an additional symmetry

$$C^a{}_{bad} \equiv 0. \tag{6.89}$$

Combining this result with the previous symmetries, it then follows that the Weyl tensor is **trace-free**, in other words, it vanishes for **any** pair of contracted indices. One can think of the Weyl tensor as that part of the curvature tensor for which all contractions vanish.

Two metrics g_{ab} and $\bar{g}_{ab}$ are said to be **conformally related** or **conformal** to each other if

$$\bar{g}_{ab} = \Omega^2 g_{ab}, \tag{6.90}$$

where $\Omega(x)$ is a non-zero differentiable function. Given a manifold with two metrics defined on it which are conformal, then it is straightforward from (6.51) and (6.52) to show that angles between vectors and ratios of magnitudes of vectors, but not lengths, are the same for each metric. Moreover, the null geodesics of one metric **coincide** with the null geodesics of the other (exercise). The metrics also possess the same Weyl tensor, i.e.

$$\bar{C}^a{}_{bcd} = C^a{}_{bcd}. \tag{6.91}$$

Any quantity which satisfies a relationship like (6.91) is called **conformally invariant** (g_{ab}, Γ^a_{bc}, and R^a_{bcd} are examples of quantities which are not conformally invariant). A metric is said to be **conformally flat** if it can be reduced to the form

$$g_{ab} = \Omega^2 \eta_{ab}, \tag{6.92}$$

where η_{ab} is a flat metric. We end this section by quoting two results concerning conformally flat metrics.

Theorem: A necessary and sufficient condition for a metric to be conformally flat is that its Weyl tensor vanishes everywhere.

Theorem: Any two-dimensional Riemannian manifold is conformally flat.

Exercises

6.1 (§6.2) Prove (6.13) by showing that $L_X\delta_b^a = 0$ in two ways: (i) using (6.17); (ii) from first principles (remembering Exercise 5.8).

6.2 (§6.2) Use (6.17) to find expressions for $L_X Z_{bc}$ and $L_X(Y^a Z_{bc})$. Use these expressions and (6.15) to check the Leibniz property in the form (6.12).

6.3 (§6.3) Establish (6.23) by assuming that the quantity defined by (6.22) has the tensor character indicated. Take the partial derivative of

$$\delta'^a_c = \frac{\partial x'^a}{\partial x'^c} = \frac{\partial x'^a}{\partial x^d}\frac{\partial x^d}{\partial x'^c}$$

with respect to x'^b to establish the alternative form (6.24).

6.4 (§6.3) Show that covariant differentiation commutes with contraction by checking that $\nabla_c \delta_b^a = 0$.

6.5 (§6.3) Assuming (6.22) and (6.25), apply the Leibniz rule to the covariant derivative of $X_a X^a$, where X^a is arbitrary, to verify (6.26).

6.6 (§6.3) Check (6.29).

6.7 (§6.4) If X, Y, and Z are vector fields, f and g smooth functions, and λ and μ constants, then show that

(i) $\nabla_X(\lambda Y + \mu Z) = \lambda\nabla_X Y + \mu\nabla_X Z$,

(ii) $\nabla_{fX+gY} Z = f\nabla_X Z + g\nabla_Y Z$,

(iii) $\nabla_X(fY) = (Xf)Y + f\nabla_X Y$.

6.8 (§6.4) Show that (6.33) leads to (6.34).

6.9 (§6.4) If s is an affine parameter, then show that, under the transformation

$$s \to \bar{s} = \bar{s}(s),$$

the parameter $\bar{s}$ will be affine only if $\bar{s} = \alpha s + \beta$, where α and β are constants.

6.10 (§6.5) Show that

$$\nabla_c\nabla_d X^a_{\ b} - \nabla_d\nabla_c X^a_{\ b} = R^a_{\ ecd}X^e_{\ b} + R^e_{\ bcd}X^a_{\ e}.$$

6.11 (§6.5) Show that

$$\nabla_X(\nabla_Y Z^a) - \nabla_Y(\nabla_X Z^a) - \nabla_{[X,\ Y]}Z^a = R^a_{\ bcd}Z^b X^c Y^d.$$

6.12 (§6.7) Prove that if a manifold is affine flat then the connection is necessarily integrable and symmetric.

6.13 (§6.8) Show that if g_{ab} is **diagonal**, i.e. $g_{ab} = 0$ if $a \neq b$, then g^{ab} is diagonal with corresponding reciprocal diagonal elements.

6.14 (§6.8) The line elements of $\mathbb{R}^3$ in Cartesian, cylindrical polar, and spherical polar coordinates are given respectively by

(i) $ds^2 = dx^2 + dy^2 + dz^2$,

(ii) $ds^2 = dR^2 + R^2 d\phi^2 + dz^2$,

(iii) $ds^2 = dr^2 + r^2 d\theta^2 + r^2\sin^2\theta\,d\phi^2$.

Find g_{ab}, g^{ab}, and g in each case.

6.15 (§6.8) Express T_{ab} in terms of T^{cd}.

6.16 (§6.9) Write down the tensor transformation law of g_{ab}. Show directly that

$$\left\{\begin{matrix} a \\ bc \end{matrix}\right\} = \tfrac{1}{2}g^{ad}(\partial_b g_{dc} + \partial_c g_{db} - \partial_d g_{bc})$$

transforms like a connection.

6.17 (§6.9) Find the geodesic equation for $\mathbb{R}^3$ in cylindrical polars. [Hint: use the results of Exercise 6.14(ii) to compute the metric connection and substitute in (6.68).]

6.18 (§6.9) Consider a 3-space with coordinates $(x^a) = (x, y, z)$ and line element

$$ds^2 = dx^2 + dy^2 - dz^2.$$

Prove that the null geodesics are given by

$$x = lu + l', \qquad y = mu + m', \qquad z = nu + n',$$

where u is a parameter and l, l', m, m', n, n' are arbitrary constants satisfying $l^2 + m^2 - n^2 = 0$.

6.19 (§**6.10**) Prove that $\nabla_c g_{ab} \equiv 0$. Deduce that $\nabla_b X_a = g_{ac} \nabla_b X^c$.

6.20 (§**6.10**) Suppose we have an arbitrary symmetric connection Γ^a_{bc} satisfying $\nabla_c g_{ab} = 0$. Deduce that Γ^a_{bc} must be the metric connection. [Hint: use the equation to find expressions for $\partial_b g_{dc}$, $\partial_c g_{db}$ and $-\partial_d g_{bc}$, as in (6.76), add the equations together, and multiply by $\frac{1}{2} g^{ad}$.]

6.21 (§**6.11**) The Minkowski line element in Minkowski coordinates

$$(x^a) = (x^0, x^1, x^2, x^3) = (t, x, y, z)$$

is given by

$$ds^2 = dt^2 - dx^2 - dy^2 - dz^2$$

(i) What is the signature?
(ii) Is the metric non-singular?
(iii) Is the metric flat?

6.22 (§**6.11**) The line element of $\mathbb{R}^3$ in a particular coordinate system is

$$ds^2 = (dx^1)^2 + (x^1)^2 (dx^2)^2 + (x^1 \sin x^2)^2 (dx^3)^2$$

(i) Identify the coordinates.
(ii) Is the metric flat?

6.23 (§**6.12**) Establish the identities (6.78) and (6.79). [Hint: choose an arbitrary point P and introduce geodesic coordinates at P.] Show that (6.78) is equivalent to $R^a_{[bcd]} \equiv 0$.

6.24 (§**6.12**) Establish the identity (6.82). [Hint: use geodesic coordinates.] Show that (6.82) is equivalent to $R_{de[ab;c]} \equiv 0$. Deduce (6.86).

6.25 (§**6.12**) Show that $G_{ab} = 0$ if and only if $R_{ab} = 0$.

6.26 (§**6.13**) Establish the identity (6.89). Deduce that the Weyl tensor is trace-free on **all** pairs of indices.

6.27 (§**6.13**) Show that angles between vectors and ratios of lengths of vectors, but not lengths, are the same for conformally related metrics.

6.28 (§**6.13**) Prove that the null geodesics of two conformally related metrics coincide. [Hint: the two classes of geodesics need not both be affinely parametrized.]

6.29 (§**6.13**) Establish (6.91).

6.30 (§**6.13**) Establish the theorem that any two-dimensional Riemann manifold is conformally flat in the case of a metric of signature 0, i.e. at any point the metric can be reduced to the diagonal form $(+1, -1)$ say. [Hint: use null curves as coordinate curves, that is, change to new co-ordinates

$$\lambda = \lambda(x^0, x^1), \qquad v = v(x^0, x^1)$$

satisfying

$$g^{ab} \lambda_{,a} \lambda_{,b} = g^{ab} v_{,a} v_{,b} = 0$$

and show that the line element reduces to the form

$$ds^2 = e^{2\mu} d\lambda dv$$

and finally introduce new coordinates $\frac{1}{2}(\lambda + v)$ and $\frac{1}{2}(\lambda - v)$.]

6.31 This final exercise consists of a long calculation which will be needed later in the book. If we take coordinates

$$x^a = (x^0, x^1, x^2, x^3) = (t, r, \theta, \phi),$$

then the four-dimensional spherically symmetric line element is

$$ds^2 = e^v dt^2 - e^\lambda dr^2 - r^2 d\theta^2 - r^2 \sin^2 d\phi^2,$$

where $v = v(t, r)$ and $\lambda = \lambda(t, r)$ are arbitrary functions of t and r.

(i) Find g_{ab}, g, and g^{ab} (see Exercise 6.13).
(ii) Use the expressions in (i) to calculate Γ^a_{bc}. [Hint: remember $\Gamma^a_{bc} = \Gamma^a_{cb}$.]
(iii) Calculate R^a_{bcd}. [Hint: use (6.39) and the symmetry relation (6.77).]
(iv) Calculate R_{ab}, R, and G_{ab}.
(v) Calculate $G^a_b (= g^{ac} G_{cb} = G_b{}^a)$.

Integration, variation, and symmetry

7.1 Tensor densities

A **tensor density** of weight W, denoted conventionally by a gothic letter, $\mathfrak{T}^{a\cdots}_{b\cdots}$, transforms like an ordinary tensor, except that in addition the Wth power of the Jacobian

$$J = \left| \frac{\partial x^a}{\partial x'^b} \right|$$

appears as a factor, i.e.

$$\mathfrak{T}'^{a\cdots}_{b\cdots} = J^W \frac{\partial x'^a}{\partial x^c} \cdots \frac{\partial x^d}{\partial x'^b} \cdots \mathfrak{T}^{c\cdots}_{d\cdots}. \tag{7.1}$$

Then, with certain modifications, we can combine tensor densities in much the same way as we do tensors. One exception, which follows from (7.1), is that the product of two tensor densities of weight W_1 and W_2 is a tensor density of weight $W_1 + W_2$. There is some arbitrariness in defining the covariant derivative of a tensor density, but we shall adhere to the definition that if $\mathfrak{T}^{a\cdots}_{b\cdots}$ is a tensor density of weight W then

$$\nabla_c \mathfrak{T}^{a\cdots}_{b\cdots} = \text{usual terms if } \mathfrak{T}^{a\cdots}_{b\cdots} \text{ were a tensor} - W\Gamma^d_{dc} \mathfrak{T}^{a\cdots}_{b\cdots}. \tag{7.2}$$

For example, the covariant derivative of a vector density of weight W is

$$\nabla_c \mathfrak{T}^a = \partial_c \mathfrak{T}^a + \Gamma^a_{bc} \mathfrak{T}^b - W\Gamma^b_{bc} \mathfrak{T}^a.$$

In the special case when $W = +1$ and $c = a$, we get the important result (check)

$$\nabla_a \mathfrak{T}^a = \partial_a \mathfrak{T}^a, \tag{7.3}$$

that is, the **covariant divergence** of a vector density of weight $+1$ is identical to its **ordinary divergence**. It can be shown that both these quantities are scalar densities of weight $+1$ (exercise).

7.2 The Levi–Civita alternating symbol

We introduce a quantity which is a generalization of the Kronecker delta δ_b^a, but which turns out to be a tensor density. The **Levi–Civita alternating symbol** ε^{abcd} is a completely anti-symmetric tensor density of weight $+1$ and contravariant rank 4, whose values in any coordinate system is $+1$ or -1 if $abcd$ is an even or odd permutation of 0123, respectively, and zero otherwise. Thus, for example, in four dimensions, if we let the coordinates range from 0 to 3 (as we shall), i.e.

$$(x^a) = (x^0, x^1, x^2, x^3),$$

then some of its values are

$$\varepsilon^{0123} = \varepsilon^{2301} = -\varepsilon^{0132} = -\varepsilon^{0321} = +1$$

and

$$\varepsilon^{0120} = \varepsilon^{0331} = \varepsilon^{0101} = 0.$$

Similarly, we can define the covariant version ε_{abcd}, which has weight -1. It can be used, in particular, to form the determinant of a second-rank density, i.e.

$$\det \mathfrak{T}^{ab} = \frac{1}{4!} \varepsilon_{abcd} \varepsilon_{efgh} \mathfrak{T}^{ae} \mathfrak{T}^{bf} \mathfrak{T}^{cg} \mathfrak{T}^{dh}.$$

Assuming this is non-zero, we can then also use it to construct the inverse of a second-rank tensor. The covariant derivatives of both ε^{abcd} and ε_{abcd} vanish identically, which from one point of view motivates the definition (7.2).

We define the **generalized Kronecker delta** by

$$\delta_{cd}^{ab} = \begin{cases} +1 & \text{for } a \neq b,\ a = c,\ b = d, \\ -1 & \text{for } a \neq b,\ a = d,\ b = c, \\ 0 & \text{otherwise,} \end{cases}$$

and similarly for higher-order tensors. They are constant tensors of the type indicated, and can be defined in terms of the Kronecker delta by the determinant relationships

$$\delta_{cd}^{ab} = \begin{vmatrix} \delta_c^a & \delta_c^b \\ \delta_d^a & \delta_d^b \end{vmatrix}$$

and

$$\delta_{def}^{abc} = \begin{vmatrix} \delta_d^a & \delta_d^b & \delta_d^c \\ \delta_e^a & \delta_e^b & \delta_e^c \\ \delta_f^a & \delta_f^b & \delta_f^c \end{vmatrix},$$

and so forth. In four dimensions they are related to products of the alternating symbols according to

$$\varepsilon^{abcd} \varepsilon_{efgh} = \delta_{efgh}^{abcd},$$

$$\varepsilon^{abcd} \varepsilon_{efgd} = \delta_{efg}^{abc},$$

$$\varepsilon^{abcd} \varepsilon_{efcd} = 2\delta_{ef}^{ab},$$

$$\varepsilon^{abcd} \varepsilon_{ebcd} = 3!\delta_e^a,$$

$$\varepsilon^{abcd} \varepsilon_{abcd} = 4!.$$

7.3 The metric determinant

If we have a Riemannian manifold with metric g_{ab}, then it transforms according to

$$g'_{ab}(x') = \frac{\partial x^c}{\partial x'^a} \frac{\partial x^d}{\partial x'^b} g_{cd}(x),$$ (7.4)

and so, taking determinants, we have

$$g' = J^2 g.$$

Hence the metric determinant g is a scalar density of weight $+2$. In the later chapters, we shall be working with metrics of **negative signature** in which case g will be negative, and so we write the last equation in the equivalent form

$$(-g') = J^2(-g).$$

Since all these terms are now positive, we can take square roots, to get

$$(-g')^{\frac{1}{2}} = J(-g)^{\frac{1}{2}}$$

and hence $(-g)^{\frac{1}{2}}$ is a **scalar density of weight $+1$**. The quantity $(-g)^{\frac{1}{2}}$ plays an important role in integration. Given any tensor $T^{a\cdots}_{b\cdots}$, we can form the product $(-g)^{\frac{1}{2}} T^{a\cdots}_{b\cdots}$ which is then a **tensor density of weight $+1$**. In particular, we can deduce an important result from equation (7.3), namely, for any **vector** T^a,

$$\nabla_a[(-g)^{\frac{1}{2}} T^a] = \partial_a[(-g)^{\frac{1}{2}} T^a].$$ (7.5)

Now, at any point, the covariant and contravariant metrics are symmetric matrices which are inverse to each other by

$$g_{ab}g^{bc} = \delta_a^c.$$

Let us digress for a moment and consider the general case of finding the derivative of a determinant of a matrix whose elements are functions of the coordinates. Consider any square matrix $A = (a_{ij})$. Then its inverse, (b^{ij}) say, is defined by

$$(b^{ij}) = \frac{1}{a}(A^{ij})' = \frac{1}{a}(A^{ji}),$$ (7.6)

where a is the determinant of A, A^{ij} is the cofactor of a_{ij}, and the prime denotes the transpose. Let us fix i, and expand the determinant a by the ith row. Then

$$a = \sum_{j=1}^{n} a_{ij} A^{ij}$$

where we have explicitly included the summation sign for clarity. If we partially differentiate both sides with respect to a_{ij}, then we get

$$\frac{\partial a}{\partial a_{ij}} = A^{ij},$$ (7.7)

since a_{ij} does not occur in any of the cofactors A^{ij} (i fixed, j runs from 1 to n). Repeating the argument for every i, as i runs from 1 to n, we see that the formula (7.7) is quite general. Let us suppose that the a_{ij} are all functions of the coordinates x^k. Then the determinant is a functional of the a_{ij}, which in turn are functions of the x^k, that is,

$$a = a(a_{ij}(x^k)).$$

Differentiating this partially with respect to x^k, using the function of a function rule and equation (7.7), we obtain

$$\frac{\partial a}{\partial x^k} = \frac{\partial a}{\partial a_{ij}} \frac{\partial a_{ij}}{\partial x^k}$$

$$= A^{ij} \frac{\partial a_{ij}}{\partial x^k}$$

$$= ab^{ji} \frac{\partial a_{ij}}{\partial x^k}$$

by equation (7.6). Applying this result to the metric determinant g and remembering that g^{ab} is symmetric, we get the useful equation

$$\partial_c g = g g^{ab} \partial_c g_{ab}. \tag{7.8}$$

We now combine this result with (6.76) (which comes directly from the vanishing of the covariant derivative of the metric) and find

$$\partial_c g = g g^{ab} (\Gamma^d_{ac} g_{db} + \Gamma^d_{bc} g_{ad})$$

$$= g \delta^a_d \Gamma^d_{ac} + g \delta^b_d \Gamma^d_{bc}$$

$$= 2g \Gamma^a_{ac}. \tag{7.9}$$

Let us compute the covariant derivative of g using (7.2). Then, since g is a scalar density of weight $+2$, we have

$$\nabla_c g = \partial_c g - 2g \Gamma^a_{ac},$$

and so by equation (7.9) it follows that

$$\nabla_c g \equiv 0. \tag{7.10}$$

This is again intimately connected with the choice of the definition (7.2). Similarly, we find from equation (7.9) that

$$\partial_c (-g)^{\frac{1}{2}} - (-g)^{\frac{1}{2}} \Gamma^a_{ac} = 0,$$

that is, by (7.2),

$$\nabla_c (-g)^{\frac{1}{2}} \equiv 0. \tag{7.11}$$

In particular, for any tensor $T^{a\cdots}_{b\cdots}$, this leads to the identity

$$\nabla_c [(-g)^{\frac{1}{2}} T^{a\cdots}_{b\cdots}] = (-g)^{\frac{1}{2}} (\nabla_c T^{a\cdots}_{b\cdots}), \tag{7.12}$$

that is, we can pull factors of $(-g)^{\frac{1}{2}}$ and g through covariant derivatives in the same way as we can with factors involving the covariant or contravariant metric.

7.4 Integrals and Stokes' theorem

Unlike tensors in general, we can add a scalar field ϕ evaluated at two different points, x_1 and x_2 say, and the resulting quantity is still a scalar, since under a coordinate transformation, the sum transforms like

$$\phi'(x_1') + \phi'(x_2') = \phi(x_1) + \phi(x_2) \tag{7.13}$$

by (5.18). Hence, we might imagine that it is possible to integrate a scalar field ϕ over some n-dimensional region Ω of a manifold M. However, it turns out that the volume element $\mathrm{d}\Omega$ is not a scalar but, as we shall see, a scalar density of weight -1. It follows that we can integrate a scalar density Φ of weight $+1$ over a region Ω,

$$\int_\Omega \Phi \, \mathrm{d}\Omega, \tag{7.14}$$

since at each point $\Phi \, \mathrm{d}\Omega$ is a scalar and can be added together by (7.13). There are analogous statements which can be made about integration over curves, surfaces, and hypersurfaces.

Consider an m-dimensional subspace of M whose parametric equation by (5.2) is

$$x^a = x^a(u^i) \quad (i = 1, 2, \ldots, m).$$

The **'volume' element** of this subspace is defined to be

$$\mathrm{d}\tau^{a_1 a_2 \cdots a_m} = \delta^{a_1 a_2 \cdots a_m}_{b_1 b_2 \cdots b_m} \frac{\partial x^{a_1}}{\partial u_1} \frac{\partial x^{a_2}}{\partial u_2} \cdots \frac{\partial x^{a_m}}{\partial u_m} \mathrm{d}u^1 \mathrm{d}u^2 \cdots \mathrm{d}u^m. \tag{7.15}$$

This element is an mth rank contravariant tensor under coordinate transformations and behaves like a scalar under arbitrary change of parameter. Hence, if $X_{a_1 a_2 \cdots a_m}$ is an mth rank covariant tensor, then $X_{a_1 a_2 \cdots a_m} \mathrm{d}\tau^{a_1 a_2 \cdots a_m}$ is a scalar under both coordinate and parameter transformations, and we can form the integral

$$\int_{\Omega_m} X_{a_1 a_2 \cdots a_m} \mathrm{d}\tau^{a_1 a_2 \cdots a_m} \tag{7.16}$$

over some region Ω_m of the subspace. The **coordinate differentials** $\mathrm{d}_i x^a$ corresponding to each parameter u_i are defined by

$$\mathrm{d}_i x^a = \frac{\partial x^a}{\partial u^i} \mathrm{d}u^i \quad \text{(no sum on } i\text{)}. \tag{7.17}$$

We now state **Stokes' theorem** for a simply connected m-dimensional subspace Ω_m bounded by the $(m-1)$-dimensional subspace $\partial\Omega_m = \Omega_{m-1}$:

$$\int_{\partial\Omega_m} X_{a_1 a_2 \cdots a_{m-1}} \mathrm{d}\tau^{a_1 a_2 \cdots a_{m-1}} = \int_{\Omega_m} \partial_{a_m} X_{a_1 a_2 \cdots a_{m-1}} \mathrm{d}\tau^{a_1 a_2 \cdots a_m}. \tag{7.18}$$

We will be particularly interested in the special case of a four-dimensional region Ω of a four-dimensional manifold M, where Ω is bounded by the

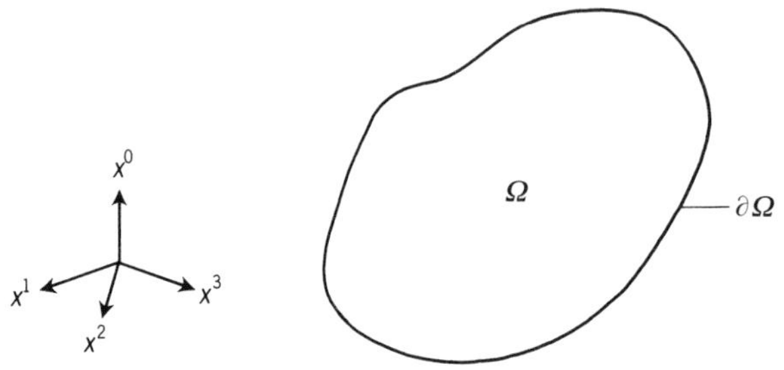

Fig. 7.1 A four-dimensional region Ω bounded by $\partial\Omega$.

hypersurface $\partial\Omega$ (Fig 7.1). Stokes's theorem then becomes the **divergence theorem** or **Gauss's theorem** for a contravariant vector density $\mathfrak{T}^a$ of weight $+1$, which we write in the form

$$\int_{\partial\Omega} \mathfrak{T}^a \, \mathrm{d}S_a = \int_{\Omega} \mathfrak{T}^a{}_{,a} \, \mathrm{d}\Omega, \qquad (7.19)$$

where

$$\mathrm{d}S_a = \frac{1}{3!} \varepsilon_{abcd} \mathrm{d}\tau^{bcd} \qquad (7.20)$$

and

$$\mathrm{d}\Omega = \frac{1}{4!} \varepsilon_{abcd} \mathrm{d}\tau^{abcd}. \qquad (7.21)$$

If we use the **coordinates** x^a as parameters then $\mathrm{d}\Omega$ is written as $\mathrm{d}^4 x$ where

$$\mathrm{d}^4 x \equiv \mathrm{d}x^0 \, \mathrm{d}x^1 \, \mathrm{d}x^2 \, \mathrm{d}x^3 \qquad (7.22)$$

and

$$\mathrm{d}S_a = (\mathrm{d}x^1 \, \mathrm{d}x^2 \, \mathrm{d}x^3, \, \mathrm{d}x^0 \, \mathrm{d}x^2 \, \mathrm{d}x^3, \, \mathrm{d}x^0 \, \mathrm{d}x^1 \, \mathrm{d}x^3, \, \mathrm{d}x^0 \, \mathrm{d}x^1 \, \mathrm{d}x^2). \qquad (7.23)$$

Note from the definition (7.21) that $\mathrm{d}^4 x$ is a **scalar density of weight** -1.

7.5 The Euler–Lagrange equations

The variational principle and with it the Euler–Lagrange equations will play an important role in this book. So, although it is something of a digression, we shall, for completeness, include a brief discussion of their derivation. Then, as a first indication of their usefulness, we shall show in the next section how they provide an efficient method for obtaining geodesics.

A **functional** may be defined as a correspondence between a real number and a function belonging to some class. Thus, a functional is a kind of function where the independent variable is itself a function. One of the basic problems in the calculus of variations is that of finding the stationary values (maxima, minima, saddle points) of the **action** I defined by

$$I[y] = \int_{x_1}^{x_2} L(y, y', x) \, \mathrm{d}x, \qquad (7.24)$$

where L is a functional of the **dynamical variable** y, its derivative $y' = \mathrm{d}y/\mathrm{d}x$, and the coordinate x, and is called the **Lagrangian**. The problem is easily generalized. In order to solve the problem, we need to make use of the following result.

Lemma: If $\int_{x_1}^{x_2} \phi(x)\eta(x)\,dx = 0$, where $\phi(x)$ is continuous and $\eta(x)$ is an **arbitrary** twice-differentiable function vanishing on the boundary, i.e. $\eta(x_1) = \eta(x_2) = 0$, then $\phi(x) \equiv 0$.

To establish this, we suppose that $\phi(x) \neq 0$ for some $x = \xi$ in the interval (x_1, x_2). To fix ideas, let us assume $\phi(\xi) > 0$. Then, by continuity, there exists a neighbourhood of ξ ($\xi_1 < \xi < \xi_2$) for which $\phi(x) > 0$. Setting

$$\eta(x) = \begin{cases} (x - \xi_1)^4(x - \xi_2)^4 & \text{for } x \in (\xi_1, \xi_2), \\ 0 & \text{otherwise,} \end{cases}$$

we find that $\eta(x)$ satisfies the conditions of the above lemma. Furthermore,

$$\int_{x_1}^{x_2} \phi(x)\eta(x)\,dx = \int_{\xi_1}^{\xi_2} \phi(x)\eta(x)\,dx > 0,$$

which produces a contradiction. Similarly, if we assume $\phi(\xi) < 0$, then again we get a contradiction, and so the result follows.

Returning to (7.24), we assume L is twice differentiable with respect to its three variables. Let us vary y by an **arbitrary small amount** and write

$$\bar{y} = y + \varepsilon\eta(x), \tag{7.25}$$

where ε is small and $\eta(x)$ satisfies the conditions of the lemma, that is, it has continuous second derivatives and vanishes at x_1 and x_2 but is otherwise arbitrary. We define a **variation** of y by

$$\delta y \equiv \bar{y} - y = \varepsilon\eta(x). \tag{7.26}$$

Differentiating (7.25) with respect to x and using the prime notation, we get

$$\bar{y}' = y' + \varepsilon\eta',$$

so that

$$\delta(y') \equiv \bar{y}' - y' = \varepsilon\eta' = (\delta y)',$$

from which we see that δ and d/dx acting on y commute. Then, working to first order in ε,

$$I[\bar{y}] = I[y + \delta y] = \int_{x_1}^{x_2} L(y + \varepsilon\eta, y' + \varepsilon\eta', x)\,dx$$

$$= \int_{x_1}^{x_2} \left(L(y, y', x) + \frac{\partial L}{\partial y}\varepsilon\eta + \frac{\partial L}{\partial y'}\varepsilon\eta' \right) dx$$

by Taylor's theorem. Thus defining the quantity

$$\delta I \equiv I[y + \delta y] - I[y],$$

we get

$$\delta I = \varepsilon \int_{x_1}^{x_2} \left(\frac{\partial L}{\partial y}\eta + \frac{\partial L}{\partial y'}\eta' \right) dx.$$

The last term can be integrated by parts, to give

$$\int_{x_1}^{x_2} \frac{\partial L}{\partial y'} \eta' \, dx = \left[\frac{\partial L}{\partial y'} \eta \right]_{x_1}^{x_2} - \int_{x_1}^{x_2} \frac{d}{dx}\left(\frac{\partial L}{\partial y'} \right) \eta \, dx.$$

The term in square brackets vanishes since $\eta(x_1) = \eta(x_2) = 0$, and hence

$$\delta I = \varepsilon \int_{x_1}^{x_2} \left[\frac{\partial L}{\partial y} - \frac{d}{dx}\left(\frac{\partial L}{\partial y'} \right) \right] \eta \, dx. \tag{7.27}$$

If $y = y(x)$ is a stationary curve, then δI must vanish to first order, and so, using the above lemma, we find that y must satisfy the **Euler–Lagrange equation** for L, that is,

$$\frac{\partial L}{\partial y} - \frac{d}{dx}\left(\frac{\partial L}{\partial y'} \right) = 0. \tag{7.28}$$

Introducing some further notation which serves as a useful abbreviation, we define the **variational derivative, functional derivative,** or **Euler–Lagrange derivative** of L by

$$\frac{\delta L}{\delta y} \equiv \frac{\partial L}{\partial y} - \frac{d}{dx}\left(\frac{\partial L}{\partial y'} \right),$$

so that (7.27) can be written as

$$\delta I = \int_{x_1}^{x_2} \frac{\delta L}{\delta y} \delta y \, dx. \tag{7.29}$$

Then, in this formalism, the **principle of stationary action** requires

$$\delta I = 0 \tag{7.30}$$

for arbitrary δy, which leads immediately by the lemma to the Euler–Lagrange equation

$$\frac{\delta L}{\delta y} = 0. \tag{7.31}$$

The argument can be generalized to n dynamical variables each of which are functions of one variable $y_1(x), \ldots, y_n(x)$ in a straightforward manner. Then the action is defined in terms of the Lagrangian by

$$I[y_1, \ldots, y_n] = \int_{x_1}^{x_2} L(y_1, \ldots, y_n, y_1', \ldots, y_n', x) \, dx \tag{7.32}$$

and the variations

$$y_i \rightarrow \bar{y}_i = y_i + \delta y_i \quad (i = 1, 2, \ldots, n),$$

where

$$\delta y_i = \varepsilon \eta_i(x), \qquad \eta_i(x_1) = \eta_i(x_2) = 0,$$

lead to

$$\delta I = \int_{x_1}^{x_2} \frac{\delta L}{\delta y_i} \delta y_i \, dx \quad \text{(summed over } i\text{)},$$

with

$$\frac{\delta L}{\delta y_i} \equiv \frac{\partial L}{\partial y_i} - \frac{\mathrm{d}}{\mathrm{d}x}\left(\frac{\partial L}{\partial y_i'}\right).$$

The principle of stationary action, $\delta I = 0$, for arbitrary independent variations δy_i, produces the Euler–Lagrange equations

$$\frac{\partial L}{\partial y_i} - \frac{\mathrm{d}}{\mathrm{d}x}\left(\frac{\partial L}{\partial y_i'}\right) = 0 \qquad (i = 1, 2, \ldots, n). \tag{7.33}$$

The further generalization to a system of m dynamical variables $y_A(x)$ ($A = 1, 2, \ldots, m$), defined on an n-dimensional manifold M, starts from the action

$$I = \int_\Omega \mathscr{L}(y_A, y_{A,b}, x^a)\,\mathrm{d}\Omega, \tag{7.34}$$

where a comma in the subscript denotes a partial derivative, i.e. $y_{A,b} = \partial_b y_A$, and the Lagrangian $\mathscr{L}$ is a scalar density of weight $+1$ and leads to the Euler–Lagrange equations

$$\frac{\delta \mathscr{L}}{\delta y_A} \equiv \frac{\partial \mathscr{L}}{\partial y_A} - \left(\frac{\partial \mathscr{L}}{\partial y_{A,b}}\right)_{,b} = 0 \quad (A = 1, 2, \ldots, m). \tag{7.35}$$

The significance of the variational principle approach is that most, if not all, physical theories may be formulated by specifying a suitable Lagrangian. The Euler–Lagrange equations can then be computed in a straightforward manner and these constitute the **field equations** of the theory.

7.6 The variational method for geodesics

We now apply the technique of the last section to finding a convenient way for computing the geodesics of a given metric. We start from the Lagrangian functional (compare with (7.32))

$$L = L(x^a, \dot{x}^a, u),$$

where u is a parameter along a timelike curve and the dot denotes differentiation with respect to u, defined in terms of the metric by

$$L = [g_{ab}(x)\dot{x}^a\dot{x}^b]^{\frac{1}{2}}. \tag{7.36}$$

It follows from (6.59) that the action is

$$\int_{P_1}^{P_2} L\,\mathrm{d}u = \int_{P_1}^{P_2} \mathrm{d}s = s, \tag{7.37}$$

where s is the interval between any two points P_1 and P_2 on a curve connecting them. The **metric geodesic** between these points P_1 and P_2 is defined as that curve joining them whose interval is stationary under small variations which vanish at the end points. In other words, we need to solve the principle of stationary action problem $\delta s = 0$. The solution consists of the Euler–Lagrange equations (7.33) in the form

$$\frac{\partial L}{\partial x^a} - \frac{d}{du}\left(\frac{\partial L}{\partial \dot{x}^a}\right) = 0. \tag{7.38}$$

In principle these equations solve the problem, but in practice there are a number of difficulties. First of all, it is much better to work where possible with L^2 rather than L to avoid square roots. Then there is the freedom in the choice of the parameter u. Finally, in the case of an indefinite metric, there is the distinction between null and non-null geodesics. Assuming $L \neq 0$ and multiplying (7.38) by $-2L$, we get

$$2L\left[\frac{d}{du}\left(\frac{\partial L}{\partial \dot{x}^a}\right) - \frac{\partial L}{\partial x^a}\right] = 0 \tag{7.39}$$

which can be rewritten as

$$\frac{d}{du}\left(\frac{\partial L^2}{\partial \dot{x}^a}\right) - \frac{\partial L^2}{\partial x^a} = 2\frac{\partial L}{\partial \dot{x}^a}\frac{dL}{du}. \tag{7.40}$$

Substituting for L^2, the left-hand side of (7.40) produces

$$\frac{d}{du}\left(\frac{\partial L^2}{\partial \dot{x}^a}\right) - \frac{\partial L^2}{\partial x^a} = \frac{d}{du}\left[\frac{\partial}{\partial \dot{x}^a}(g_{bc}\dot{x}^b\dot{x}^c)\right] - \frac{\partial}{\partial x^a}(g_{bc}\dot{x}^b\dot{x}^c)$$

$$= \frac{d}{du}(2g_{ab}\dot{x}^b) - (\partial_a g_{bc})\dot{x}^b\dot{x}^c$$

$$= 2g_{ab}\ddot{x}^b + 2\partial_c g_{ab}\dot{x}^b\dot{x}^c - \partial_a g_{bc}\dot{x}^b\dot{x}^c$$

$$= 2g_{ab}\ddot{x}^b + 2\dot{x}^b\dot{x}^c[\tfrac{1}{2}(\partial_c g_{ba} + \partial_b g_{ca} - \partial_a g_{bc})]$$

$$= 2g_{ab}\ddot{x}^b + 2\dot{x}^b\dot{x}^c\{bc, a\},$$

where we have used symmetry, interchange of dummy indices, and (6.62). If we again assume that $L \neq 0$, then the right-hand side of (7.40) produces

$$2\frac{\partial L}{\partial \dot{x}^a}\frac{dL}{du} = 2\frac{\partial}{\partial \dot{x}^a}(g_{bc}\dot{x}^b\dot{x}^c)^{\frac{1}{2}}\frac{d}{du}\left(\frac{ds}{du}\right)$$

$$= 2(g_{bc}\dot{x}^b\dot{x}^c)^{-\frac{1}{2}}g_{ad}\dot{x}^d\frac{d^2s}{du^2}$$

$$= 2\left(\frac{d^2s}{du^2}\Big/\frac{ds}{du}\right)g_{ab}\dot{x}^b.$$

Equating these two results and dividing by 2 gives the equation (6.61). Multiplying through by g^{ad} and using (6.64) leads to

$$\ddot{x}^a + \Gamma^a_{bc}\dot{x}^b\dot{x}^c = (\ddot{s}/\dot{s})\dot{x}^a. \tag{7.41}$$

If we choose the parameter $u = s$, then the right-hand side vanishes, giving

$$\ddot{x}^a + \Gamma^a_{bc}\dot{x}^b\dot{x}^c = 0. \tag{7.42}$$

and hence s is an affine parameter. It follows from (7.41) that any other affine parameter is related to s by

$$\bar{s} = \alpha s + \beta, \tag{7.43}$$

where α and β are constants. A similar argument applies to spacelike geodesics (exercise).

In the case of an indefinite metric, the interval ds between neighbouring points on a curve may be zero. A null geodesic is a geodesic whose interval between any of its two points is zero. It follows from (7.36) that L vanishes and so the argument given above breaks down. However, it is possible to modify the argument (we shall not do it) to show that the general equations of a null geodesic are

$$\ddot{x}^a + \Gamma^a_{bc}\dot{x}^b\dot{x}^c = \lambda(u)\dot{x}^a,$$

where $\lambda(u)$ is some function of the curve's parameter u and where the tangent vector $\dot{x}^a$ satisfies $g_{ab}\dot{x}^a\dot{x}^b = 0$. As before, if the geodesic equations do not possess a right-hand side, that is, $\lambda = 0$, then the parameter u is called affine. Any other parameter $\bar{u}$ will be affine if it is related to u by

$$\bar{u} = \alpha u + \beta, \tag{7.44}$$

where α and β are constants.

Summarizing, if we define the quantity K by

$$2K \equiv g_{ab}\dot{x}^a\dot{x}^b = \alpha, \tag{7.45}$$

where α is a constant, and if we take u to be an **affine** parameter, then the most useful form of the geodesic equations is (exercise)

$$\frac{\partial K}{\partial x^a} - \frac{d}{du}\left(\frac{\partial K}{\partial \dot{x}^a}\right) = 0, \tag{7.46}$$

where

$$2K = \alpha = \begin{cases} 0, \\ +1, \\ -1, \end{cases} \tag{7.47}$$

depending on whether the tangent vector is null, or has positive or negative length, respectively, and where in the last two cases we take u to be the distance parameters s and σ. This is the approach we shall adopt in our ensuing work. It is possible, by (7.42), to read off directly from (7.46) the components of the connection Γ^a_{bc}, and this proves to be a very efficient way of calculating Γ^a_{bc}.

7.7 Isometries

Tensor calculus is largely concerned with how quantities change under coordinate transformations. It is of particular interest when a quantity does not change, i.e. remains invariant, under coordinate transformations. For example, coordinate transformations which leave a metric invariant are of importance since they contain information about the **symmetries** of a Riemannian manifold. Just as in an ordinary Euclidean space, there are two sorts of transformations: **discrete** ones, like reflections, and **continuous** ones, like translations and rotations. In most applications, these latter types are the more important ones and they can in principle be obtained systematically by obtaining the so-called Killing vectors of a metric, which we now discuss below.

A metric g_{ab} is **form-invariant** or simply **invariant** under the transformation $x^a \rightarrow x'^a$ if

$$g'_{ab}(y) = g_{ab}(y) \quad \text{for all coordinates } y^c, \tag{7.48}$$

that is, the transformed metric $g'_{ab}(x')$ is the **same** function of its argument x'^c as the original metric $g_{ab}(x)$ is of its argument x^c. Then a transformation leaving g_{ab} form-invariant is called an **isometry**. Since g_{ab} is a covariant tensor it transforms according to (7.4), or equivalently (interchanging primes and unprimes as we are free to do)

$$g_{ab}(x) = \frac{\partial x'^c}{\partial x^a} \frac{\partial x'^d}{\partial x^b} g'_{cd}(x').$$

Then, using (7.48), $x^a \rightarrow x'^a$ will be an isometry if

$$g_{ab}(x) = \frac{\partial x'^c}{\partial x^a} \frac{\partial x'^d}{\partial x^b} g_{cd}(x'). \tag{7.49}$$

It will be convenient to consider all quantities appearing in this equation to be functions of x using $x'^a = x'^a(x)$. In general, the condition (7.49) is very complicated, but it may be greatly simplified if we consider the special case of an **infinitesimal** coordinate transformation

$$x^a \rightarrow x'^a = x^a + \varepsilon X^a(x) \tag{7.50}$$

where ε is small and arbitrary and X^a is a vector field. Differentiating (7.50) gives

$$\frac{\partial x'^a}{\partial x^b} = \delta_b^a + \varepsilon \partial_b X^a,$$

and so, substituting in (7.49) and using Taylor's theorem, we get

$$g_{ab}(x) = (\delta_a^c + \varepsilon \partial_a X^c)(\delta_b^d + \varepsilon \partial_b X^d) g_{cd}(x^e + \varepsilon X^e)$$
$$= (\delta_a^c + \varepsilon \partial_a X^c)(\delta_b^d + \varepsilon \partial_b X^d)[g_{cd}(x) + \varepsilon X^e \partial_e g_{cd}(x) + \cdots]$$
$$= g_{ab}(x) + \varepsilon[g_{ad} \partial_b X^d + g_{bd} \partial_a X^d + X^e \partial_e g_{ab}] + O(\varepsilon^2).$$

Working to first order in ε and subtracting $g_{ab}(x)$ from each side, it follows that the quantity in square brackets must vanish. This quantity is simply the Lie derivative of g_{ab} with respect to X by (6.17), namely,

$$L_X g_{ab} = X^e \partial_e g_{ab} + g_{ad} \partial_b X^d + g_{bd} \partial_a X^d. \qquad (7.51)$$

Now we can replace ordinary derivatives by covariant derivatives in any expression for a Lie derivative and so, using (6.73) and (6.57), the condition for an infinitesimal isometry becomes

$$L_X g_{ab} = \nabla_b X_a + \nabla_a X_b = 0. \qquad (7.52)$$

These are called **Killing's equations** and any solution of them is called a **Killing vector field** X^a. In the language of §6.2, equation (7.52) states that the metric is 'dragged into itself' by the vector field X^a. We have thus established the following important result.

Theorem: An infinitesimal isometry is generated by a Killing vector $X^a(x)$ satisfying $L_X g_{ab} = 0$

It proves sufficient to restrict attention to infinitesimal transformations because it can be shown that it is possible to build up any finite transformation with non-zero Jacobian (i.e. a continuous transformation) by an integration process involving an infinite sequence of infinitesimal transformations.

Exercises

7.1 (§**7.1**) Write down the expression for the covariant derivative of a scalar density Φ of weight $+1$.

7.2 (§**7.3**) Denoting the transformation matrices by

$$J_{ab} = \left(\frac{\partial x^a}{\partial x'^b}\right), \qquad J^{ab} = \left(\frac{\partial x'^a}{\partial x^b}\right),$$

use the argument of §7.3 to show that

$$\partial_c J = J J^{ab} \partial_c J_{ab},$$

where $J = \det(J_{ab})$ is the Jacobian. Hence show from first principles that if $\mathfrak{T}^a$ is a vector density of weight $+1$ then so is $\partial_a \mathfrak{T}^a$.

7.3 (§**7.3**) Start from the assumption that, for an arbitrary vector field T^a,

$$\nabla_a[(-g)^{\frac{1}{2}} T^a] = \partial_a[(-g)^{\frac{1}{2}} T^a],$$

and show that this leads directly to the result

$$\nabla_a[(-g)^{\frac{1}{2}}] = \partial_a[(-g)^{\frac{1}{2}}] - \Gamma^b_{ba}(-g)^{\frac{1}{2}}$$

(which is consistent with the definition in Exercise 7.1)

7.4 (§**7.4**) Show that, for any vector field T^a, the divergence theorem in four dimensions can be written in the form

$$\int_{\partial\Omega} T^a(-g)^{\frac{1}{2}} dS_a = \int_{\Omega} \nabla_a T^a(-g)^{\frac{1}{2}} d^4 x.$$

7.5 (§**7.5**) Find the Euler–Lagrange equations for the Lagrangians
(i) $L(y, y', x) = y^2 + y'^2$,
(ii) $L(y_1, y_2, y'_1, y'_2, x) = xy_1^3 + y_1 y_2 + y_1(y_1'^2 + y_2'^2)$.

7.6 (§**7.6**) Trace the variational argument which leads to the equations for a spacelike geodesic. Defining K by (7.45) and (7.47), show that (7.40) can be written in the form (7.46). [Hint: if u is affine, then $dL/du = 0$.]

7.7 (§**7.7**) Use (7.45), (7.46), and (7.47) to find the geodesic equations of the spherically symmetric line element given in Exercise 6.31. Use the equations to read off directly the components Γ^a_{bc} and check them with those obtained in Exercise 6.31(ii). [Hint: remember $\Gamma^a_{bc} = \Gamma^a_{cb}$.]

7.8 (§**7.7**) Find all Killing vector solutions of the metric

$$g_{ab} = \begin{bmatrix} x^2 & 0 \\ 0 & x \end{bmatrix},$$

where $(x^a) = (x^0, x^1) = (x, y)$.

7.9 (§**7.7**) Deduce (7.52) from (7.51).

7.10 (§**7.7**) Find all the Killing vectors X^a of the three-dimensional Euclidean line element

$$ds^2 = dx^2 + dy^2 + dz^2.$$

[Hint: deduce from Killing's equations that $\partial_b X_a + \partial_a X_b = 0$, differentiate with respect to x^c, permute the indices to show that $\partial_b \partial_c X_a = 0$ and integrate to get $X^a = \omega^a_b x^b + t^a$, where $\omega_{ab} = -\omega_{ba}$ and t^a are constants of integration, usually termed **parameters**.]

Denoting the six independent constants of integration by $\lambda_1, \lambda_2, \lambda_3, \lambda_4, \lambda_5, \lambda_6$, write the general solution for X^a in the form

$$\lambda_1 X^{1a} + \lambda_2 X^{2a} + \lambda_3 X^{3a} + \lambda_4 X^{4a} + \lambda_5 X^{5a} + \lambda_6 X^{6a}$$

Find expressions for the vector fields X^a ($a = 1, 2, \ldots, 6$), and hence, or otherwise, find all values of $[X^\alpha, X^\beta]$. Interpret the six Killing vector fields in terms of geometrical transformations.

7.11 (§**7.7**) Show that if X^a and Y^a are Killing vectors then so is any linear combination $\lambda X^a + \mu Y^a$, where λ and μ are constants.

7.12 (§**7.7**) Consider the following operator identity:

$$L_u L_v - L_v L_u = L_{[u, v]}.$$

(i) Check it holds when applied to an arbitrary scalar function f.
(ii) Check it holds when applied to an arbitrary contravariant vector field m^a. [Hint: use the Jacobi identity.]
(iii) Deduce that the identity holds when applied to a covariant vector field p_a. [Hint: let $f = m^a p_a$, where m^a is arbitrary.]

Use the identity to prove that if u and v are vector fields then so is their commutator $[u, v]$.

Given that $\partial/\partial x$ and $-y\partial/\partial x + x\partial/\partial y$ are Killing vector fields, find another.

7.13 (§**7.7**) Express $(\nabla_c \nabla_b - \nabla_b \nabla_c)X_a$ in terms of the Riemann tensor. Use this result to prove that any Killing vector satisfies

$$g^{bc}\nabla_b \nabla_a X_c - R_{ab}X^b = 0.$$

7.14 (§**7.7**) By making use of the identity

$$R^a{}_{bcd} + R^a{}_{dbc} + R^a{}_{cdb} \equiv 0$$

or otherwise, prove that a Killing vector satisfies

$$\nabla_c \nabla_b X_a = R_{abcd}X^d.$$

C. General Relativity

Special relativity revisited

8

8.1 Minkowski space-time

As we saw in Chapter 2, special relativity discards the old Newtonian picture in which absolute time is split off from three-dimensional Euclidean space. Instead, we introduce a four-dimensional continuum called space-time in which an event has coordinates (t, x, y, z) and where the square of the infinitesimal interval ds between infinitesimally separated events satisfies the Minkowski line element (2.13). The essence of special relativity lies in the special Lorentz transformations, and the significance of the Minkowski line element is that it is invariant under such transformations. We now use the language of Part B to formulate this more precisely.

Minkowski space-time, or simply **flat space**, is defined as a four-dimensional manifold endowed with a **flat** metric of signature -2. Then, by definition, since the metric is flat, there exists a special coordinate system covering the whole manifold in which the metric is diagonal, with diagonal elements equal to ± 1. From now on, we shall use the convention that lower case latin indices run from 0 to 3. The special coordinate system is called a **Minkowski coordinate** system and is written

$$(x^a) = (x^0, x^1, x^2, x^3) = (t, x, y, z). \tag{8.1}$$

We adopt the sign convention in which the **Minkowski line element** takes the form

$$ds^2 = dt^2 - dx^2 - dy^2 - dz^2. \tag{8.2}$$

We write this in tensorial form as

$$ds^2 = \eta_{ab} dx^a dx^b, \tag{8.3}$$

where from now on we will always take η_{ab} to denote the **Minkowski metric**

$$\eta_{ab} \equiv \begin{bmatrix} 1 & 0 & 0 & 0 \\ 0 & -1 & 0 & 0 \\ 0 & 0 & -1 & 0 \\ 0 & 0 & 0 & -1 \end{bmatrix} = \text{diag}(1, -1, -1, -1). \tag{8.4}$$

If we use some other general coordinate system then we shall write the metric in the form

$$ds^2 = g_{ab}\,dx^a dx^b.$$

For example, in spherical polar coordinates,

$$(x^a) = (t, r, \theta, \phi),$$

where, as usual,

$$x = r\sin\theta\cos\phi, \qquad y = r\sin\theta\sin\phi, \qquad z = r\cos\theta,$$

the line element becomes

$$ds^2 = dt^2 - dr^2 - r^2\,d\theta^2 - r^2\sin^2\theta\,d\phi^2$$

and the metric is

$$g_{ab} = \text{diag}(1, -1, -r^2, -r^2\sin^2\theta).$$

One of the main results of Part B is the theorem of §6.11, which states that a necessary and sufficient condition for a metric to be flat is that its Riemann tensor vanishes. In Minkowski coordinates, the metric η_{ab} is constant and so the connection Γ^a_{bc} vanishes in this coordinate system, from which it is clear that the Riemann curvature tensor vanishes. However, in a general coordinate system, the connection components will not necessarily vanish. For example, in spherical polar coordinates, we find that Γ^a_{bc} has non-vanishing components

$$\left.\begin{array}{ll} \Gamma^1_{22} = -r, & \Gamma^1_{33} = r\sin^2\theta, \\[4pt] \Gamma^2_{12} = r^{-1}, & \Gamma^2_{33} = -\sin\theta\cos\theta, \\[4pt] \Gamma^3_{13} = r^{-1}, & \Gamma^3_{23} = \cot\theta, \end{array}\right\} \tag{8.5}$$

but if we compute the Riemann tensor we will again find

$$R^a{}_{bcd} = 0,$$

as required by the theorem.

8.2 The null cone

In Minkowski space-time, the square of the length or norm of a vector is defined as usual by

$$X^2 = g_{ab}X^a X^b = X_a X^a. \tag{8.6}$$

The vector is said to be

$$\left.\begin{array}{lll} \textbf{timelike} & \quad \text{if } X^2 > 0, \\[4pt] \textbf{spacelike} & \quad \text{if } X^2 < 0, \\[4pt] \textbf{null} \text{ or } \textbf{lightlike} & \quad \text{if } X^2 = 0. \end{array}\right\} \tag{8.7}$$

Two vectors X^a and Y^a are orthogonal if their **inner product** vanishes, that is,

$$g_{ab} X^a Y^a = 0,$$

from which it follows that a null vector is orthogonal to itself.

The set of all null vectors at a point P of a Minkowski manifold forms a double cone called the **null cone** or **light cone**. In Minkowski coordinates, the null vectors X^a at P satisfy

$$\eta_{ab} X^a X^b = 0,$$

that is,

$$(X^0)^2 - (X^1)^2 - (X^2)^2 - (X^3)^2 = 0, \tag{8.8}$$

which is the equation of a double cone. This null cone lies in the tangent space T_P at P, but since it is easy to show that the tangent space is itself a Minkowski manifold (by (8.8)) we can identify the tangent space with the underlying manifold and regard the null cone as lying in the manifold. We will not be able to do this when we go on to consider non-flat manifolds. If we define the timelike vector T^a in Minkowski coordinates by $T^a = (1, 0, 0, 0)$, then a timelike or null vector X^a is said to be

future-pointing if $\eta_{ab} X^a T^b > 0$,

past-pointing if $\eta_{ab} X^a T^b < 0$.

The future-pointing vectors all lie inside or on one sheet of the cone called the **future sheet** and past-pointing vectors lie inside or on the **past sheet** (Fig. 8.1).

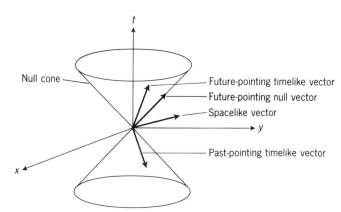

Fig. 8.1 The null cone with one dimension (the z-direction) suppressed.

8.3 The Lorentz group

The Lorentz transformations are defined as those linear homogeneous transformations

$$x^a \rightarrow x'^a = L^a{}_b x^b \tag{8.9}$$

of Minkowski coordinates which leave the Minkowski metric η_{ab} **invariant**. From (8.9),

$$\frac{\partial x'^a}{\partial x^b} = L^a{}_b,$$

and, substituting in the transformation formula for a metric (7.4) (with primes and unprimes interchanged), we get (exercise)

$$L^a_{\ c}L^b_{\ d}\eta_{ab} = \eta_{cd} \tag{8.10}$$

since the metric remains invariant. We see from (7.49) that Lorentz transformations are **isometries**. If follows immediately from (8.10) that Lorentz transformations preserve lengths and innner products of vectors. The Lorentz transformations form a group called the **Lorentz group** L. The identity element of the group is δ^a_b and the inverse element is given by the inverse matrix. The matrix $L^a_{\ b}$ is invertible, because if we take determinants of each side of (8.10) we get

$$(\det L^a_{\ b})^2 = 1 \quad \Rightarrow \quad \det L^a_{\ b} = \pm 1,$$

and so the matrix is non-singular. If we set $c = d = 0$ in (8.10), we also find that

$$(L^0_{\ 0})^2 - [(L^1_{\ 0})^2 + (L^2_{\ 0})^2 + (L^3_{\ 0})^2] = 1,$$

from which it follows that $(L^0_{\ 0})^2 \geqslant 1$ and so either $L^0_{\ 0} \geqslant 1$ or $L^0_{\ 0} \leqslant -1$. We divide Lorentz transformations into four separate classes depending on whether $\det L^a_{\ b} = \pm 1$ and $L^0_{\ 0} \geqslant 1$ or $L^0_{\ 0} \leqslant -1$. If $\det L^a_{\ b} = +1$, then $L^a_{\ b}$ is called **proper** or **orientation preserving**. An example of an improper Lorentz transformation is the discrete transformation

$$t' = t, \qquad x' = -x, \qquad y' = y, \qquad z' = z,$$

which reverses the x-direction. If $L^0_{\ 0} \geqslant 1$, then $L^a_{\ b}$ is called **orthocronous** or **time-orientation preserving**. An example of a non-orthocronous Lorentz transformation is the discrete transformation

$$t' = -t, \qquad x' = x, \qquad y' = y, \qquad z' = z,$$

which reverses the t-direction. The proper orthochronous transformations, denoted by $L^\uparrow_+$ (read 'L arrow plus') from a **subgroup** of L. Clearly, $L^\uparrow_+$ contains the identity, whereas the other three subsets do not and hence are not subgroups.

In fact, $L^\uparrow_+$ is a six-parameter continuous group of transformations. We can interpret the parameters physically by considering the transformation actively as transforming one inertial frame S into another one at rest with respect to an inertial frame S' in general position (see Fig. 2.20). Then two parameters correspond to the two Euler rotations required to line up the x-axis of S with the velocity of S', one parameter corresponds to a boost from S to a frame at rest relative to S' (and this parameter depends on the velocity of S' relative to S), and the final three parameters correspond to the three Euler rotations required to rotate the frame into the same orientation that S' has. Another subgroup of L is the ordinary three-dimensional rotation group.

The **Poincaré group** P consists of those linear inhomogeneous transformations which leave η_{ab} invariant. A Poincaré transformation is made up of a Lorentz transformation together with an arbitrary translation (in space and time), i.e.

$$x^a \rightarrow x'^a = L^a_{\ b}x^b + t^a. \tag{8.11}$$

The Lorentz group L is a proper subgroup of P and the translations form an invariant (normal) subgroup of P. The Poincaré group P is a ten-parameter group, consisting of six Lorentz parameters plus four translation parameters. The continuous Poincaré transformations constitute the full set of isometries of the Minkowski metric. Physically, a Poincaré transformation maps one inertial frame S into another inertial frame S' in general position.

8.4 Proper time

A **timelike world-line** or **timelike curve** is defined as a curve whose tangent vector is everywhere timelike. If, in particular, the curve is a geodesic, it is called a **timelike geodesic**. Timelike curves represent tracks on which material particles or observers can travel. From §8.2, we see that the velocity tangent vector to a timelike curve at any point P must lie within the null cone emanating from P (Fig. 8.2). This is a manifestation of the special relativity result that material particles travel with speeds always less than the speed of light. Spacelike and null curves and geodesics are defined in an analogous manner to timelike ones.

At any point P, we define the **null cone** or **light cone** which consists of all null geodesics passing through P. This coincides with the null cone of null vectors passing through P. Then the null cone divides space-time into three distinct regions namely future, past, and elsewhere (Fig. 8.3). Any point in the **future** or **past** may be reached by a future-directed or past-directed timelike geodesic, respectively. Any point in the region exterior to the null cone, called **elsewhere**, can be reached by a geodesic which is everywhere spacelike. This is an invariant division of events which all observers agree upon. This follows because of the invariance of η_{ab} under a Lorentz transformation, which means that null cones get mapped onto null cones. Moreover, events to the future of P get mapped into events which are still to the future of P under an orthochronous Lorentz transformation. A similar result holds for past events. However, non-orthochronous Lorentz transformations reverse the past and future.

Since Γ^a_{bc} vanishes in Minkowski coordinates, the equations for a non-null geodesic (7.42) reduce to

$$\frac{d^2 x^a}{du^2} = 0 \tag{8.12}$$

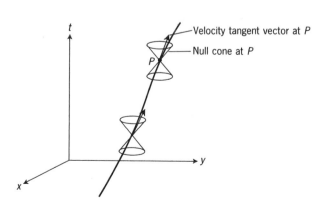

Fig. 8.2 World-line of material particle.

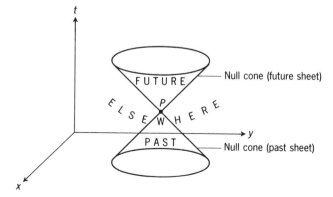

Fig. 8.3 Invariant classification of events relative to P.

for some affine parameter u, where the tangent vector satisfies

$$\eta_{ab}\frac{\mathrm{d}x^a}{\mathrm{d}u}\frac{\mathrm{d}x^b}{\mathrm{d}u} = k. \tag{8.13}$$

The geodesic is timelike or spacelike depending on whether $k > 0$ or $k < 0$, respectively. In the case when $k > 0$, we introduce a new parameter

$$u \to \bar{u} = \bar{u}(u)$$

satisfying

$$\left(\frac{\mathrm{d}\bar{u}}{\mathrm{d}u}\right)^2 = k.$$

It follows from (8.13) that the new tangent vector $\mathrm{d}x^a/\mathrm{d}\bar{u}$ has **unit** length. The parameter $\bar{u}$ is called the **proper time** and is denoted by τ. Thus, in relativistic units, from (8.3) and (8.13), the proper time satisfies

$$\mathrm{d}\tau^2 = \mathrm{d}s^2. \tag{8.14}$$

This shows that τ is an **affine parameter** along timelike geodesics.

In non-relativistic units the equation for the proper time becomes

$$\mathrm{d}\tau^2 = \frac{1}{c^2}\,\mathrm{d}s^2, \tag{8.15}$$

which checks dimensionally since s is a distance parameter. Let us see how proper time τ relates to coordinate time t for any observer whose velocity at time t is v, where

$$v = \left(\frac{\mathrm{d}x}{\mathrm{d}t}, \frac{\mathrm{d}y}{\mathrm{d}t}, \frac{\mathrm{d}z}{\mathrm{d}t}\right).$$

From (8.15) and (3.13), we have

$$\mathrm{d}\tau^2 = \frac{1}{c^2}\,\mathrm{d}s^2 = \frac{1}{c^2}(c^2\,\mathrm{d}t^2 - \mathrm{d}x^2 - \mathrm{d}y^2 - \mathrm{d}z^2)$$

$$= \mathrm{d}t^2\left\{1 - \frac{1}{c^2}\left[\left(\frac{\mathrm{d}x}{\mathrm{d}t}\right)^2 + \left(\frac{\mathrm{d}y}{\mathrm{d}t}\right)^2 + \left(\frac{\mathrm{d}z}{\mathrm{d}t}\right)^2\right]\right\}$$

$$= \mathrm{d}t^2\left(1 - \frac{v^2}{c^2}\right)$$

So the proper time between t_0 and t_1, is given by

$$\tau = \int_{t_0}^{t_1}\left(1 - \frac{v^2}{c^2}\right)^{\frac{1}{2}}\mathrm{d}t, \tag{8.16}$$

in agreement with (3.17).

8.5 An axiomatic formulation of special relativity

We are now in a position to give a completely precise formulation of special relativity which will prove useful when we wish to generalize to the general theory. We do this by stating two sets of postulates or axioms.

Axiom I. Space and time are represented by a four-dimensional manifold endowed with a symmetric affine connection Γ^a_{bc} and a metric tensor g_{ab} satisfying the following:

(i) $\nabla_c g_{ab} = 0$;
(ii) along any timelike world-line, a parameter τ is defined by
$$d\tau^2 = g_{ab} dx^a dx^b;$$
(iii) $R^a{}_{bcd} = 0$.

Axiom II. There exist privileged classes of curves in the manifold singled out as follows:

(i) ideal clocks travel along timelike curves and measure the parameter τ (called the 'proper time');
(ii) free particles travel along timelike geodesics;
(iii) light rays travel along null geodesics.

The first axiom defines the **geometry** of the theory and the second axiom puts in the **physics**. Thus, the first axiom states that Γ^a_{bc} is the metric connection (by I(i)) and that the metric is flat (by I(iii)) and defines a formal parameter whose physical significance is given in the second axiom. The first part of the second axiom makes physical the distinction between space and time in the manifold. In canonical (Minkowski) coordinates, it distinguishes the coordinate x^0 from the other three as the 'time' coordinate. More precisely, it states that it is the proper time τ which a clock measures in accordance with the clock hypothesis. The remainder of Axiom II singles out the privileged curves that free particles and light rays travel along.

Looking at this theory from a purely axiomatic viewpoint, one can ask, Is there any a priori reason for singling out timelike and null geodesics as trajectories for material particles and photons or light rays, or could one make some other choice (say, spacelike geodesics)? In Newtonian theory, free particles travel in straight lines, by Newton's first law. It would seem natural, therefore, to take geodesics as the analogue of straight lines. The significance of timelike geodesics is that their choice, unlike the case of spacelike geodesics, is consistent with **causality**. As we have seen, Minkowski space-time admits the Poincaré group as its invariance group. Hence, if two neighbouring events P and Q of the history of a free particle occur on a timelike geodesic at proper times τ and $\tau + d\tau$, respectively, then an orthochronous Poincaré transformation preserves the fact that Q occurs **after** P. This is consistent with causality, since we say that the arrival of the particle at Q is **caused** by its having previously been at P.

Null geodesics possess a special property which makes them natural candidates for light signals. The equation of a null geodesic in Minkowski coordinates is

$$\frac{d^2 x^a}{du^2} = 0, \tag{8.17}$$

where

$$\eta_{ab} \frac{dx^a}{du} \frac{dx^b}{du} = 0 \qquad (8.18)$$

for an affine parameter u. Integrating (8.17), we get

$$\frac{dx^a}{du} = k^a, \qquad (8.19)$$

where the components of k^a are constants of integration. Substituting in (8.18), we obtain

$$\eta_{ab} k^a k^b = 0, \qquad (8.20)$$

and so k^a is a null vector. Let us define the 3-velocity $\boldsymbol{v}$ along the null geodesic by

$$\boldsymbol{v} = (v^1, v^2, v^3) = \left(\frac{dx^1}{dx^0}, \frac{dx^2}{dx^0}, \frac{dx^3}{dx^0} \right) = \left(\frac{k^1}{k^0}, \frac{k^2}{k^0}, \frac{k^3}{k^0} \right), \qquad (8.21)$$

using (8.19) and the fact that $k^0 \neq 0$ (why?). Writing (8.20) out fully, we find

$$(k^0)^2 - (k^1)^2 - (k^2)^2 - (k^3)^2 = 0,$$

and hence it follows from (8.21) that $\boldsymbol{v}^2 = 1$. Thus, null geodesics have associated with them a characteristic velocity of magnitude 1. Furthermore, this property is preserved under a Poincaré transformation, and so they seem natural candidates for encoding the constancy of the velocity of light.

8.6 A variational principle approach to classical mechanics

We met an introduction to relativistic mechanics in Chapter 4. We shall now look for a formulation which rests on a variational principle. The importance of the variational formulation of a physical theory is that it is often very simple and elegant and, moreover, it is one method which lends itself easily to generalization. Indeed, most current theories use the variational approach as their starting point. We start by summarizing the variational formulation of a classical system moving under a conservative force.

A mechanical system is described by n **generalized coordinates** x^a ($a = 1, 2, \ldots, n$) which are functions of time t, n **generalized velocities** $\dot{x}^a$, the **kinetic energy** $T = \frac{1}{2} g_{ab} \dot{x}^a \dot{x}^b$, and the **potential energy** $V(x)$ which gives rise to n **generalized forces** $F_a = -\partial V/\partial x^a$. The **Lagrangian** L is defined to be

$$L \equiv T - V.$$

Then the principle of stationary action is

$$\delta S = \delta \int_{t_1}^{t_2} L \, dt = 0$$

and this leads to the Euler–Lagrange equations

$$\frac{\partial L}{\partial x^a} - \frac{d}{dt} \left(\frac{\partial L}{\partial \dot{x}^a} \right) = 0.$$

A straightforward calculation leads to the equations of motion

$$\ddot{x}^a + \Gamma^a_{bc} \dot{x}^b \dot{x}^c = F^a, \qquad (8.22)$$

where Γ^a_{bc} is the metric connection of g_{ab}. If there are no external forces, then the above equations can be thought of as defining **geodesics** on an n-dimensional Riemannian manifold, with metric g_{ab} called **configuration space**. We define **generalized momenta** $p_a \equiv \partial L/\partial \dot{x}^a$ and the **Hamiltonian** H by

$$H \equiv p_a \dot{x}^a - L.$$

If H is time-independent, then it can be shown to be equal to the total energy E of the system.

As an example of this formalism, let us consider the simple case of a **free particle** moving in three dimensions with velocity $\boldsymbol{u}$. Adopting Cartesian coordinates, we have

$$(x^a) = (x^1, x^2, x^3) = (x, y, z).$$

Then

$$T = \tfrac{1}{2}mu^2 = \tfrac{1}{2}m(\dot{x}^2 + \dot{y}^2 + \dot{z}^2),$$

from which we find

$$g_{ab} = \mathrm{diag}(m, m, m) = m\delta_{ab}$$

By assumption, $V = 0$, and so

$$L = T = \tfrac{1}{2}mu^2, \tag{8.23}$$

giving generalized momenta

$$p_x = \frac{\partial L}{\partial \dot{x}} = m\dot{x}, \qquad p_y = m\dot{y}, \qquad p_z = m\dot{z}.$$

The Euler–Lagrange equations are

$$\frac{\mathrm{d}}{\mathrm{d}t}(m\dot{x}) = 0, \qquad \frac{\mathrm{d}}{\mathrm{d}t}(m\dot{y}) = 0, \qquad \frac{\mathrm{d}}{\mathrm{d}t}(m\dot{z}) = 0,$$

which are just the three components of Newton's second law. The Hamiltonian is

$$H = \boldsymbol{p} \cdot \boldsymbol{u} - L = m(\dot{x}^2 + \dot{y}^2 + \dot{z}^2) - T = \tfrac{1}{2}mu^2 = T = E.$$

In general, if we consider a system with no forces acting, then the Lagrangian reduces to

$$T = \tfrac{1}{2}g_{ab}\dot{x}^a\dot{x}^b.$$

This Lagrangian is identical to the quantity K defined in (7.45) of §7.6. In that section, we saw that (if we work with affine parameters) this gives the same Euler–Lagrange equations as the Lagrangian (7.37), namely, as

$$\frac{\mathrm{d}s}{\mathrm{d}t} = (g_{ab}\dot{x}^a\dot{x}^b)^{\frac{1}{2}},$$

does. Thus, for convenience, we may take the action S for a free particle to be

$$S = \int_{t_1}^{t_2} \frac{\mathrm{d}s}{\mathrm{d}t}\,\mathrm{d}t = \int_{t_1}^{t_2} \mathrm{d}s \tag{8.24}$$

8.7 A variational principle approach to relativistic mechanics

We now consider a free particle in relativistic mechanics moving on a curve

$$x^a = x^a(\tau),$$

where τ is the proper time. Since τ is an affine parameter, we assume from (8.24) of the last section that the action can be written as

$$S = -\alpha \int_{\tau_1}^{\tau_2} \mathrm{d}s, \qquad (8.25)$$

where α is a constant to be determined. Working in Minkowski coordinates, we can write the action as

$$S = -\alpha \int_{\tau_1}^{\tau_2} (\eta_{ab} \dot{x}^a \dot{x}^b)^{\frac{1}{2}} \, \mathrm{d}\tau,$$

where a dot denotes differentiation with respect to τ. The Lagrangian is therefore

$$L = -\alpha(\eta_{ab} \dot{x}^a \dot{x}^b)^{\frac{1}{2}}$$

and the Euler–Lagrange equations

$$\frac{\partial L}{\partial x^a} - \frac{\mathrm{d}}{\mathrm{d}\tau}\left(\frac{\partial L}{\partial \dot{x}^a}\right) = 0$$

produce

$$\frac{\mathrm{d}}{\mathrm{d}\tau}\left[\alpha(\eta_{cd}\dot{x}^c\dot{x}^d)^{-\frac{1}{2}}\eta_{ab}\dot{x}^b\right] = 0. \qquad (8.26)$$

Since

$$\eta_{ab}\dot{x}^a\dot{x}^b = \eta_{ab}\frac{\mathrm{d}x^a}{\mathrm{d}\tau}\frac{\mathrm{d}x^b}{\mathrm{d}\tau} = \frac{\mathrm{d}s^2}{\mathrm{d}\tau^2} = 1$$

in relativistic units, the field equations (8.26) reduce to $\ddot{x}^a = 0$, which are the standard geodesic equations in Minkowski coordinates.

Instead of using the proper time τ as our time parameter, let us use instead the coordinate time t and see how various quantities are defined in terms of time and space coordinates. The equation of the world-line of the particle is now

$$x = x(t), \qquad y = y(t), \qquad z = z(t),$$

and it has a 3-velocity $\boldsymbol{u}$ defined by

$$\boldsymbol{u} = (u_1, u_2, u_3) = \left(\frac{\mathrm{d}x}{\mathrm{d}t}, \frac{\mathrm{d}y}{\mathrm{d}t}, \frac{\mathrm{d}z}{\mathrm{d}t}\right).$$

Using

$$\mathrm{d}s^2 = \eta_{ab}\mathrm{d}x^a\mathrm{d}x^b$$
$$= \mathrm{d}t^2 - \mathrm{d}x^2 - \mathrm{d}y^2 - \mathrm{d}z^2$$
$$= \mathrm{d}t^2(1 - u^2),$$

we can write the action (8.25) as

$$S = -\alpha \int_{t_1}^{t_2} (1 - u^2)^{\frac{1}{2}} \, \mathrm{d}t,$$

where the new Lagrangian (which we shall also write as L) is

$$L = -\alpha(1-u^2)^{\frac{1}{2}} = -\alpha + \tfrac{1}{2}\alpha u^2 + \cdots$$

for small velocities. Comparing this with the classical expression (8.23), namely $\tfrac{1}{2}mu^2$, we may identify α with the mass of the particle as $u \to 0$. Note that the additive constant $-\alpha$ in the Lagrangian is unimportant (see Exercise 8.9). Thus α is equal to the **rest mass** m_0 of the particle. Hence, we have

$$L = -m_0(1-u^2)^{\frac{1}{2}}. \tag{8.27}$$

We define the 3-momentum $\boldsymbol{p}$ by (check)

$$\boldsymbol{p} = \left(\frac{\partial L}{\partial u_1}, \frac{\partial L}{\partial u_2}, \frac{\partial L}{\partial u_3}\right) = m_0(1-u^2)^{-\frac{1}{2}}\boldsymbol{u}. \tag{8.28}$$

Comparing this with the classical relationship $\boldsymbol{p} = m\boldsymbol{u}$, we define the **relativistic mass** m by (see (4.11))

$$m = m_0(1-u^2)^{-\frac{1}{2}}.$$

Using the Hamiltonian to define the **energy** E (see (4.17)), we find

$$E = H = \boldsymbol{p}\cdot\boldsymbol{u} - L = m_0(1-u^2)^{-\frac{1}{2}} = m \tag{8.29}$$

after some simple algebra. We have thus regained the results of (4.19) in relativistic units.

8.8 Covariant formulation of relativistic mechanics

We finish this discussion of relativistic mechanics by giving a full **4-dimensional** or **covariant** formulation of the variational principle. The action S is defined as

$$S = -m_0 \int_{\tau_1}^{\tau_2} (g_{ab}\dot{x}^a\dot{x}^b)^{\frac{1}{2}} \, d\tau,$$

where g_{ab} is a flat metric and is used for raising and lowering indices. The **4-velocity** u^a is defined by

$$u^a \equiv \frac{dx^a}{d\tau} = \dot{x}^a, \tag{8.30}$$

and the **4-acceleration** a^b by

$$a^b \equiv \frac{du^b}{d\tau} = \frac{d^2x^b}{d\tau^2} = \ddot{x}^b. \tag{8.31}$$

The covariant **4-momentum** p_a is defined by

$$p_a \equiv -\frac{\partial L}{\partial \dot{x}^a},$$

from which we find that its contravariant form is given by

$$p^a = g^{ab} p_b = m_0 u^a. \tag{8.32}$$

If a particle is acted on by a force, then the four-dimensional version of Newton's second law becomes

$$f^a = \frac{dp^a}{d\tau}, \tag{8.33}$$

where f^a is called the **4-force**. If there is no external force acting, then

$$\frac{dp^a}{d\tau} = 0 \quad \Rightarrow \quad p^a = l^a, \tag{8.34}$$

where l^a is a constant 4-vector. This is the **conservation of 4-momentum** law and generalizes to an isolated system of n particles with 4-momenta $p_i{}^a$ ($i = 1, 2, \ldots, n$)

$$\sum_{i=1}^{n} p_i{}^a = l^a,$$

where l^a is a constant 4-vector. Finally, we define the **angular momentum tensor** l^{ab} of the particle by

$$l^{ab} = x^a p^b - x^b p^a. \tag{8.35}$$

If we now assume that m_0 is a **scalar**, then it follows that all the quantities have the tensor character indicated under a general coordinate transformation. If, in particular, we restrict attention to Minkowski coordinates we can relate these four-dimensional quantities to the three-dimensional ones of the last section and Chapter 4. We can then consider how the four-dimensional quantities transform under a **Lorentz transformation** and so obtain the transformation law for the three-dimensional quantities (exercise). Thus, in particular, we can confirm the transformation equations (4.21) for the energy and momentum of a particle.

We have considered the main ingredients of special relativistic mechanics, but we shall not pursue the topic further. We shall, rather, concentrate on our main task — that of establishing the general theory.

Exercises

8.1 (§8.1) Check (8.5) and show that the Riemann tensor vanishes.

8.2 (§8.2) Show that a timelike vector cannot be orthogonal to a null vector or to another timelike vector. Show that two null vectors are orthogonal if and only if they are parallel.

8.3 (§8.2) The vectors T, X, Y, and Z have components

$$T^a = (1, 0, 0, 0), \quad X^a = (0, 1, 0, 0), \quad Y^a = (0, 0, 1, 0),$$

$$Z^a = (0, 0, 0, 1).$$

Show that the only non-vanishing inner products between the vectors are

$$T^2 = -X^2 = -Y^2 = -Z^2 = 1.$$

Define the following:

$$L^a = \frac{1}{\sqrt{2}}(T^a + Z^a), \qquad N^a = \frac{1}{\sqrt{2}}(T^a - Z^a),$$

$$M^a = \frac{1}{\sqrt{2}}(X^a + iY^a), \qquad \bar{M}^a = \frac{1}{\sqrt{2}}(X^a - iY^a),$$

where $i = \sqrt{-1}$. Treating M^a and $\bar{M}^a$ as vectors, show that all four vectors are null and the only non-vanishing inner products are

$$L^a N_a = -M^a \bar{M}_a = 1.$$

8.4 (§8.3) (i) Check that (8.9) leads to (8.10), assuming invariance.
(ii) Show that the Lorentz transformations form a group.
(iii) Show that the Poincaré transformations form a group.

8.5 (§8.3) Show that a Killing vector X_a satisfies the equation $\partial_b \partial_c X_a = 0$ in flat space in Minkowski coordinates. [Hint: use Exercise 7.10 or Exercise 7.14.] Deduce that the Killing vectors are given by

$$X_a = \omega_{ab} x^b + t_a,$$

where $\omega_{ab} = -\omega_{ba}$ and t_a are arbitrary parameters (constants of integration). How many parameters are there in
(a) an n-dimensional manifold?
(b) Minkowski space-time?
What do the parameters correspond to physically in Minkowski space-time?

8.6 (§8.4) Prove that the proper time is an affine parameter along timelike geodesics.

8.7 (§8.6) Establish the equation of motion (8.22).

8.8 (§8.6) Consider two masses m_1 and m_2 suspended on the ends of a rope passing over a frictionless pulley. Show that the Lagrangian can be written in the form

$$L = \tfrac{1}{2}(m_1 + m_2)\dot{x}^2 + m_1 g x + m_2 g(l - x),$$

where the mass m_1 is a distance x below the horizontal and l is a constant. Find the Euler–Lagrange equation of motion. Define the generalized momentum for the system and hence obtain the Hamiltonian.

8.9 (§8.7) If L is a Lagrangian, then show that the Lagrangians L_1 and L_2, where (i) $L_1 = \lambda L$ and (ii) $L_2 = L + \mu$, with λ and μ constants, possess the same field equations as L. Show also that if $L \neq 0$ then the Lagrangians (iii) $L_3 = L^2$ and (iv) $L_4 = L^{\frac{1}{2}}$ give rise to the same field equations.

8.10 (§8.8) Show that, in Minkowski space-time in Minkowski coordinates, $u^a = (u^0, u^1, u^2, u^3) = \gamma(1, \boldsymbol{u})$, where $\gamma = (1 - u^2)^{-\frac{1}{2}}$. Show also that $p^a = (E, \boldsymbol{p})$. By considering the invariant $p_a p^a$, deduce that (see (4.20))

$$E^2 - p^2 = m_0{}^2.$$

Use the four-dimensional version of Newton's second law to identify the 4-force as

$$f^a = \gamma(\boldsymbol{u} \cdot \boldsymbol{F}, \boldsymbol{F}),$$

where $\boldsymbol{F}$ is the force acting on the particle. Show also that

$$\frac{\mathrm{d}p^a}{\mathrm{d}\tau} = \gamma\left(\frac{\mathrm{d}E}{\mathrm{d}t}, \frac{\mathrm{d}\boldsymbol{p}}{\mathrm{d}t}\right)$$

and give a physical interpretation of the zero component of the four-dimensional Newton's law.

8.11 (§8.8)
(i) Use the tensor transformation law on the 4-velocity u^a to find the transformation properties of $\boldsymbol{u}$ under a special Lorentz transformation between two frames in standard configuration moving with velocity v. Show in particular, that $\gamma'/\gamma = \beta(1 - u_x v)$, where $\beta = (1 - v^2)^{-\frac{1}{2}}$.
(ii) Find the transformation properties of E and $\boldsymbol{p}$ under a special Lorentz transformation.
(iii) Find the transformation properties of $\boldsymbol{F}$ under a special Lorentz transformation. Are forces still absolute quantities in special relativity?
(iv) A particle moves parallel to the x-axis under the influence of a force $\boldsymbol{F} = (F, 0, 0)$. What is the force in a frame co-moving with the particle?

9 The principles of general relativity

9.1 The role of physical principles

We are at last ready to embark on our central task, namely, that of extending special relativity to a theory which incorporates gravitation. In this chapter, we shall undertake a detailed consideration of the physical principles which guided Einstein in his search for the general theory. There is a school of thought that considers this an unnecessary process, but rather argues that it is sufficient to state the theory and investigate its consequences. There seems little doubt, however, that consideration of these physical principles helps give insight into the theory and promotes understanding. The mere fact that they were important to Einstein would seem sufficient to justify their inclusion. If nothing else, it will help us to understand how one of the greatest achievements of the human mind came about. Many physical theories today start by specifying a Lagrangian from which everything else flows. Indeed, we could adopt the same attitude with general relativity, but in so doing we would miss out on gaining some understanding of how the framework of general relativity is different again from the framework of Newtonian theory or special relativity. Moreover, if we discover limitations in the theory, then there is more chance of rescuing it by investigating the physical basis of the theory rather than simply tinkering with the mathematics. It is perhaps significant that Einstein devoted much of his later life to an attempt to unify general relativity and electromagnetism by various mathematical devices, but without success.

There are five principles which, explicitly or implicitly, guided Einstein in his search. Their names are:

(1) Mach's principle
(2) principle of equivalence
(3) principle of covariance
(4) principle of minimal gravitational coupling
(5) correspondence principle.

The status of these principles has been the source of much controversy. For example, the principle of covariance is considered by some authors (e.g. Bondi, Fock) to be empty, whereas there are others (e.g. Anderson) who believe it possible to derive general relativity more or less solely from this principle. There is fairly general agreement that the principle of equivalence is the key principle. One source of confusion arises from the fact that their formulation differs quite markedly from author to author. Since some of the principles are more of a philosophical nature, this is perhaps not so surprising. We shall attempt to make some precise formulations of them in the hope that we can ultimately check the principles out against the theory. We now discuss the principles in turn.

9.2 Mach's principle

The essence of the first two principles comes from understanding the nature of Newton's laws more precisely. Do Newton's laws hold in all frames of reference? As we have seen before, they are stated only for a privileged class of frames called **inertial** frames. So the question arises as to what form they take in other, non-inertial, reference frames.

We shall investigate the status of Newton's second law for a non-inertial frame S' being uniformly accelerated relative to an inertial frame S with acceleration a. For simplicity, we shall assume the frames are in standard configuration with the acceleration along the common axis (Fig. 9.1). Assuming that the observers initialize their clocks when they meet, then the relationship between the frames is given by

$$x = x' + s, \qquad y = y', \qquad z = z', \qquad t = t'. \tag{9.1}$$

Letting a dot denote differentiation with respect to t (or t', which is the same by the last equation), then we find from the first equation that

$$\dot{x} = \dot{x}' + \dot{s}$$

and, differentiating again,

$$\ddot{x} = \ddot{x}' + \ddot{s} = \ddot{x}' + a, \tag{9.2}$$

by assumption. Consider a particle of mass m moving along the x-axis under the influence of a force $F = (F, 0, 0)$. Then Newton's second law becomes $F = m\ddot{x}$, which by (9.2) gives

$$F = m\ddot{x}' + ma.$$

From the point of view of the observer S', this equation can be rewritten in a standard form with the term mass times **acceleration relative to S'** on the right-hand side, to give

$$F - ma = m\ddot{x}'. \tag{9.3}$$

Thus, compared to S, observer S' detects a **reduction** of the force on the particle by an amount ma. This additional force is called an **inertial force**. Other well-known inertial forces are **centrifugal** and **Coriolis** forces arising in

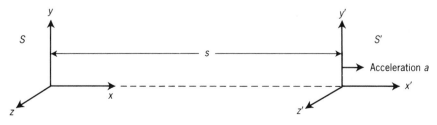

Fig. 9.1 Position of S and S' at time t.

a frame rotating relative to an inertial frame (exercise). Notice that all inertial forces have the **mass as a constant of proportionality** in them. The status of inertial forces is again a controversial one. One school of thought describes them as **apparent** or **fictitious** forces which arise in non-inertial frames of reference (and which can be eliminated mathematically by putting the terms back on the right-hand side). We shall adopt the attitude that if you judge them by their effects then they are very real forces. For, after all, inertial forces cause astronauts to black-out in rocket ships and flywheels to break under centrifugal effects. Is it enough to describe these as being due to apparent forces or reference frame effects? There must be some interaction going on to cause such dramatic effects. The question arises, What is the physical origin of inertial forces? Newtonian theory makes no attempt to answer this question; the Machian viewpoint, as we shall see, does.

Let us ask another fundamental question. If Newton's laws only hold in inertial frames, then how do we detect inertial frames? Newton realized that this was a fundamental question and attempted to answer it by devising an ingenious thought experiment — the famous **bucket experiment**. He first of all postulated the existence of absolute space: 'Absolute space, in its own nature, without relation to anything external, remains always similar and immovable.' Thus he sees absolute space as the backcloth against which all motion is observed. An inertial observer then becomes an observer at rest or in uniform motion relative to absolute space. Inertial forces arise in the manner described above only when an observer is in **absolute acceleration** relative to absolute space. The bucket experiment is a device for detecting such motion. More precisely, the experiment determines whether or not a system is in **absolute rotation** relative to absolute space.

The experiment consists of suspending a bucket containing water by a rope in an inertial frame. The rope is twisted and the bucket is released. The motion divides into four phases:

B1 At first, the bucket rotates, but the water does not, its surface remaining **flat**.

B2 The frictional effects between the bucket and the water eventually communicate the rotation to the water. The centrifugal forces cause the water to pile up round the edges of the bucket and the surface becomes **concave** (Fig. 9.2). The faster the water rotates, the more concave the surface becomes.

B3 Eventually the bucket will slow down and stop, but the water will continue rotating for a while, its surface remaining concave.

B4 Finally, the water will return to rest with a flat surface.

Fig. 9.2 The bucket and water in absolute rotation.

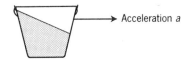

Acceleration a

Fig. 9.3 Inclination of surface of water in absolute linear acceleration.

Newton's explanation of this experiment is that the curvature of the water surface in B2 and B3 arises from centrifugal effects due to the rotation of the **water** relative to absolute space. This curvature is not directly connected to local considerations such as the bucket's rotation since in B1 the surface is flat when the bucket is rotating and in B3 curved when the bucket is at rest. In this way, Newton gave a prescription for determining whether a system is in absolute rotation or not. Similar arguments apply to systems which are linearly accelerated relative to absolute space. Here, the surface becomes inclined at angle to the horizontal (Fig. 9.3) (see Exercise 9.1(ii)). In simple terms, all observers should be equipped with a bucket of water. Then an observer will be inertial if and only if the surface of the water is flat.

We now turn to the view which was proposed by Mach in 1893, although it grew out of similar ideas arrived at earlier by Bishop Berkeley. This is a semi-philosophical view, the starting point of which is that there is no meaning to the concept of motion, but only to that of **relative** motion. For example, a body in an otherwise empty universe cannot be said to be in motion according to Mach, since there is nothing to which the body's motion can be referred. Moreover, in a populated universe, it is the interaction between all the matter in the universe (over and above the usual gravitational interaction) which is the source of the inertial effects we have discussed above. In our universe, the bulk of the matter resides in what is called the 'fixed stars'. Then, from Mach's viewpoint, an inertial frame is a frame in some privileged state of motion relative to the average motion of the fixed stars. Thus, it is the fixed stars through their masses, distribution, and motion which **determine** a local inertial frame. This is Mach's principle in essence. Returning to the bucket experiment, Newton gives no reason why the surface curves up when it is in rotation relative to absolute space. Mach, however, says that the curvature stems from the fact that the water is in rotation **relative** to the fixed stars. One way of seeing the difference between the two viewpoints is to ask what would happen if the bucket was fixed and the **universe** (i.e. the fixed stars) rotated. Since all motion is relative, it follows from the Machian viewpoint that the surfaces of the water would be curved, whereas in Newtonian theory no such effect would be detected. Hence, Mach sees all matter coupled together in such a way that inertial forces have their physical origin in matter. The bucket has very little effect on the water's motion because its mass is so small. On the other hand, the fixed stars contain most of the matter in the universe and this counteracts the fact that they are a very long way away.

There is one very outstanding and simple fact that lends support to the Machian viewpoint. Consider a pendulum set swinging at the North Pole (Fig. 9.4). According to Newton, the pendulum swings in a frame which is not rotating relative to absolute space. In this frame the Earth is rotating under the pendulum. An observer fixed on the Earth will see the pendulum rotating. The time taken for the pendulum to swing through 360° is therefore the time taken for the Earth to rotate through 360° with respect to **absolute space**. We can also measure how long the Earth takes to rotate through 360° relative to the **fixed stars**. The remarkable fact is that, within the limits of experimental accuracy, the two times are the **same**. In other words, the fixed stars are not rotating relative to absolute space, from which it follows that **inertial frames are those in which the fixed stars are not rotating**. In Newtonian theory, there is nothing a priori to predict this, it is simply a **coincidence**. Whenever we find coincidences in a physical theory, we should be highly suspicious of the theory — it is usually saying that something fundamental is going on. From the Machian viewpoint, it is the fixed stars which determine the inertial frames and the result is precisely what we would expect.

Can one say anything more precise about the interaction postulated by Mach? Since inertial forces involve the mass of the body experiencing them, it would seem likely for reasons of reciprocity that the effect of the stars should be due to their masses and proportional to them. On the other hand, inertial forces are unaffected (at least to the accuracy of experiment) by local masses

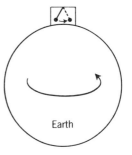

Fig. 9.4 Pendulum swinging in non-rotating frame.

such as the Earth or the Sun. Accordingly the influence of the distant bodies preponderates. So we would not expect inertial effects to vary appreciably from place to place.

Consider the motion of a particle in an otherwise empty universe. Then, according to Mach, since there are no other masses in existence, the particle cannot experience any inertial effects. Now introduce another particle of tiny mass. It is inconceivable that the introduction of this very small mass would restore the inertial properties of the first particle to its customary magnitude — its effect can only be slight. This implies that the magnitude of an inertial force on a body is determined by the mass of the universe and its distribution. If, in particular, the universe were not isotropic, then inertial effects would not be isotropic. For example, if there were a preponderance of matter in a particular direction, then inertial effects would be direction-dependent (as illustrated schematically in Fig. 9.5).

Experiments were carried out separately by Hughes and Drever around 1960 which established that mass is isotropic to at least 1 part in 10^{18}. The Hughes–Drever experiment has been called the most precise null experiment ever performed. This null result can be interpreted in two ways. Either Mach's principle is untenable or the universe is highly isotropic. There is evidence from other sources to suggest that our universe is indeed highly isotropic on the large scale.

In Newtonian theory, the gravitational potential ϕ at a point a distance r from the origin due to a particle of mass m situated at the origin is $\phi = -Gm/r$, where G is Newton's gravitational constant. The potential at any point can only depend on the properties of the body itself. However, from the Machian point of view, the mass m of the body depends on the state of the universe. Hence, the ratio of these two effects, namely G, contains **information about the universe**. In particular, if the universe was in a different state at any earlier epoch, then the 'constant' G would have a different value. An evolutionary universe would require $G = G(t)$, i.e. a function of epoch. Again, if the universe did not present the same aspect from every point (except for local irregularities), G would vary from point to point. A fully Machian theory should essentially allow one to calculate G from a knowledge of the structure of the universe.

What is the current status of Mach's principle? The biggest limitation of the principle is that it does not give a quantitative relation for the interaction of matter. Similarly, it can be argued that Mach's ideas do not really

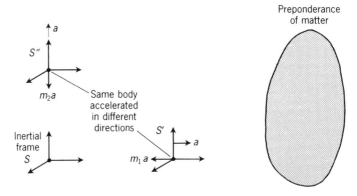

Fig. 9.5 Direction-dependent inertial effects in an anisotropic universe ($m_1 \neq m_2$).

contribute to an understanding of why there appears to be such a fundamental distinction between unaccelerated and accelerated motion in nature, i.e. it does not explain why the interaction should be velocity-independent but acceleration-dependent. Some critics claim that Mach only replaced Newton's absolute space by the distant stars and learnt nothing new thereby. The principle was considered to be of great importance to Einstein who attempted to incorporate it into his general theory. This, as we shall see, he only partially succeeded in doing (however, a more recent alternative theory to general relativity, called the Brans–Dicke theory, claims to be more fully Machian).

We finish this section by trying to make more precise the statements of Mach's principle which are relevant to the formulation of general relativity. Thinking in terms of the axiomatic formulation of the last chapter, let us refer to the privileged paths which particles and light rays travel on as the 'geometry' of the universe. The first statement tries to incorporate the essential part of Mach's ideas.

M1. The matter distribution determines the geometry.

The next statement refers to the belief that it is impossible to talk about motion or geometry in an empty universe, so that there should be no solution corresponding to an empty universe.

M2. If there is no matter then there is no geometry.

The final statement refers to a universe containing just one body, then, since there is nothing for it to interact with, it should not possess any inertial properties.

M3. A body in an otherwise empty universe should possess no inertial properties.

9.3 Mass in Newtonian theory

Up to now, we have talked rather glibly about the mass m of a body. Even in Newtonian theory, we can ascribe three masses to any body which describe quite different properties. Their names, notation, and general description are:

(1) **inertial mass** m^I, which is a measure of the body's resistance to change in motion;
(2) **passive gravitational mass** m^P, which is a measure of its reaction to a gravitational field;
(3) **active gravitational mass** m^A, which is a measure of its source strength for producing a gravitational field.

We shall discuss each of these in turn.

Inertial mass m^I is the quantity occurring in Newton's second law, which we met in Chapter 4. It is at any one time a measure of a body's resistance to change in motion and is also called the body's **inertia**. Newton's second law, stated more precisely, is

$$F = \frac{\mathrm{d}(m^{\mathrm{I}}v)}{\mathrm{d}t}, \tag{9.4}$$

or

$$F = m^{\mathrm{I}}a \tag{9.5}$$

for constant inertial mass m^{I}. Note that, a priori, m^{I} has **nothing** directly to do with gravitation. The next two masses, however, do.

Passive gravitational mass m^{P} measures a body's response to being placed in a gravitational field. Let the gravitational potential at some point be denoted by ϕ. Then, if m^{P} is placed at this point, it will experience a force on it given by

$$F = -m^{\mathrm{P}} \operatorname{grad} \phi. \tag{9.6}$$

Active gravitational mass m^{A} measures the strength of the gravitational field produced by the body itself. If m^{A} is placed at the origin, then the gravitational potential at any point distant r from the origin is given by

$$\phi = -\frac{Gm^{\mathrm{A}}}{r}. \tag{9.7}$$

We shall now see how these three masses are related in the Newtonian framework. We start from the observational result that if we neglect non-fundamental forces, like air resistance, then two bodies dropped from the same height will reach the ground together. In other words they suffer the **same** acceleration irrespective of their internal composition. This empirical result is attributed to Galileo in his famous Pisa experiments (Fig. 9.6).

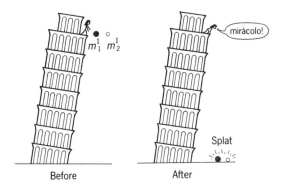

Fig. 9.6 Galileo's Pisa experiments.

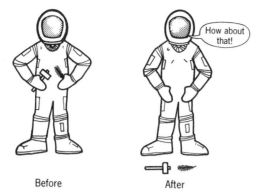

Before   After

Fig. 9.7 The moon landing 'experiment'.

Of course, you would not get this result with a hammer and a feather, say, because the air resistance would slow down the fall of the feather. It would be possible on the Moon, however, since the Moon has no atmosphere. Indeed, readers may remember the incident on one of the Moon landings when an astronaut tried this 'experiment' and confirmed the anticipated result (Fig. 9.7).

Let us assume that two particles of inertial masses m_1^I and m_2^I and passive gravitational masses m_1^P and m_2^P are dropped from the same height in a gravitational field. Then, from (9.5) and (9.6), we have

$$m_1^I \boldsymbol{a}_1 = \boldsymbol{F}_1 = -m_1^P \operatorname{grad} \phi,$$
$$m_2^I \boldsymbol{a}_2 = \boldsymbol{F}_2 = -m_2^P \operatorname{grad} \phi.$$

The observational result is $\boldsymbol{a}_1 = \boldsymbol{a}_2$, from which we get

$$m_1^I / m_1^P = m_2^I / m_2^P.$$

Repeating this experiment with other bodies, we see that the ratio m^I/m^P for any body is equal to a universal constant, α say. By a suitable choice of units, we can, without loss of generality, take $\alpha = 1$, from which we obtain the result

$$\text{inertial mass} = \text{passive gravitational mass.} \qquad (9.8)$$

This equality is one of the best attested results in physics and has been verified to 1 part in 10^{12} (see §15.7).

In order to relate passive gravitational mass to active gravitational mass, we make use of the observation that nothing can be shielded from a gravitational field. All matter is both acted upon by a gravitational field and is itself a source of a gravitational field. Consider two isolated bodies situated at points Q and R moving under their mutual gravitational interaction (Fig. 9.8). The gravitational potential due to each body is, by (9.7),

$$\phi_1 = -\frac{Gm_1^A}{r}, \qquad \phi_2 = -\frac{Gm_2^A}{r}.$$

Q F_1 F_2 R

Body 1   Body 2

Fig. 9.8 The mutual gravitational interaction of two isolated bodies.

The force which each body experiences is, by (9.6),

$$F_1 = -m_1^P \, \text{grad}_Q \, \phi_2, \qquad F_2 = -m_1^P \, \text{grad}_R \, \phi_1.$$

If we take Q to be the origin, then the gradient operators are

$$\text{grad}_R = \hat{r} \frac{\partial}{\partial r} = -\text{grad}_Q,$$

so that

$$F_1 = \frac{Gm_1^P m_1^A}{r^2} \hat{r}, \quad F_2 = -\frac{Gm_2^P m_2^A}{r^2} \hat{r}.$$

But, by Newton's third law, $F_1 = -F_2$, and so we conclude

$$m_1^P / m_1^A = m_2^P / m_2^A.$$

Using the same argument as before, we see that

> passive gravitational mass = active gravitational mass. (9.9)

This is why in Newtonian theory we can simply refer to the **mass** m of a body, where

$$m = m^I = m^P = m^A.$$

9.4 The principle of equivalence

We define a **gravitational test particle** to be a test particle which experiences a gravitational field but does **not** itself alter the field or contribute to the field. We wish to embody the empirical result of the Pisa experiments into a principle.

> **P1.** The motion of a gravitational test particle in a gravitational field is independent of its mass and composition.

This is known as the **strong** form of the principle of equivalence, and we are going to build general relativity on this principle. Notice the difference in its status in the two theories. In Newtonian theory, it is an observational result—another **coincidence**. It could be possible, for example, that if we looked closer (with an accuracy greater than 1 in 10^{12}) then different bodies would possess different accelerations when placed in a gravitational field. This would not upset Newtonian theory, which could accommodate such a result. In general relativity, it forms an essential hypothesis of the theory, and if it falls then so does the theory.

Next, we wish to make explicit the assumption that matter both responds to, and is a source of, a gravitational field. However, we have seen in special relativity that matter and energy are equivalent, so the statement about the gravitational field applies to energy as well. We incorporate this result into a statement which is known as the **weak** form of the principle of equivalence.

> **P2.** The gravitational field is coupled to everything.

Thus, no body can be shielded from a gravitational field. However, it is possible to remove gravitational effects locally from our theory and so regain special relativity. This we do by considering a frame of reference which is in **free fall**, i.e. co-moving with a gravitational test particle. If, in particular, we choose a freely falling frame which is not rotating, then we regain the concept of an **inertial frame**, at least locally. We mean here by 'locally' that observations are confined to a region over which the **variation** of the gravitational field is unobservably small. In such inertial frames, test particles remain at rest or move in straight lines with uniform velocity. This leads to the following statement of the principle of equivalence.

> **P3.** There are no local experiments which can distinguish non-rotating free fall in a gravitational field from uniform motion in space in the absence of a gravitational field.

Notice that once again we have encoded our principle as a statement of impossibility.

Einstein noticed one other **coincidence** in Newtonian theory which proved to be of great importance in formulating a statement of the principle of equivalence. All inertial forces are proportional to the mass of the body experiencing them. There is one other force which behaves in the same way, that is, the force of **gravitation**. For, if we drop two bodies in the Earth's gravitational field, then they experience forces $m_1 g$ and $m_2 g$, respectively. This coincidence suggested to Einstein that the two effects should be considered as arising from the same origin. Thus he suggested that we treat gravitation as an **inertial** effect as well, in other words it is an effect which arises from not using an inertial frame. Comparing the force mg of a falling body with the inertial force ma of (9.3) suggests the following version of the principle of equivalence.

> **P4.** A frame linearly accelerated relative to an inertial frame in special relativity is locally identical to a frame at rest in a gravitational field.

Fig. 9.9 Case 1: The lift in an accelerated rocket ship.

These last two versions of the principle of equivalence can be vividly clarified by considering the famous thought (in German, *gedänken*) experiments of Einstein called the **lift experiments**.

We consider an observer confined to a lift, or more precisely a room with no windows in it or other means of communication with the outside world. The observer is allowed equipment to carry out simple dynamical experiments. The object of the exercise is to try and determine the observer's state of motion. We consider four states of motion (Figs. 9.9–9.12).

Case 1. The lift is placed in a rocket ship in a part of the universe far removed from gravitating bodies. The rocket is accelerated forward with a constant acceleration g relative to an inertial observer. The observer in the lift releases a body from rest and (neglecting the influence of the lift, etc.) sees it fall to the floor with acceleration g.

Case 2. The rocket motor is switched off so that the lift undergoes uniform motion relative to the inertial observer. A released body is found to remain at rest relative to the observer.

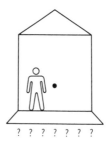

Fig. 9.10 Case 2: The lift in an unaccelerated rocket ship.

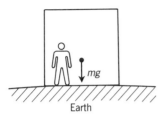

Fig. 9.11 Case 3: The lift placed on the Earth's surface.

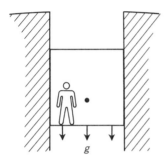

Fig. 9.12 Case 4: The lift dropped down an evacuated lift shaft.

Case 3. The lift is next placed on the surface of the Earth, whose rotational and orbital motions are ignored. A released body is found to fall to the floor with acceleration g.

Case 4. Finally, the lift is placed in an evacuated lift shaft and allowed to fall freely towards the centre of the Earth. A released body is found to remain at rest relative to the observer.

Clearly, from the point of view of the observer, cases 1 and 3 are indistinguishable, as required by P4, and cases 2 and 4 are indistinguishable, as required by P3. Let us trace the argument that shows that these requirements lead to the concept of a non-flat, i.e. a curved space-time.

In special relativity, in a coordinate system adapted to an inertial frame, namely, Minkowski coordinates, the equation for a test particle is

$$\frac{d^2x^a}{d\tau^2} = 0.$$

If we use a non-inertial frame of reference, then this is equivalent to using a more general coordinate system. In this case, the equation becomes

$$\frac{d^2x^a}{d\tau^2} + \Gamma^a_{bc}\frac{dx^b}{d\tau}\frac{dx^c}{d\tau} = 0,$$

where Γ^a_{bc} is the metric connection of g_{ab}, which is still a flat metric but not the Minkowski metric η_{ab}. The additional terms involving Γ^a_{bc} which appear are precisely the **inertial force** terms we have encountered before. Then the principle of equivalence requires that the gravitational forces, **as well as** the inertial forces, should be given by an appropriate Γ^a_{bc}. In this case, we can no longer take space-time to be flat, for otherwise there would be no distinction from the non-gravitational case. The **simplest** generalization is to keep Γ^a_{bc} as the metric connection, but now take it to be the metric connection of a **non-flat** metric. If we are to interpret the Γ^a_{bc} as force terms, then it follows that we should regard the g_{ab} as **potentials.** The field equations of Newtonian gravitation consist of second-order partial differential equations in the gravitational potential ϕ. In an analogous manner, we would expect general relativity also to involve second-order partial differential equations in the potentials g_{ab}. The remaining task which will allow us to build a relativistic theory of gravitation is to choose a likely set of second-order partial differential equations.

9.5 The principle of general covariance

Recall the principle of special relativity, namely, that all **inertial** observers are equivalent. As we have seen in the last section, general relativity attempts to include non-inertial observers into its area of concern in order to cope with gravitation. Einstein argued that all observers, whether inertial or not, should be capable of discovering the laws of physics. If this were not true, then we would have little chance in discovering them since we are bound to the Earth whose motion is almost certainly not inertial. Thus, Einstein proposed the following as the logical completion of the principle of special relativity.

Principle of general relativity: All observers are equivalent.

Observers are intimately tied up with their reference systems or coordinate systems. So, if any observer can discover the laws of physics, then any old coordinate system should do. The situation is somewhat different in special relativity, where, because the metric is flat and the connection integrable, there exists a **canonical** or preferred coordinate system; namely, Minkowski coordinates. In a curved space-time, that is, a manifold with a non-flat metric, there is no canonical coordinate system. This is just another statement of the non-existence of a global inertial observer. However, the statement needs to be treated with caution, because in many applications, there will be preferred coordinate systems. For example, many problems possess symmetries and the simplest thing to do is to adapt the coordinate system to the underlying symmetry. It is not so much that any coordinate system will do, but rather that the theory should be invariant under a coordinate transformation. Thus, the full import of the principle of general relativity is contained in the following statement.

> **Principle of general covariance:** The equations of physics should have tensorial form.

Some authors argue that this statement is empty, because it is possible to formulate any physical theory in tensorial form. (Of course, this realization only came **after** the advent of general relativity.) Whether or not this is the case, it was clearly of central importance to Einstein, as is evident from the name he gave it. We shall make use of it in the form of the principle of general covariance, which is why we undertook our major digression in Part B to learn the language of tensors.

9.6 The principle of minimal gravitational coupling

The principles we have discussed so far do not tell us how to obtain field equations of systems in general relativity when the corresponding equations are known in special relativity. The principle of minimal gravitational coupling is a **simplicity principle** or Occam's razor that essentially says we should not add unnecessary terms in making the transition from the special to the general theory. For example, we shall later meet the conservation law

$$\partial_b T^{ab} = 0 \tag{9.10}$$

in special relativity in Minkowski coordinates. The simplest generalization of this to the general theory is to take the tensor equation

$$\nabla_b T^{ab} = 0. \tag{9.11}$$

However, we could equally well take

$$\nabla_b T^{ab} + g^{be} R^a{}_{bcd} \nabla_e T^{cd} = 0 \tag{9.12}$$

since $R^a{}_{bcd} = 0$ in special relativity and (9.12) again reduces to (9.10) in Minkowski coordinates. We therefore adopt the following principle.

> **Principle of minimal gravitational coupling:** No terms explicitly containing the curvature tensor should be added in making the transition from the special to the general theory.

The principle was not stated by Einstein but was used implicitly. Unfortunately, it is rather vague and ambiguous and needs to be used with care.

9.7 The correspondence principle

As we stated from the outset, we are engaged with modelling, and together with any model should go its range of validity. Then any new theory must be consistent with any acceptable earlier theories within their range of validity. General relativity must agree on the one hand with special relativity in the absence of gravitation and on the other hand with Newtonian gravitational theory in the limit of weak gravitational fields and low velocities (compared with the speed of light). This gives rise to a **correspondence principle**, as indicated in Fig. 9.13, where arrows indicate directions of increased specialization.

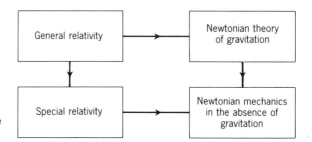

Fig. 9.13 The correspondence principle for general relativity.

Exercises

9.1 (§9.2)
(i) A pendulum is suspended from the roof of a car moving in a straight line with uniform acceleration a. Find the angle the pendulum makes with the vertical. Explain what is happening from the viewpoint of an inertial observer external to the car and a non-inertial observed fixed in the car.
(ii) A bucket of water is located in the car as well. Find the angle which the surface of the water makes with the horizontal.
(iii) A bucket of water slides freely under gravity down a slope of fixed angle α to the horizontal. What is the angle of inclination of the surface of the water relative to the base of the bucket?

9.2 (§9.2)
(i) Consider a body rotating relative to an inertial frame about a fixed point O with angular velocity $\boldsymbol{\omega}$ in Newtonian theory. The velocity $\boldsymbol{v}$ of any point P in the body with position vector $\overrightarrow{OP} = \boldsymbol{r}$ is given by

$$\boldsymbol{v} = \boldsymbol{\omega} \times \boldsymbol{r}$$

Let $\boldsymbol{i}, \boldsymbol{j}, \boldsymbol{k}$ denote unit vectors in the inertial frame S and $\boldsymbol{i'}, \boldsymbol{j'}, \boldsymbol{k'}$ denote unit vectors in a frame S' fixed in the body, where both origins are at O. If $\boldsymbol{u} = \boldsymbol{u}(t)$ is a general vector with components

$$\boldsymbol{u} = u'_1 \boldsymbol{i'} + u'_2 \boldsymbol{j'} + u'_3 \boldsymbol{k'}$$

in S', show, by differentiating this equation, that

$$\left[\frac{d\boldsymbol{u}}{dt}\right]_S = \left[\frac{d\boldsymbol{u}}{dt}\right]_{S'} + \boldsymbol{\omega} \times \boldsymbol{u}.$$

(ii) Consider a non-inertial frame S' moving arbitrarily relative to an inertial frame S, where the position of the origin O' of S' relative to the origin O of S is $s(t)$ and its angular velocity is $\omega(t)$. A particle of constant mass m situated at a point with position vectors r and r' relative to S and S', respectively, is acted on by a force F. Show that S' can write the equation of motion of the particle in the form

$$F - [ma + 2m\omega \times \dot{r}' + m\omega \times (\omega \times r')$$
$$+ m\dot{\omega} \times r'] = m\ddot{r}',$$

where a is the acceleration of O' relative to O and a dot denotes differentiation with respect to time in the frame of S'. What are the quantities in square brackets? Interpret these quantities physically.

9.3 (§**9.3**) Fill in the details that lead to the equalities (9.8) and (9.9).

9.4 (§**9.3**) Write down the equations of motion for an isolated system of three bodies of inertial masses m_1^I, m_2^I and m_3^I. Eliminate the internal forces from these equations and demonstrate that if two of the bodies are rigidly bound to form a composite system then the inertial mass is additive.

9.5 (§**9.4**) In the lift experiments, explain the motion of the released body from the point of view of: case (1) an inertial observer; case (2), an inertial observer who initially sees the rocket moving away with constant velocity v; case (4), an observer at rest on the surface of the Earth.

9.6 (§**9.4**) Consider a sphere of non-interacting particles falling towards the Earth's surface. Taking into account the different accelerations of particles in the sphere, what is the ensuing shape of the enclosing volume?

9.7 (§**9.4**) Find the geodesic equations for $\mathbb{R}^3$ in cylindrical polar coordinates (see Exercise 6.17). Interpret the terms occurring which involve Γ^a_{bc}.

9.8 (§**9.4**) What is the path of a free particle
(i) in an inertial frame?
(ii) in the presence of a uniform gravitational field?
Use the principle of equivalence and the particle theory of light to find the path of a light ray in the above two cases and hence deduce light bending in a gravitational field.

9.9 (§**9.6**) Write down a generalization of (9.10) to a curved space which involves a term quadratic in the Riemann tensor.

9.10 (§**9.6**) An anti-symmetric tensor F_{ab} satisfies the equation in special relativity in Minkowski coordinates

$$\partial_{[a} F_{bc]} = 0.$$

Write down the simplest generalization to a curved space-time and show that it is identical to the original equation.

9.11 (§**9.7**) Write down the correspondence principle for the transition from special relativity (in non-relativistic units) to Newtonian theory in the absence of gravitation. Express this transition as a limit involving the speed of light. Draw a sequence of diagrams to indicate what happens to the null cone in this limit. What happens to the three regions defined by the null cone in special relativity? What happens to the concept of simultaneity in the limit?

10 The field equations of general relativity

10.1 Non-local lift experiments

The considerations of the last chapter led us to conclude that, locally, i.e. neglecting variations in the gravitational field, we can regain special relativity. However, in a non-local situation, we require a non-flat metric which may be thought of as the potentials of the gravitational field. Correspondence with Newtonian theory then suggests that we require second-order field equations in these potentials, and, moreover, from the principle of covariance, these equations must be tensorial in character. In this chapter, we shall pursue the Newtonian correspondence further and reformulate Newtonian theory in such a way that it leads naturally to the particular set of field equations of general relativity.

We return to the lift experiments and consider performing the following **non-local** experiments. We assume that the observer's equipment is sufficiently sensitive to detect variations in the gravitational field. The four experiments take the same form as before, but this time the observer releases two bodies, whose mutual interactions we ignore (Figs. 10.1–10.4).

Case 1. From the point of view of the observer in the lift, the two bodies fall to the ground parallel to each other.

Case 2. The bodies remain at rest relative to the observer.

Case 3. The two bodies fall towards the centre of the Earth and hence fall on paths which **converge**.

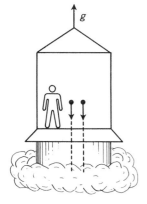

Fig. 10.1 Case 1: The lift in an accelerated rocket ship.

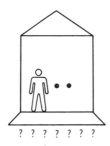

Fig. 10.2 Case 2: The lift in an unaccelerated rocket ship.

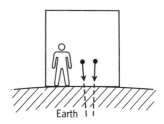

Fig. 10.3 Case 3: The lift placed on the earth's surface.

Case 4. The bodies appear to the observer to move closer together, because they are falling on lines which converge towards the centre of the earth.

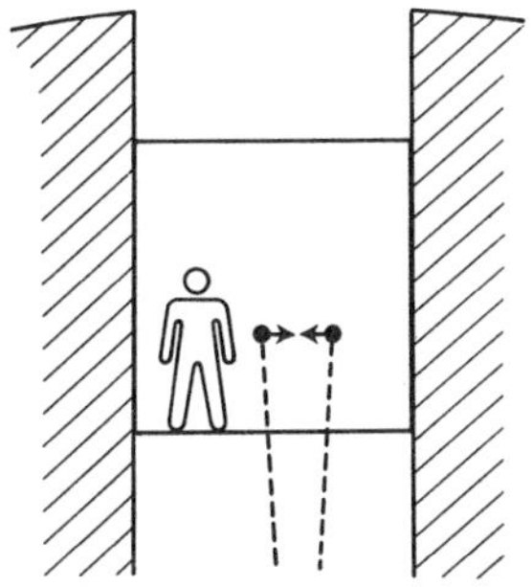

It follows that the observer can distinguish the **uniform** inertial field of case 1 from the Earth's **non-uniform** gravitational field of case 3. Again, in free fall, bodies travel on geodesics in a gravitational field which **converge** (or diverge), as in case 4. The point of these thought experiments is that the presence of a genuine gravitational field, as distinct from an inertial field, is verified by the observation of the **variation** of the field rather than by the observation of the field itself. We shall see that in general relativity this variation is described by the Riemann tensor through the equation of geodesic deviation.

Fig. 10.4 Case 4: The lift dropped down an evacuated lift shaft.

10.2 The Newtonian equation of deviation

The non-local lift experiments reveal that we should focus our attention on two neighbouring test particles in free fall in a gravitational field. We look at this motion first of all in Newtonian theory using the tensor apparatus of Part B. We introduce Cartesian coordinates

$$(x^\alpha) = (x^1, x^2, x^3) = (x, y, z),$$

where, for the rest of this chapter, Greek indices run from 1 to 3, and then the line element of Euclidean 3-space $\mathbb{R}^3$ is

$$d\sigma^2 = dx^2 + dy^2 + dz^2,$$

from which we obtain the Euclidean metric

$$g_{\alpha\beta} = \delta_{\alpha\beta} = \text{diag}(1, 1, 1) \tag{10.1}$$

We therefore raise and lower indices with the three-dimensional Kronecker delta. This means that in Newtonian theory there is really no distinction between raised and lowered indices, but we will retain the notation in order to help us compare results later with the general theory. We consider the paths of two neighbouring gravitational test particles of unit mass travelling **in vacuo** in a gravitational field whose potential is ϕ.

Let the particles travel on curves C_1 and C_2 so that they reach the points P and Q at time t (Fig. 10.5). If we use the time t as the parameter along the curves, then the parametric equations of C_1 are

$$x^\alpha = x^\alpha(t) \tag{10.2}$$

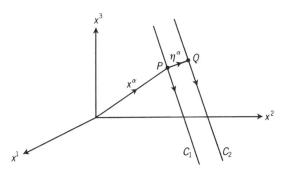

Fig. 10.5 Freely falling gravitational test particles at time t.

and those of C_2 are

$$x^\alpha = x^\alpha(t) + \eta^\alpha(t), \tag{10.3}$$

when η^α is a small **connecting vector** which connects points on the two curves with equal values of t. Since the particles have unit mass, the equation of motion of the first particle, by (9.5) and (9.6), can be written in the tensor form

$$1\ddot{x}^\alpha = -1\partial^\alpha\phi, \tag{10.4}$$

where a dot denotes differentiation with respect to time and

$$\partial^\alpha\phi = \delta^{\alpha\beta}\partial_\beta\phi = \left(\frac{\partial\phi}{\partial x}, \frac{\partial\phi}{\partial y}, \frac{\partial\phi}{\partial z}\right) = (\text{grad }\phi)_P. \tag{10.5}$$

Similarly, omitting the unit masses, the equation of motion of the second particle is

$$\ddot{x} + \ddot{\eta}^\alpha = -(\partial^\alpha\phi)_Q. \tag{10.6}$$

Since η^α is small, we may expand the term on the right-hand side by Taylor's theorem (exercise), to obtain

$$-(\partial^\alpha\phi)_Q = -(\partial^\alpha\phi)_P - (\eta^\beta\partial_\beta\partial^\alpha\phi)_P \tag{10.7}$$

to first order. Subtracting (10.4) from (10.6), we get

$$\ddot{\eta}^\alpha = -\eta^\beta\partial_\beta\partial^\alpha\phi. \tag{10.8}$$

If we define the tensor $K^\alpha{}_\beta$ by

$$K^\alpha{}_\beta = K_\beta{}^\alpha = \partial^\alpha\partial_\beta\phi, \tag{10.9}$$

then the equation of motion (10.8) of the connecting vector η^α, which we call the **Newtonian equation of deviation**, becomes

$$\ddot{\eta}^\alpha + K^\alpha{}_\beta\eta^\beta = 0. \tag{10.10}$$

This equation is intimately connected with the Newtonian field equations in empty space, namely, Laplace's equation (4.6), which can be written (exercise)

$$K^\alpha{}_\alpha = 0. \tag{10.11}$$

In other words, **the tensor $K^\alpha{}_\beta$ is trace-free**. We now search for a relativistic generalization of these equations.

10.3 The equation of geodesic deviation

Following the axioms of §8.5, we assume that free test particles in general relativity travel on timelike geodesics. We therefore consider a 2-surface S

ruled by a **congruence of timelike geodesics**, that is, a family of geodesics such that exactly one of the curves goes through every point of S. The parametric equation of S is given by

$$x^a = x^a(\tau, v), \qquad (10.12)$$

where τ is the **proper time** along the geodesics and v labels distinct geodesics. We define two vector fields on S by

$$v^a = \frac{dx^a}{d\tau} \qquad (10.13)$$

and

$$\xi^a = \frac{dx^a}{dv}. \qquad (10.14)$$

Then v^a is the **tangent vector** to the timelike geodesic at each point and ξ^a is a **connecting vector** connecting two neighbouring curves in the congruence (Fig. 10.6). The commutator of v^a and ξ^a satisfies

$$
\begin{aligned}
[v, \xi]^a &= v^b \partial_b \xi^a - \xi^b \partial_b v^a \\
&= \frac{dx^b}{d\tau} \frac{\partial}{\partial x^b}\left(\frac{dx^a}{dv}\right) - \frac{dx^b}{dv} \frac{\partial}{\partial x^b}\left(\frac{dx^a}{d\tau}\right) \\
&= \frac{d^2 x^a}{d\tau\, dv} - \frac{d^2 x^a}{dv\, d\tau} \\
&= 0 \qquad (10.15)
\end{aligned}
$$

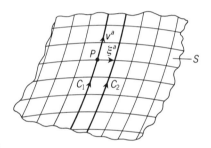

Fig. 10.6 The vectors v^a and ξ^a at a point P in S.

since the mixed partial derivatives commute. (It can be shown that the vanishing of the commutator is a necessary and sufficient condition for the vector fields to be **surface-forming**, which means that the congruences generated by the two vectors knit together to form a 2-surface.) By (6.15), the commutator is also equal to the Lie derivative $L_v \xi^a$. We now use the result which allows us to replace partial derivatives by covariant derivatives in an expression for a Lie derivative

$$
\begin{aligned}
0 &= L_v \xi^a \\
&= v^b \partial_b \xi^a - \xi^b \partial_b v^a \\
&= v^b \nabla_b \xi^a - \xi^b \nabla_b v^a \\
&= \nabla_v \xi^a - \nabla_\xi v^a. \qquad (10.16)
\end{aligned}
$$

Taking the covariant derivative of this equation with respect to v^a, we find

$$\nabla_v \nabla_v \xi^a = \nabla_v \nabla_\xi v^a. \qquad (10.17)$$

The equation we are seeking derives from the identity (Exercise 6.11)

$$\nabla_X(\nabla_Y Z^a) - \nabla_Y(\nabla_X Z^a) - \nabla_{[X,Y]} Z^a = R^a{}_{bcd} Z^b X^c Y^d \qquad (10.18)$$

If we set $X^a = Z^a = v^a$ and $Y^a = \xi^a$, then the second term on the left vanishes, because v^a is tangent to an affinely parametrized geodesic, and so, by (6.36),

$$\nabla_v v^a = 0. \qquad (10.19)$$

The third term vanishes by (10.15), since the covariant derivative of any tensor with respect to the zero tensor is zero. Thus, (10.18) becomes

$$\nabla_v \nabla_\xi v^a - R^a{}_{bcd} v^b v^c \xi^d = 0. \tag{10.20}$$

By definition,

$$\frac{D^2 \xi^a}{D\tau^2} = \nabla_v \nabla_v \xi^a,$$

and so, using (10.17), equation (10.20) becomes the promised **equation of geodesic deviation**

$$\frac{D^2 \xi^a}{D\tau^2} - R^a{}_{bcd} v^b v^c \xi^d = 0. \tag{10.21}$$

The absolute derivative along the curve is the tensorial analogue of the time derivative along the curve in (10.10). However, this is not quite the form we want to compare with (10.10), because it involves the 4-vector ξ^a, which has four pieces of information in it. We are really only interested in the spatial information in this equation.

We extract this by first introducing a **projection operator** $h^a{}_b$, defined by

$$h^a{}_b \equiv \delta^a{}_b - v^a v_b, \tag{10.22}$$

which projects tensors into the 3-space orthogonal to v^a at any point P of S. It possesses the following properties (exercise) which establish it as a projection operator:

(a) $h^a{}_b v^b = 0,$

(b) $w^a v_a = 0 \quad \Leftrightarrow \quad h^a{}_b w^b = w^a,$

(c) $h^a{}_b h^b{}_c = h^a{}_c,$ $\qquad\qquad\qquad\qquad$ (10.23)

(d) $h^a{}_a = 3,$

(e) $h_{ab} = h_{ba}.$

We thus define the **orthogonal connecting vector** η^a by

$$\eta^a \equiv h^a{}_b \xi^b. \tag{10.24}$$

We need one more result which follows from the fact that v^a is a **unit** tangent vector, since

$$v^a v_a = g_{ab} v^a v^b = g_{ab} \frac{dx^a}{d\tau} \frac{dx^a}{d\tau} = 1. \tag{10.25}$$

Taking the covariant derivative with respect to ξ^a, we get the result

$$\nabla_\xi (v^a v_a) = v^a (\nabla_\xi v_a) + v_a (\nabla_\xi v^a) = 2 v_a \nabla_\xi v^a = 0, \tag{10.26}$$

since the covariant derivative of 1 is zero (why?). Then

$$\frac{D\xi^a}{D\tau} = \nabla_v \xi^a$$

$$= \nabla_v(\eta^a + v^a v_b \xi^b)$$

$$= \nabla_v \eta^a + (\nabla_v v^a)v_b \xi^b + v^a(\nabla_v v_b)\xi^b + v^a v_b(\nabla_v \xi^b)$$

$$= \frac{D\eta^a}{D\tau} + v^a v_b(\nabla_\xi v^b)$$

$$= \frac{D\eta^a}{D\tau}, \qquad (10.27)$$

using (10.24), (10.19), (10.16), and (10.26). In addition,

$$R^a{}_{bcd} v^b v^c \xi^d = R^a{}_{bcd} v^b v^c(\eta^d + v^d v_e \xi^e) = R^a{}_{bcd} v^b v^c \eta^d, \qquad (10.28)$$

since $R^a{}_{bcd}$ is anti-symmetric on c and d (see Exercise 5.11(ii)). So, finally, (10.21) can be written in terms of η^a, using (10.27) and (10.28), to give

$$\frac{D^2 \eta^a}{D\tau^2} - R^a{}_{bcd} v^b v^c \eta^d = 0. \qquad (10.29)$$

We have now written the equation of geodesic deviation in terms of the orthogonal connecting vector. However, this is still a four-dimensional equation and so, in the next section, we shall show how to extract the three-dimensional information.

10.4 The Newtonian correspondence

At any point P on the curve C_1, we introduce an orthogonal frame of three unit spacelike vectors

$$e_\alpha{}^a = (e_1{}^a, e_2{}^a, e_3{}^a),$$

which are all orthogonal to v^a and where α is a bold label running from 0 to 3. We define

$$e_0{}^a \equiv v^a,$$

and then, remembering (10.25), we have the following set of **orthonormality relations**

$$\left.\begin{aligned} e_0{}^a e_{0a} = -e_1{}^a e_{1a} = -e_2{}^a e_{2a} = -e_3{}^a e_{3a} = 1, \\ e_0{}^a e_{1a} = e_0{}^a e_{2a} = e_0{}^a e_{3a} = e_1{}^a e_{2a} = e_1{}^a e_{3a} = e_2{}^a e_{3a} = 0 \end{aligned}\right\} \qquad (10.30)$$

The four vectors $e_i{}^a$ ($i = 0, 1, 2, 3$) are said to form a **frame** or **tetrad** (**vierbein**, in German) at P, and the orthonormality relations (10.30) can be succinctly summarized as

$$e_i{}^a e_{ja} = \eta_{ij} \qquad (10.31)$$

where η_{ij} is the Minkowski metric, that is,

$$\eta_{ij} = \text{diag}(1, -1, -1, -1).$$

Treating $e_i{}^a$ as a 4×4 matrix at P, we can define its inverse (called the **dual basis**) $e^i{}_a$ by requiring

$$e_i{}^a e^i{}_a = \delta_i^j \qquad (10.32)$$

where δ_i^j is the Kronecker delta, or the identity matrix in matrix terms. We have introduced the frame notation merely as a convenience so far, but it turns out that frames possess a powerful formalism of their own (which is outside the scope of this book, but see §19.1). For example, in exactly the same way that we raise and lower **tensor** indices with the metric g_{ab}, we can raise and lower **frame** indices (**i, j,** ...) with the **frame metric** η_{ij}. Let us multiply (10.32) by $e^i{}_b$ and write it in the form

$$(e^i{}_b e_i{}^a) e^i{}_a = e^i{}_b,$$

from which it should be clear that the quantity in parentheses must be the tensorial Kronecker delta, namely,

$$e^i{}_b e_i{}^a = \delta_b^a. \qquad (10.33)$$

The physical interpretation of the frame is as follows: $e_0{}^a = v^a$ is the 4-velocity of an observer whose world-line is C_1, and the three spacelike vectors $e_{\boldsymbol{\alpha}}{}^a$ are rectangular coordinate vectors (such as the usual Cartesian basis $\boldsymbol{i}, \boldsymbol{j}$, and $\boldsymbol{k}$, for example) at P, where the bold Greek indices run from 1 to 3. So far, the frame has only been defined at P, but we now propagate the frame along C_1 by parallel propagation, i.e.

$$\frac{D}{D\tau}(e_i{}^a) = 0 \qquad (10.34)$$

In the same way as we can get the Cartesian components of a three-dimensional vector by taking the scalar product of it with $\boldsymbol{i}, \boldsymbol{j}$, and $\boldsymbol{k}$, then we define the **spatial frame components** of the orthogonal connecting vector η^a by

$$\eta^{\boldsymbol{\alpha}} = e^{\boldsymbol{\alpha}}{}_a \eta^a \qquad (10.35)$$

This is the precise analogue of the vector η^{α} of §10.2. Note that

$$\eta^0 = e^0{}_a \eta^a = 0 \qquad (10.36)$$

by (10.24) and (10.23a). We represent the various quantities schematically in Fig. 10.7.

To find the spatial part of (10.29) we contract it with $e^{\boldsymbol{\alpha}}{}_a$, and then, using (10.34), we find

$$\frac{D^2 \eta^{\boldsymbol{\alpha}}}{D\tau^2} - R^a{}_{bcd} e^{\boldsymbol{\alpha}}{}_a v^b v^c \eta^d = 0. \qquad (10.37)$$

Using (10.33), (10.36), and (10.35), we get

$$\eta^d = \delta_c^d \eta^c = e_i{}^d e^i{}_c \eta^c = e_0{}^d e^0{}_c \eta^c + e_{\boldsymbol{\beta}}{}^d e^{\boldsymbol{\beta}}{}_c \eta^c = e_{\boldsymbol{\beta}}{}^d \eta^{\boldsymbol{\beta}},$$

and so, substituting in (10.37), we can write the spatial part of the equation of

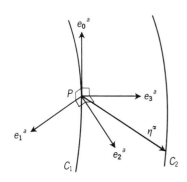

Fig. 10.7 The frame and the orthogonal connecting vector at P.

geodesic deviation as

$$\frac{\mathrm{D}^2\eta^a}{\mathrm{D}\tau^2} + K^\alpha{}_\beta\eta^\beta = 0, \tag{10.38}$$

where

$$K^\alpha{}_\beta = -R^a{}_{bcd}e^\alpha{}_a v^b v^c e^d{}_\beta. \tag{10.39}$$

Equation (10.38) is the analogue of (10.10) which we have been seeking.

10.5 The vacuum field equations of general relativity

We saw that the vacuum field equations of Newtonian theory could be expressed as the vanishing of the trace of $K^\alpha{}_\beta$. In an analogous manner, let us investigate the vanishing of the trace of (10.39), namely,

$$R^a{}_{bcd}e^\alpha{}_a v^b v^c e_\alpha{}^d = 0. \tag{10.40}$$

We do this by introducing a special coordinate system at P adapted to the frame, so that in this coordinate system

$$e_0{}^a \overset{*}{=} (1, 0, 0, 0), \quad e_1{}^a \overset{*}{=} (0, 1, 0, 0) \quad e_2{}^a \overset{*}{=} (0, 0, 1, 0), \quad e_3{}^a \overset{*}{=} (0, 0, 0, 1)$$

or more succinctly

$$e_i{}^a \overset{*}{=} \delta_i{}^a.$$

Then (10.40) reduces to

$$R^\alpha{}_{00\alpha} \overset{*}{=} 0.$$

But, since the Riemann tensor is anti-symmetric on its last pair of indices, it follows that, in any coordinate system,

$$R^0{}_{000} \equiv 0,$$

and we can combine the last two equations to give

$$R^a{}_{00a} \overset{*}{=} 0.$$

Then

$$0 \overset{*}{=} R^a{}_{00a} \overset{*}{=} R^a{}_{bca}\delta_0{}^b\delta_0{}^c \overset{*}{=} R^a{}_{bca}v^b v^c = -R^a{}_{bac}v^b v^c = -R_{bc}v^b v^c,$$

by (6.83). Now $R_{bc}v^b v^c$ is a scalar, and hence if it vanishes in one coordinate system then it must vanish in all coordinate systems. Moreover, since it vanishes for **all** observers, that is, for all timelike vectors v^a at P, then it follows that (exercise) $[R_{ab}]_P = 0$. Finally, since P is arbitrary, our analogy suggests that the vacuum field equations of general relativity should be

$$R_{ab} = 0. \tag{10.41}$$

By Exercise 6.25, the vanishing of the Ricci tensor is equivalent to the vanishing of the Einstein tensor, so that we can write alternatively

$$G_{ab} = 0. \tag{10.42}$$

Equations (10.41) or (10.42) are the equations which Einstein proposed should serve as the **vacuum field equations of general relativity**.

10.6 The story so far

Our arrival at the vacuum field equations of general relativity has involved rather a long story. This is not so surprising when you consider that it took Einstein over ten years of endeavour to move from the formulation of the special theory (1905) to a final formulation of the general theory (1916). It might be helpful, therefore, to outline again the main points of the argument.

1. The principle of equivalence reveals that if we freefall in a gravitational field then we can eliminate gravity locally and regain special relativity.
2. It also states that locally we cannot distinguish a gravitational field from a (uniform accelerative) inertial field and consequently we should regard gravitation as an inertial force.
3. Following special relativity, we assume that free test particles travel on timelike geodesics. Then inertial forces arise in the geodesic equations in the terms involving the metric connection of a flat metric. In order to include the extra effect of gravitation in the metric connection, we generalize the metric to being curved.
4. The metric then plays the role of the potentials of the theory and, in analogy with Newtonian theory, we seek a set of second-order partial differential equations for the potentials as field equations of the theory. Moreover, by the covariance principle, these equations must be tensorial.
5. If we now take non-local effects into account, then a genuine gravitational field can be observed by the variation in the field rather than by an observation of the field itself. This variation causes test particles to travel on timelike geodesics which converge (or diverge), and the convergence is described by the Riemann tensor through the equation of geodesic deviation.
6. The Riemann tensor is a tensor which involves second partial derivatives of the metric and so we might expect the field equations of the theory to involve the Riemann tensor. The fact that the Newtonian vacuum field equations involve the vanishing of a contracted tensor suggests that we might consider a contraction of the Riemann tensor. There is only one meaningful contraction (why?), namely, the Ricci tensor, and its vanishing is equivalent to the vanishing of the Einstein tensor.

10.7 The full field equations of general relativity

For completeness, we introduce briefly the full field equations which hold in the presence of fields other than gravitation. As we shall see, these fields are described by the **energy–momentum tensor** T^{ab}. Now the equivalence of mass and energy from special relativity suggests that **all** forms of energy act as sources for the gravitational field; indeed this is the content of the weak form of the principle of equivalence P2. We therefore take T^{ab} as a **source** term in the field equations. In special relativity in Minkowski coordinates, the energy–momentum tensor satisfies the conservation equations (see Chapter 12)

$$\partial_b T^{ab} = 0.$$

The principle of minimal gravitational coupling suggests the general relativistic generalization

$$\nabla_b T^{ab} = 0.$$

However, we know that the covariant derivative of the Einstein tensor vanishes through the contracted Bianchi identities (6.86):

$$\nabla_b G^{ab} \equiv 0.$$

The last two equations suggest that the two tensors are proportional and one can write consistently

$$G^{ab} = \kappa T^{ab}, \tag{10.43}$$

where κ is a constant of proportionality called the **coupling constant**. Note that this equation is in line with Mach's principle in the form M1 since the matter (T^{ab}) determines the geometry (G^{ab}) and hence is the source of inertial effects. The constant κ is then determined by the correspondence principle, since this equation must reduce to Poisson's equation (4.5) in the appropriate limit. We shall see in §12.10 that this is given in non-relativistic units by

$$\kappa = 8\pi G/c^2. \tag{10.44}$$

The equations (10.43) subject to (10.44) constitute the **full field equations** of general relativity. We shall, for the most part, work in **relativistic units**, in which we can take both $c = 1$ and $G = 1$, and then the coupling constant is simply

$$\kappa = 8\pi. \tag{10.45}$$

At this stage, we shall define the **theory of general relativity** to consist of the axioms of special relativity as stated in §8.5 except that I(iii) is now replaced by equation (10.43) subject to (10.44). However, before we consider further the significance of the field equations, we shall look at, in the next chapter, an alternative derivation based on a mathematical principle rather than physical principles, namely, the variational principle, and follow this up with an investigation of the right-hand side of (10.43), namely, the energy–momentum tensor.

Exercises

10.1 (§**10.2**) Taylor's theorem in three dimensions can be written

$$f(x + h) = f(x) + \sum_1^\infty \frac{(h \cdot \nabla)^n}{n!} f(x),$$

where

$$x = x\boldsymbol{i} + y\boldsymbol{j} + z\boldsymbol{k},$$

$$h = h_1\boldsymbol{i} + h_2\boldsymbol{j} + h_3\boldsymbol{k},$$

$$\nabla = \boldsymbol{i}\frac{\partial}{\partial x} + \boldsymbol{j}\frac{\partial}{\partial y} + \boldsymbol{k}\frac{\partial}{\partial z}.$$

Write out the first three terms of the expansion.

10.2 (§**10.2**) (i) Use Exercise 10.1 to verify (10.7).
(ii) Verify that Laplace's equation can be written in the form (10.11).

10.3 (§**10.3**) Verify the properties (10.23) of the projection operator $h^a{}_b$.

10.4 (§**10.3**) If $v^a = dx^a/d\tau$ is the tangent vector to a timelike geodesic parametrized by the proper time, and ζ^a is an arbitrary vector field, show that

(i) $\nabla_v v^a = 0$,
(ii) $\nabla_v v_a = 0$,
(iii) $v_a \nabla_\zeta v^a = 0$,
(iv) $v^a \nabla_\zeta v_a = 0$,
(v) $L_v h^a{}_b = 0$.

10.5 (§**10.3**) Show that a Killing vector X^a satisfies the equation of geodesic deviation

$$\frac{D^2 X^a}{Du^2} - R^a{}_{bcd}\frac{dx^b}{du}\frac{dx^c}{du}X^d = 0$$

along any geodesic $x^a = x^a(u)$. [Hint: use Exercise 7.14.]

10.6 (§**10.4**) Show that if a frame $e_i{}^a$ is parallelly propagated along C then so is its dual frame $e^i{}_a$.

10.7 (§**10.4**) If η^{ij} is the inverse of η_{ij}, then show that

$$g_{ab} = \eta_{ij}e^i{}_a e^j{}_b \quad \text{and} \quad g^{ab} = \eta^{ij}e_i{}^a e_j{}^b.$$

If $(x^a) = (t, r, \theta, \phi)$ and

$$e_0{}^a = (A^{-\frac{1}{2}}, 0, 0, 0), \qquad e_1{}^a = (0, A^{\frac{1}{2}}, 0, 0),$$

$$e_2{}^a = (0, 0, 1/r, 0), \qquad e_3{}^a = (0, 0, 0, 1/r\sin\theta),$$

where $A = A(r)$ is an arbitrary function, then find g^{ab}, g_{ab} and the line element ds^2.

10.8 (§**10.5**) If at some point P, the symmetric tensor R_{ab} satisfies

$$R_{ab}v^a v^b = 0$$

for an **arbitrary** timelike vector v^a, then deduce that R_{ab} must vanish at P. [Hint: let $v^a = t^a + \lambda s^a$, where $t^a t_a = 1$, $s^a s_a = -1$, $t^a s_a = 0$, $0 \leqslant \lambda < 1$, λ arbitrary, and consider a special coordinate system in which $t^a \overset{*}{=} \delta_0^a$ and $s^a \overset{*}{=} \delta_1^a, \delta_2^a, \delta_3^a$ in turn.]

10.9 (§**10.6**) What principles are used in each of the six steps outlined in §10.6?

10.10 (§**10.7**) What principles are used in the transition to the full theory?

General relativity from a variational principle 11

11.1 The Palatini equation

Many tensor identities are best derived using the technique of geodesic coordinates, where we choose an arbitrary point P at which $\Gamma^a_{bc} \overset{*}{=} 0$. Then, in particular, covariant derivatives reduce to ordinary derivatives at the point P. The Riemann tensor (6.39) reduces to

$$R^a{}_{bcd} \overset{*}{=} \partial_c \Gamma^a_{bd} - \partial_d \Gamma^a_{bc}. \tag{11.1}$$

We now contemplate a variation of the connection Γ^a_{bc} to a new connection $\bar{\Gamma}^a_{bc}$:

$$\Gamma^a_{bc} \to \bar{\Gamma}^a_{bc} = \Gamma^a_{bc} + \delta\Gamma^a_{bc}. \tag{11.2}$$

Then $\delta\Gamma^a_{bc}$, being the difference of two connections, is a tensor of type $(1, 2)$. This variation results in a change in the Riemann tensor:

$$R^a{}_{bcd} \to \bar{R}^a{}_{bcd} = R^a{}_{bcd} + \delta R^a{}_{bcd},$$

where

$$\delta R^a{}_{bcd} \overset{*}{=} \partial_c(\delta\Gamma^a_{bd}) - \partial_d(\delta\Gamma^a_{bc})$$
$$\overset{*}{=} \nabla_c(\delta\Gamma^a_{bd}) - \nabla_d(\delta\Gamma^a_{bc}),$$

since partial derivative commutes with variation and is equivalent to covariant derivative in geodesic coordinates. Now both $\delta R^a{}_{bcd}$, being the difference of two tensors, and the quantities on the right-hand side of the last equation are tensors, and so by our fundamental result (if a tensor equation holds in one coordinate system it must hold in all coordinate systems) we can deduce the **Palatini equation**

$$\delta R^a{}_{bcd} = \nabla_c(\delta\Gamma^a_{bd}) - \nabla_d(\delta\Gamma^a_{bc}) \tag{11.3}$$

at the point P. Since P is an arbitrary point the result holds quite generally. Contraction on a and c gives the useful result

$$\delta R_{bd} = \nabla_a(\delta\Gamma^a_{bd}) - \nabla_d(\delta\Gamma^a_{ba}). \tag{11.4}$$

11.2 Differential constraints on the field equations

The variational principle proceeds from the specification of a Lagrangian density $\mathscr{L}$ which is assumed to be a functional of the metric g_{ab} and its first and possibly higher derivatives, that is,

$$\mathscr{L} = \mathscr{L}(g_{ab}, \partial_c g_{ab}, \partial_d \partial_c g_{ab}, \dots). \tag{11.5}$$

$\mathscr{L}$ is required to be a scalar density of weight $+1$ so that we can form the action integral

$$I = \int_\Omega \mathscr{L} \, \mathrm{d}\Omega \tag{11.6}$$

over some region Ω of the manifold. The principle of stationary action then states that, if we make arbitrary variations of the g_{ab} which vanish on the boundary $\partial \Omega$ of Ω, then I must be stationary. Writing this out formally using the variational notation of Chapter 7, we obtain

$$g_{ab} \to g_{ab} + \delta g_{ab} \quad \Rightarrow \quad I \to I + \delta I \quad \text{with } \delta I = 0, \tag{11.7}$$

where

$$\delta I = \int_\Omega \mathscr{L}^{ab} \, \delta g_{ab} \, \mathrm{d}\Omega \tag{11.8}$$

and $\mathscr{L}^{ab}$ is the Euler–Lagrange derivative

$$\mathscr{L}^{ab} \equiv \frac{\delta \mathscr{L}}{\delta g_{ab}}. \tag{11.9}$$

The field equations are then

$$\mathscr{L}^{ab} = 0. \tag{11.10}$$

Since δI is the difference between two scalars it must itself be a scalar, and hence from (11.8) it follows that $\mathscr{L}^{ab}$ is a symmetric tensor density of weight $+1$. We shall consider the details of the calculation of $\mathscr{L}^{ab}$ in later sections. However, before we do this we shall derive some very important differential constraints on the field equations which hold **whether or not the field equations hold** and which follow simply from the fact that $\mathscr{L}$ **is a density**. In general relativity, these will turn out to be the contracted Bianchi identities.

The idea is to generate a 'variation' in the g_{ab} which is brought about simply by carrying out a change of coordinates in Ω. Then, since I remains

invariant, it follows that δI must be **identically** zero,

$$\delta I \equiv 0. \qquad (11.11)$$

We consider an infinitesimal change of coordinates (7.50) in Ω

$$x^a \rightarrow x'^a = x^a + \varepsilon X^a(x), \qquad (11.12)$$

where X^a is a smooth vector field which vanishes on the boundary of Ω. Performing a similar calculation to that of §7.7, we find (exercise)

$$\delta g_{ab} = g'_{ab}(x) - g_{ab}(x) = -\mathscr{L}_{\varepsilon X} g_{ab} = -\varepsilon(\nabla_b X_a + \nabla_a X_b). \qquad (11.13)$$

Hence, combining this with (11.8) and (11.11), we obtain

$$0 \equiv \delta I = -2\varepsilon \int_\Omega \mathscr{L}^{ab}(\nabla_b X_a) \, d\Omega,$$

since $\mathscr{L}^{ab}$ is symmetric by the definition (11.9). We now use a standard trick, called **integration by parts**, to write the integral as a difference of two terms, namely (check),

$$0 \equiv 2\varepsilon \int_\Omega (\nabla_b \mathscr{L}^{ab}) X_a \, d\Omega - 2\varepsilon \int_\Omega \nabla_b[\mathscr{L}^{ab} X_a] \, d\Omega. \qquad (11.14)$$

The term in square brackets is a vector density of weight $+1$, and hence by (7.3) its covariant derivative can be replaced by an ordinary derivative. Then the divergence theorem (7.19) gives

$$2\varepsilon \int_\Omega \partial_b[\mathscr{L}^{ab} X_a] \, d\Omega = 2\varepsilon \int_{\partial\Omega} \mathscr{L}^{ab} X_a \, dS_b, \qquad (11.15)$$

which converts the last term in (11.14) to a surface integral. But, by assumption, X^a vanishes on $\partial\Omega$, and hence this term must vanish. Thus (11.14) reduces to

$$\int_\Omega (\nabla_b \mathscr{L}^{ab}) X_a \, d\Omega \equiv 0, \qquad (11.16)$$

and, since Ω is **arbitrary**, we must conclude (exercise)

$$(\nabla_b \mathscr{L}^{ab}) X_a \equiv 0. \qquad (11.17)$$

Finally, since X^a is arbitrary, we obtain the promised **differential identities**

$$\nabla_b \mathscr{L}^{ab} \equiv 0. \qquad (11.18)$$

11.3 A simple example

Let us use the following notation: a **gothicized** tensor is to represent the corresponding tensor multiplied by $(-g)^{\frac{1}{2}}$. Thus, for example,

$$\mathfrak{g}_{ab} = (-g)^{\frac{1}{2}} g_{ab} \quad \text{and} \quad \mathfrak{T}_{ab} = (-g)^{\frac{1}{2}} T_{ab}.$$

Then all tensors in gothic type will be tensor densities of weight $+1$.

The simplest scalar density that we can make out of g_{ab} alone is $(-g)^{\frac{1}{2}}$ itself, namely,

$$\mathscr{L}(g_{ab}) = (-g)^{\frac{1}{2}}, \qquad (11.19)$$

where $(-g)^{\frac{1}{2}}$ is to be regarded as a functional of the dynamical variable g_{ab}. Recalling (7.7), we write

$$\frac{\partial g}{\partial g_{ab}} = gg^{ab}, \tag{11.20}$$

and so

$$\frac{\partial(-g)^{\frac{1}{2}}}{\partial g_{ab}} = \frac{1}{2}\frac{(-g)}{(-g)^{\frac{1}{2}}}g^{ab} = \frac{1}{2}(-g)^{\frac{1}{2}}g^{ab} = \frac{1}{2}\mathfrak{g}^{ab}, \tag{11.21}$$

from which we deduce that

$$\mathscr{L}^{ab} = \frac{\partial\mathscr{L}}{\partial g_{ab}} = \frac{1}{2}\mathfrak{g}^{ab}.$$

Clearly, $\mathfrak{g}^{ab} = 0$ cannot serve as field equations. The identities (11.18) become

$$\nabla_b \mathfrak{g}^{ab} \equiv 0, \tag{11.22}$$

which is trivially satisfied, since both g^{ab} and $(-g)^{\frac{1}{2}}$ have vanishing covariant derivatives by (6.73) and (7.11).

11.4 The Einstein Lagrangian

The Lagrangian (11.19) clearly cannot serve as the Lagrangian of a physical theory. However, it turns out that the next most complicated scalar which can be built out of g_{ab} and its derivatives — and it is very much more complicated — is the curvature scalar R. The resulting Lagrangian

$$\mathscr{L}_G = (-g)^{\frac{1}{2}}R \tag{11.23}$$

is called the **Einstein Lagrangian**, where the label G denotes that it is the Lagrangian for gravitation. We shall employ the notation of a comma for partial differentiation, otherwise we end up writing terms like $\partial\mathscr{L}/\partial(\partial_c g_{ab})$. Then, explicitly,

$$
\begin{aligned}
\mathscr{L}_G &= (-g)^{\frac{1}{2}}g^{cd}R_{cd} \\
&= \mathfrak{g}^{cd}R^e_{ced} \\
&= \mathfrak{g}^{cd}(\Gamma^e_{cd,e} - \Gamma^e_{ce,d} + \Gamma^f_{cd}\Gamma^e_{fe} - \Gamma^f_{ce}\Gamma^e_{fd}) \\
&= \mathfrak{g}^{cd}\{[\tfrac{1}{2}g^{ef}(g_{cf,d} + g_{df,c} - g_{cd,f})]_{,e} \\
&\quad - [\tfrac{1}{2}g^{ef}(g_{cf,e} + g_{ef,c} - g_{ce,f})]_{,d} \\
&\quad + [\tfrac{1}{2}g^{fh}(g_{ch,d} + g_{dh,c} - g_{cd,h})][\tfrac{1}{2}g^{ei}(g_{fi,e} + g_{ei,f} - g_{fe,i})] \\
&\quad - [\tfrac{1}{2}g^{fh}(g_{ch,e} + g_{eh,c} - g_{ce,h})][\tfrac{1}{2}g^{ei}(g_{fi,d} + g_{di,f} - g_{fd,i})]\}.
\end{aligned}
\tag{11.24}
$$

We must think of this as a functional of g_{ab} and its first and second derivatives, namely,

$$\mathscr{L}_G = \mathscr{L}_G(g_{ab}, g_{ab,c}, g_{ab,cd}),$$

where we regard g^{ab} and g (and therefore $\mathfrak{g}^{ab}$) as functions of g_{ab}. Note that we could equally well regard $\mathscr{L}_G$ as a functional of one of g^{ab}, $\mathfrak{g}^{ab}$, or $\mathfrak{g}_{ab}$ and their corresponding first and second derivatives. In the case where g_{ab} are the dynamical variables, the Euler–Lagrange derivative is a generalization of (7.35) and becomes

$$\frac{\delta \mathcal{L}_G}{\delta g_{ab}} = \frac{\partial \mathcal{L}_G}{\partial g_{ab}} - \left(\frac{\partial \mathcal{L}_G}{\partial g_{ab,c}}\right)_{,c} + \left(\frac{\partial \mathcal{L}_G}{\partial g_{ab,cd}}\right)_{,cd}. \tag{11.25}$$

Following the procedure of the last section, we would expect next to calculate actual expressions for each of these terms. For example (exercise),

$$\frac{\partial \mathcal{L}_G}{\partial g_{ab,cd}} = (-g)^{\frac{1}{2}}\left[\tfrac{1}{2}(g^{ac}g^{bd} + g^{ad}g^{bc}) - g^{ab}g^{cd}\right]. \tag{11.26}$$

The calculation of the remaining terms, though straightforward, is, unfortunately, absolutely horrendous and we shall not pursue it further. Instead, we will exploit the variational formalism in the next section and show how this indirect approach leads to a more tractable calculation. However, had we proceeded, then we would have found (exercise for the completely dedicated reader!)

$$\mathcal{L}_G^{ab} = \frac{\delta \mathcal{L}_G}{\delta g_{ab}} = -(-g)^{\frac{1}{2}}G^{ab}, \tag{11.27}$$

and so the Euler–Lagrange equations lead to the vacuum field equations

$$-(-g)^{\frac{1}{2}}G^{ab} = 0, \tag{11.28}$$

that is, the vanishing of the Einstein tensor. In addition the identities (11.18) become

$$\nabla_b[-(-g)^{\frac{1}{2}}G^{ab}] \equiv 0 \quad \Rightarrow \quad \nabla_b G^{ab} \equiv 0, \tag{11.29}$$

that is, the contracted Bianchi identities.

11.5 Indirect derivation of the field equations

The approach depends on exploiting the δ notation fully. It can be shown (exercise) that δ behaves much like a derivative when applied to sums, differences, and products. For example, let us see what happens when we apply δ to the tensor δ_c^a. The variation

$$g_{ab} \rightarrow g_{ab} + \delta g_{ab}$$

induces a variation in g^{ab}, which we write

$$g^{ab} \rightarrow g^{ab} + \delta g^{ab}. \tag{11.30}$$

Then

$$\delta_c^a = g^{ab}g_{bc} \rightarrow (g^{ab} + \delta g^{ab})(g_{bc} + \delta g_{bc})$$

$$= \delta_c^a + \delta g^{ab}g_{bc} + g^{ab}\delta g_{bc} + O(\delta^2)$$

But, since δ_c^a is a constant tensor, it cannot change and therefore

$$\delta g^{ab} g_{bc} + g^{ab} \delta g_{bc} = 0 \tag{11.31}$$

to first order, or, multiplying through by g^{cd},

$$\delta g^{ad} = -g^{ab} g^{cd} \delta g_{bc}. \tag{11.32}$$

Compare this with the corresponding relationship between partial derivatives.

Starting from I written in the form

$$I = \int_\Omega \mathfrak{g}^{ab} R_{ab} \, d\Omega,$$

we carry out a variation and use the Leibniz rule for products, to get

$$\delta I = \int_\Omega (\delta \mathfrak{g}^{ab} R_{ab} + \mathfrak{g}^{ab} \delta R_{ab}) \, d\Omega. \tag{11.33}$$

We now use the Palatini equation in the form (11.4), so that the second term on the right-hand side becomes

$$\int_\Omega \mathfrak{g}^{ab} \delta R_{ab} \, d\Omega = \int_\Omega \mathfrak{g}^{ab} [\nabla_c \delta \Gamma_{ab}^c - \nabla_b \delta \Gamma_{ac}^c] \, d\Omega$$

$$= \int_\Omega [\nabla_c (\mathfrak{g}^{ab} \delta \Gamma_{ab}^c) - \nabla_b (\mathfrak{g}^{ab} \delta \Gamma_{ac}^c)] \, d\Omega$$

$$= \int_\Omega \partial_c (\mathfrak{g}^{ab} \delta \Gamma_{ab}^c - \mathfrak{g}^{ac} \delta \Gamma_{ab}^b) \, d\Omega,$$

since the covariant derivative of $\mathfrak{g}^{ab}$ vanishes identically and the quantities in parentheses are vector densities of weight $+1$. Using the same argument as we did in §11.2, this can be converted to a surface integral by the divergence theorem, which vanishes because the variations are assumed to vanish on the surface of Ω. Hence, (11.33) reduces to

$$\delta I = \int_\Omega R_{ab} \delta \mathfrak{g}^{ab} \, d\Omega$$

$$= \int_\Omega R_{ab} \delta [(-g)^{\frac{1}{2}} g^{ab}] \, d\Omega$$

$$= \int_\Omega [R_{ab} g^{ab} \delta(-g)^{\frac{1}{2}} + R_{ab} (-g)^{\frac{1}{2}} \delta g^{ab}] \, d\Omega$$

$$= \int_\Omega (-g)^{\frac{1}{2}} (\tfrac{1}{2} R g^{cd} - R_{ab} g^{ac} g^{bd}) \delta g_{cd} \, d\Omega$$

$$= -\int_\Omega (-g)^{\frac{1}{2}} (R^{cd} - \tfrac{1}{2} R g^{cd}) \delta g_{cd} \, d\Omega$$

$$= \int_\Omega [-(-g)^{\frac{1}{2}} G^{ab}] \delta g_{ab} \, d\Omega, \tag{11.34}$$

where we have used (11.31) and the result (exercise)

$$\delta(-g)^{\frac{1}{2}} = \tfrac{1}{2}(-g)^{\frac{1}{2}} g^{ab} \delta g_{ab}. \tag{11.35}$$

Using (11.8), we again get the vacuum field equation in the form (11.28) and the contracted Bianchi identities (11.29) as the corresponding differential constraints on the field equations.

11.6 An equivalent Lagrangian

The resulting field equations are second order in the partial derivatives. This is at first sight rather surprising since by (11.25) we might expect the last term to produce fourth-order equations. However, it turns out, as we have seen in (11.26), that $\partial\mathscr{L}_G/\partial g_{ab,cd}$ only involves undifferentiated g_{ab}'s and $\partial\mathscr{L}_G/\partial g_{ab,c}$ only involves once differentiated g_{ab}'s (exercise). In this section, we make the second-order nature of the equations more evident by showing that

$$\mathscr{L}_G = \bar{\mathscr{L}}_G + Q^a{}_{,a}, \tag{11.36}$$

where $\bar{\mathscr{L}}_G$ depends on the metric and its first derivatives only. It can be shown that in applying the variational principle argument to such an equation the divergence $Q^a{}_{,a}$ can be discarded (by converting to a vanishing surface integral), and hence it follows that $\mathscr{L}_G$ and $\bar{\mathscr{L}}_G$ give rise to the **same** field equations. However, $\bar{\mathscr{L}}_G$ is no longer a scalar density. We sketch the argument below.

The Einstein Lagrangian

$$\mathscr{L}_G = (-g)^{\frac{1}{2}} R$$

$$= g^{ab} R_{ab}$$

$$= g^{ab}(\Gamma^c_{ab,c} - \Gamma^c_{ac,b} + \Gamma^c_{ab}\Gamma^d_{cd} - \Gamma^d_{ac}\Gamma^c_{bd})$$

$$= g^{ab}\Gamma^c_{ab,c} - g^{ab}\Gamma^c_{ac,b} - \bar{\mathscr{L}}_G, \tag{11.37}$$

where

$$\bar{\mathscr{L}}_G = g^{ab}(\Gamma^d_{ac}\Gamma^c_{bd} - \Gamma^c_{ab}\Gamma^d_{cd}). \tag{11.38}$$

Integrating the first two terms in (11.37) by parts, we get

$$\mathscr{L}_G = -g^{ab}{}_{,c}\Gamma^c_{ab} + g^{ab}{}_{,b}\Gamma^c_{ac} - \bar{\mathscr{L}}_G + Q^a{}_{,a}, \tag{11.39}$$

where

$$Q^a = g^{bc}\Gamma^a_{bc} - g^{ab}\Gamma^c_{bc}. \tag{11.40}$$

From the fact that the covariant derivative of g^{ab} vanishes, we find (exercise)

$$g^{ab}{}_{,c} = \Gamma^d_{dc}g^{ab} - \Gamma^a_{dc}g^{db} - \Gamma^b_{dc}g^{ad}. \tag{11.41}$$

Substituting in (11.39) and simplifying, we obtain the result (11.36).

Once again, we could consider $\bar{\mathscr{L}}_G$ as a functional of one of g_{ab}, g^{ab}, $\mathfrak{g}_{ab}$, or $\mathfrak{g}^{ab}$ and their corresponding first derivatives. For example, let us choose the $\mathfrak{g}^{ab}$ as the dynamical variables. Then

$$\bar{\mathscr{L}}_G = \bar{\mathscr{L}}_G(\mathfrak{g}^{ab}, \mathfrak{g}^{ab}{}_{,c}),$$

from which it can be shown that

$$\frac{\partial\bar{\mathscr{L}}_G}{\partial\mathfrak{g}^{ab}} = \Gamma^d_{ac}\Gamma^c_{bd} - \Gamma^c_{ab}\Gamma^d_{cd} \tag{11.42}$$

and

$$\frac{\partial\bar{\mathscr{L}}_G}{\partial\mathfrak{g}^{ab}{}_{,c}} = \Gamma^c_{ab} - \tfrac{1}{2}\delta^c_a\Gamma^d_{bd} - \tfrac{1}{2}\delta^c_b\Gamma^d_{ad}. \tag{11.43}$$

The Euler–Lagrange equations

$$\mathscr{\bar{L}}_G^{ab} = \frac{\partial \mathscr{L}_G}{\partial g^{ab}} - \left(\frac{\partial \mathscr{L}_G}{\partial g^{ab}{}_{,c}}\right)_c = 0 \qquad (11.44)$$

then lead to

$$\mathscr{\bar{L}}_G^{ab} = -\Gamma^c_{ab,c} + \tfrac{1}{2}\Gamma^d_{bd,a} + \tfrac{1}{2}\Gamma^d_{ad,b} - \Gamma^c_{ab}\Gamma^d_{cd} + \Gamma^d_{ac}\Gamma^c_{bd}. \qquad (11.45)$$

If we use the result (exercise)

$$[\ln(-g)^{\frac{1}{2}}]_{,a} = \Gamma^d_{ad}, \qquad (11.46)$$

then

$$\Gamma^d_{ad,b} = [\ln(-g)^{\frac{1}{2}}]_{,ab} = [\ln(-g)^{\frac{1}{2}}]_{,ba} = \Gamma^d_{bd,a},$$

and so (11.45) gives

$$\mathscr{\bar{L}}_G^{ab} = -(\Gamma^c_{ab,c} - \Gamma^d_{ad,b} + \Gamma^c_{ab}\Gamma^d_{cd} - \Gamma^d_{ac}\Gamma^c_{bd}) = -R_{ab}.$$

The field equations are correspondingly $R_{ab} = 0$.

11.7 The Palatini approach

The Palatini approach is very elegant and is based on the idea of treating both the metric **and** the connection separately as dynamical variables in the Einstein Lagrangian. To be specific, let us choose $\mathscr{L}_G$ as a functional of g^{ab} and a symmetric connection Γ^a_{bc} and its derivatives, i.e.

$$\mathscr{L}_G = \mathscr{L}_G(g^{ab}, \Gamma^a_{bc}, \Gamma^a_{bc,d}),$$

where

$$\mathscr{L}_G = g^{ab} R_{ab}$$
$$= g^{ab}(\Gamma^c_{ab,c} - \Gamma^d_{ad,b} + \Gamma^c_{ab}\Gamma^d_{cd} - \Gamma^d_{ac}\Gamma^c_{bd}), \qquad (11.47)$$

so that the Ricci tensor depends on Γ^a_{bc} and its derivatives only. Then, if we carry out a variation with respect to g^{ab} only,

$$\delta I = \int_\Omega \delta g^{ab} R_{ab}\, d\Omega$$

and the principle of stationary action gives immediately the vacuum field equations $R_{ab} = 0$.

We next carry out a variation with respect to Γ^a_{bc}, so that

$$\delta I = \int_\Omega g^{ab}\, \delta R_{ab}\, d\Omega$$

$$= \int_\Omega g^{ab}[\nabla_c(\delta\Gamma^c_{ab}) - \nabla_b(\delta\Gamma^c_{ac})]\, d\Omega$$

by the corollary of the Palatini equation (11.4). Integrating by parts and discarding the divergence term by the usual argument, we get

$$\delta I = \int_\Omega [\nabla_b g^{ab}\, \delta\Gamma^c_{ac} - \nabla_c g^{ab}\, \delta\Gamma^c_{ab}]\, d\Omega$$

$$= \int_\Omega [(\delta^b_c \nabla_d g^{ad} - \nabla_c g^{ab})\, \delta\Gamma^c_{ab}]\, d\Omega.$$

Since δI vanishes for arbitrary volumes Ω, the integrand must vanish, i.e.

$$(\delta_c^b \nabla_d g^{ad} - \nabla_c g^{ab}) \delta \Gamma_{ab}^c = 0.$$

The variations $\delta \Gamma_{ab}^c$ are arbitrary, but **symmetric** in a and b, and so only the symmetric part of the expression in brackets vanishes, i.e.

$$\tfrac{1}{2}\delta_c^b \nabla_d g^{ad} + \tfrac{1}{2}\delta_c^a \nabla_d g^{bd} - \nabla_c g^{ab} = 0. \tag{11.48}$$

Manipulating this equation, one can show in turn (exercise) that the covariant derivatives of g^{ab}, $(-g)^{\frac{1}{2}}$, g^{ab}, and g_{ab} vanish. Finally, by Exercise 6.20, if

$$\nabla_c g_{ab} = 0,$$

then it follows that Γ_{bc}^a is necessarily the metric connection

$$\Gamma_{bc}^a = \tfrac{1}{2}g^{ad}(g_{bd,c} + g_{cd,b} - g_{bc,d})$$

   To summarize, the Palatini approach starts from the Einstein Lagrangian (11.47) considered as a functional of a metric and an arbitrary symmetric connection and its derivatives. Variation with respect to the metric produces the vacuum field equations of general relativity, and variation with respect to the connection reveals that the connection is necessarily the metric connection.

11.8 The full field equations

So far, we have been concerned with the vacuum field equations. To obtain the full field equations, we assume that there are other fields present beside the gravitational field, which can be described by an appropriate Lagrangian density $\mathscr{L}_M$—the matter Lagrangian. The action is then

$$I = \int_\Omega (\mathscr{L}_G + \kappa \mathscr{L}_M)\, d\Omega, \tag{11.49}$$

where κ is the coupling constant. Both Lagrangians are to be considered as functionals of the metric and its derivatives, and so, varying with respect to g_{ab} (say), we obtain

$$\frac{\delta \mathscr{L}_G}{\delta g_{ab}} = -(-g)^{\frac{1}{2}} G^{ab} \tag{11.50}$$

and

$$\frac{\delta \mathscr{L}_M}{\delta g_{ab}} = (-g)^{\frac{1}{2}} T^{ab}, \tag{11.51}$$

where the latter equation defines the **energy–momentum tensor** T^{ab} for the fields present. Dividing through by $(-g)^{\frac{1}{2}}$, the field equations become

$$G^{ab} = \kappa T^{ab}, \tag{11.52}$$

in agreement with (10.43). In the next chapter, we shall investigate the right-hand side of this equation and look at the definition of the energy–momentum tensor for various important fields.

Exercises

11.1 (§11.2) Show that, under an infinitesimal change of coordinates

$$x^a \rightarrow x'^a = x^a + \varepsilon X^a(x),$$

the transformed metric satisfies

$$g'_{ab}(x) - g_{ab}(x) = -\varepsilon(\nabla_b X_a + \nabla_a X_b)$$

to first order in ε.

11.2 (§11.4) Show that

$$\frac{\partial \mathscr{L}_G}{\partial g_{ab,cd}} = (-g)^{\frac{1}{2}} [\tfrac{1}{2}(g^{ac}g^{bd} + g^{ad}g^{bc}) - g^{ab}g^{cd}].$$

11.3 (§11.4) Show that

$$\frac{\partial g^{cd}}{\partial g_{ab}} = -\tfrac{1}{2}(g^{ac}g^{bd} + g^{bc}g^{ad})$$

11.4 (§11.4) Check that $\partial \mathscr{L}_G/\partial g_{ab,c}$ depends only on g_{ab} and its first derivatives. [Hint: consider (11.24).]

11.5 (§11.5) If y_A are dynamical variables and $L_1 = L_1(y_A)$ and $L_2 = L_2(y_A)$, then show from first principles that

(i) $\delta(\lambda L_1 + \mu L_2) = \lambda \delta L_1 + \mu \delta L_2$, where λ and μ are constants,

(ii) $\delta(L_1 L_2) = L_1 \delta L_2 + L_2 \delta L_1$.

11.6 (§11.5) Show that

(i) $g_{ab}\delta g^{ab} = -g^{ab}\delta g_{ab}$,

(ii) $\delta g = g g^{ab}\delta g_{ab}$, (compare this with (7.8))

(iii) $\delta(-g)^{\frac{1}{2}} = \tfrac{1}{2}(-g)^{\frac{1}{2}}g^{ab}\delta g_{ab}$.

11.7 (§11.5) Show that, if we regard g^{ab}, g_{ab}, and g^{ab}, respectively, as dynamical variables, then

(i) $\dfrac{\delta \mathscr{L}_G}{\delta g^{ab}} = R_{ab}$,

(ii) $\dfrac{\delta \mathscr{L}_G}{\delta g_{ab}} = -R^{ab}$,

(iii) $\dfrac{\delta \mathscr{L}_G}{\delta g^{ab}} = (-g)^{\frac{1}{2}}G_{ab}$.

What differential constraints do each of these quantities satisfy?

11.8 (§11.5) (i) If $\int_\Omega \Phi \, d\Omega = 0$, where Ω is arbitrary, then prove that $\Phi = 0$. [Hint: choose an arbitrary point P where $\Phi(P) > 0$, say, use continuity to show that there is a region surrounding P where Φ remains positive, deduce that $\int_\Omega \Phi \, d\Omega > 0$ for a suitable Ω and derive a contradiction; then complete the proof.]

(ii) If $W^a X_a = 0$, where X_a is arbitrary, then show that $W^a = 0$. [Hint: take $X_a \overset{*}{=} (1, 0, 0, 0)$, etc.]

11.9 (§11.6) If the Lagrangians $L(y, y', x)$ and $\bar{L}(y, y', x)$ differ by a divergence, i.e.

$$L = \bar{L} + \frac{dQ(y, y', x)}{dx},$$

then show that L and $\bar{L}$ give rise to the same field equation.

11.10 (§11.6) (i) Establish the results (11.41) and (11.46).

(ii) Establish the result (11.36) for the Einstein Lagrangian.

(iii) Establish the result (11.42) and (11.43).

11.11 (§11.7) Show that, if

$$\tfrac{1}{2}\delta_c^b \nabla_d g^{ad} + \tfrac{1}{2}\delta_c^a \nabla_d g^{bd} - \nabla_c g^{ab} = 0$$

for an arbitrary connection, then

(i) $\nabla_c g^{ab} = 0$,

(ii) $\nabla_c(-g)^{\frac{1}{2}} = 0$,

(iii) $\nabla_c g^{ab} = 0$,

(iv) $\nabla_c g_{ab} = 0$,

and deduce that the connection is necessarily the metric connection.

11.12 (§11.7) Use the variational principle approach to find the field equations of the theory (considered by A. S. Eddington)

$$\mathscr{L} = (-g)^{\frac{1}{2}} R^{abcd} R_{abcd}.$$

11.13 (§11.8) Find the energy–momentum tensor for the Lagrangian

$$\mathscr{L} = (-g)^{\frac{1}{2}}(\phi_{,a}\phi_{,b}g^{ab} + m_0^2\phi^2),$$

where $\phi = \phi(x)$ is a scalar field.

The energy–momentum tensor

12

12.1 Preview

Our programme for this chapter is to look at the three most important energy–momentum tensors in general relativity, namely, the energy–momentum tensors for incoherent matter or dust, a perfect fluid, and the electromagnetic field. In passing, we shall encounter a tensor formulation of Maxwell's equations governing the electromagnetic field. Again, our treatment will not be exhaustive or complete, but will be sufficient for generating the explicit expressions for the three tensors, and these expressions will be essentially all that we require in future chapters. We shall also look more carefully at the Newtonian limit and discuss the calculation for determining the coupling constant.

12.2 Incoherent matter

We start by considering the simplest kind of matter field, namely, that of **non-interacting incoherent matter** or **dust**. Such a field may be characterized by two quantities, the **4-velocity** vector field of flow

$$u^a = \frac{dx^a}{d\tau},$$

where τ is the proper time along the world-line of a dust particle (Fig. 12.1), and a scalar field

$$\rho_0 = \rho_0(x)$$

describing the **proper density** of the flow, that is, the density which would be measured by an observer moving with the flow (a co-moving observer). The simplest second-rank tensor we can construct from these two quantities is

$$T^{ab} = \rho_0 u^a u^b, \tag{12.1}$$

and this turns out to be the energy–momentum tensor for the matter field.

Let us investigate this tensor in special relativity in Minkowski coordinates. Then, by Exercise 8.10, the 4-velocity is

$$u^a = \gamma(1, \boldsymbol{u}), \tag{12.2}$$

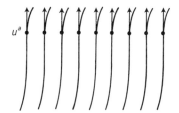

Fig. 12.1 The world-lines of dust particles.

where $\gamma = (1 - u^2)^{-\frac{1}{2}}$. The proper time is defined by

$$d\tau^2 = ds^2$$
$$= \eta_{ab}\, dx^a\, dx^b$$
$$= dt^2 - dx^2 - dy^2 - dz^2$$
$$= dt^2(1 - u^2)$$
$$= \gamma^{-2}\, dt^2. \tag{12.3}$$

Then the zero–zero component of T^{ab} is

$$T^{00} = \rho_0 \frac{dx^0}{d\tau} \frac{dx^0}{d\tau} = \rho_0 \frac{dt^2}{d\tau^2} = \gamma^2 \rho_0, \tag{12.4}$$

by (12.3). This quantity has a simple physical interpretation. First of all, in special relativity, the mass of a body in motion is greater than its rest mass by a factor γ, by (4.11). In addition, if we consider a moving three-dimensional volume element, then its volume decreases by a factor γ through the Lorentz contraction. Thus, from the point of view of a fixed as opposed to a co-moving observer, the density increases by a factor γ^2. Hence, if a field of material of proper density ρ_0 flows past a fixed observer with velocity $\boldsymbol{u}$, then the observer will measure a density

$$\rho = \gamma^2 \rho_0. \tag{12.5}$$

The component T^{00} may therefore be interpreted as the **relativistic energy density** of the matter field since the only contribution to the energy of the field is from the motion of the matter.

The components of T^{ab} can be written, using (12.2) and (12.5), in the form

$$T^{ab} = \rho \begin{bmatrix} 1 & u_x & u_y & u_z \\ u_x & u_x^2 & u_x u_y & u_x u_z \\ u_y & u_x u_y & u_y^2 & u_y u_z \\ u_z & u_x u_z & u_y u_z & u_z^2 \end{bmatrix}. \tag{12.6}$$

We now show that the equations governing the force-free motion of a matter field of dust can be written in the following very succinct way:

$$\partial_b T^{ab} = 0. \tag{12.7}$$

Using (12.6), in the case when $a = 0$, this equation becomes (exercise)

$$\frac{\partial \rho}{\partial t} + \frac{\partial}{\partial x}(\rho u_x) + \frac{\partial}{\partial y}(\rho u_y) + \frac{\partial}{\partial z}(\rho u_z) = 0.$$

This is precisely the classical **equation of continuity**

$$\frac{\partial \rho}{\partial t} + \mathrm{div}\,(\rho \boldsymbol{u}) = 0. \tag{12.8}$$

In classical fluid dynamics, this expresses the conservation of matter with density ρ moving with velocity $\boldsymbol{u}$. Since matter is the same as energy in special relativity, it follows that the **conservation of energy** equation for dust is $\partial_b T^{0b} = 0$. The equations corresponding to $a = \alpha$ $(\alpha = 1, 2, 3)$ are similarly found to be (exercise)

$$\frac{\partial}{\partial t}(\rho\boldsymbol{u}) + \frac{\partial}{\partial x}(\rho u_x \boldsymbol{u}) + \frac{\partial}{\partial y}(\rho u_y \boldsymbol{u}) + \frac{\partial}{\partial z}(\rho u_z \boldsymbol{u}) = \boldsymbol{0}.$$

Combining this with (12.8), the equation can be written as (exercise)

$$\rho\left[\frac{\partial\boldsymbol{u}}{\partial t} + (\boldsymbol{u}\cdot\nabla)\boldsymbol{u}\right] = \boldsymbol{0}. \tag{12.9}$$

Comparing this with the **Navier–Stokes equation of motion** for a perfect fluid in classical fluid dynamics, namely,

$$\rho\left[\frac{\partial\boldsymbol{u}}{\partial t} + (\boldsymbol{u}\cdot\nabla)\boldsymbol{u}\right] = -\operatorname{grad}p + \rho X, \tag{12.10}$$

where p is the pressure in the fluid and X is the body force per unit mass, we see that (12.9) is simply this equation in the absence of pressure and external forces.

We have seen that the requirement that the energy–momentum tensor has zero divergence in special relativity is equivalent to demanding conservation of energy and conservation of momentum in the matter field—hence the name **energy–momentum** tensor. Moreover, (12.7) is known as the energy–momentum **conservation law**. If we use a non-flat metric in special relativity, then (12.7) is replaced by its covariant counterpart

$$\nabla_b T^{ab} = 0. \tag{12.11}$$

We now make the transition to general relativity and once again define the energy–momentum tensor for incoherent matter by (12.1), and, using the principle of minimal gravitational coupling, retain (12.11) as the statement of the conservation law.

12.3 Perfect fluid

A **perfect fluid** is characterized by three quantities: a **4-velocity** $u^a = dx^a/d\tau$; a **proper density** field $\rho_0 = \rho_0(x)$; and a **scalar pressure** field $p = p(x)$. In the limit as p vanishes, a perfect fluid reduces to incoherent matter. This suggests that we take the energy–momentum tensor for a perfect fluid to be of the form

$$T^{ab} = \rho_0 u^a u^b + pS^{ab} \tag{12.12}$$

for some symmetric tensor S^{ab}. The only second-rank tensors which are associated with the fluid are $u^a u^b$ and the metric g^{ab}, and so the simplest assumption we can make is

$$S^{ab} = \lambda u^a u^b + \mu g^{ab}, \tag{12.13}$$

where λ and μ are constants. Proceeding as we did in the last section, we investigate the conservation law $\partial_b T^{ab} = 0$ in special relativity in Minkowski coordinates and demand that it reduces in an appropriate limit to the continuity equation (12.11) and the Navier–Stokes equation (12.10) in the absence of body forces. This requirement leads to $\lambda = 1$ and $\mu = -1$. Then (12.12) and (12.13) give

$$T^{ab} = (\rho_0 + p)u^a u^b - pg^{ab}, \tag{12.14}$$

which we take as the definition for the energy–momentum tensor of a perfect fluid. If we use a non-flat metric in special relativity, then we again take the covariant form (12.11) for the conservation law. In the full theory, we also take (12.14) as the definition of a perfect fluid and (12.11) as the conservation equations.

In addition, p and ρ are related by an **equation of state** governing the particular sort of perfect fluid under consideration. In general, this is an equation of the form $p = p(\rho, T)$, where T is the absolute temperature. However, we shall only be concerned with situations in which T is effectively constant so that the equation of state reduces to

$$p = p(\rho).$$

12.4 Maxwell's equations

In this section, we wish to reformulate Maxwell's equations for the electromagnetic field in tensorial form. We start by rewriting them in special relativity in Minkowski coordinates. Working in Heavyside–Lorentz units with $c = 1$, we find that **Maxwell's equations in vacuo** for the electromagnetic field split up into two pairs of equations, namely, the **source equations**

$$\text{div } \boldsymbol{E} = \rho \tag{12.15}$$

$$\text{curl } \boldsymbol{B} - \frac{\partial \boldsymbol{E}}{\partial t} = \boldsymbol{j}, \tag{12.16}$$

and the **internal equations**

$$\text{div } \boldsymbol{B} = 0 \tag{12.17}$$

$$\text{curl } \boldsymbol{E} + \frac{\partial \boldsymbol{B}}{\partial t} = \boldsymbol{0}, \tag{12.18}$$

where $\boldsymbol{E}$ is the electric field, $\boldsymbol{B}$ is the magnetic induction, ρ is the charge density, and $\boldsymbol{j}$ is the current density. In simple physical terms: (12.15) is the differential form of Gauss's law relating the flux through a closed surface to the enclosed charge; (12.16) is a generalized Ampère's law relating the magnetic field to a flow of current (where the term involving E is Maxwell's displacement current added in part to produce wave equations for E and B); (12.17) is the statement that magnetic monopoles do not exist; and (12.18) is essentially Faraday's law of induction. The quantities ρ and j cannot be prescribed independently because, differentiating (12.15) with respect to t, we get (remembering that $\partial/\partial t$ commutes with $\partial/\partial x$, $\partial/\partial y$, and $\partial/\partial z$)

$$\operatorname{div}\frac{\partial E}{\partial t} = \frac{\partial\rho}{\partial t},$$

and taking the divergence of (12.16) gives

$$-\operatorname{div}\frac{\partial E}{\partial t} = \operatorname{div}j.$$

Thus, ρ and j must satisfy the **equation of continuity**

$$\frac{\partial\rho}{\partial t} + \operatorname{div}j = 0. \tag{12.19}$$

If we interpret j as a convection current, i.e. $j = \rho u$, where u is the velocity field of the material with charge density ρ, then (12.19) is identical to (12.8), the continuity equation of fluid dynamics.

In order to write these equations in tensorial form, we define an anti-symmetric tensor F^{ab}, called the **electromagnetic field tensor** or **Maxwell tensor**, by

$$F^{ab} = \begin{bmatrix} 0 & E_x & E_y & E_z \\ -E_x & 0 & B_z & -B_y \\ -E_y & -B_z & 0 & B_x \\ -E_z & B_y & -B_x & 0 \end{bmatrix} \tag{12.20}$$

and the **current density** or **source** 4-vector j^a by

$$j^a = (\rho, j). \tag{12.21}$$

Then (exercise) the source equations and internal equations can be written in the form

$$\partial_b F^{ab} = j^a, \tag{12.22}$$

$$\partial_a F_{bc} + \partial_c F_{ab} + \partial_b F_{ca} = 0. \tag{12.23}$$

The anti-symmetry of F_{ab} means that (12.23) can be written more succinctly as

$$\partial_{[a} F_{bc]} = 0. \tag{12.24}$$

The continuity equation (12.19) becomes

$$\partial_a j^a = 0. \tag{12.25}$$

Let us be clear what we have done so far. We have merely shown that, given the definitions (12.20) and (12.21), Maxwell's equations (12.15)–(12.18) can be written **formally** as (12.22) and (12.23). We have treated F^{ab} and j^a as tensors, but the only justification for doing this is knowing their transformation properties under Lorentz transformations. Before the advent of special relativity, their transformation properties were in fact unclear. Indeed, from one point of view, it was precisely the desire to make Maxwell's equations Lorentz-covariant that led to the development of special relativity. The approach we shall adopt is to propose (12.20) and (12.21) as an **ansatz** (working hypothesis) and from these definitions work out their transformation properties. The ultimate justification then, as always, lies in comparing the predictions with observation and there are a host of experiments which support the ansatz.

12.5 Potential formulation of Maxwell's equations

Rather than working with the fields E and B directly, it is usually more convenient to work in terms of the potentials. The **scalar potential** ϕ and the **vector potential** A are defined by

$$E = -\operatorname{grad} \phi - \frac{\partial A}{\partial t}, \tag{12.26}$$

$$B = \operatorname{curl} A. \tag{12.27}$$

If we define the **4-potential** by

$$\phi^a = (\phi, A), \tag{12.28}$$

then we find that (12.26) and (12.27) are equivalent to (exercise)

$$F_{ab} = \partial_b \phi_a - \partial_a \phi_b. \tag{12.29}$$

The 4-potential is not defined uniquely by this equation since we may perform a **gauge transformation**

$$\phi_a \to \bar{\phi}_a = \phi_a + \partial_a \psi, \tag{12.30}$$

where ψ is an arbitrary scalar field. Although a gauge transformation alters the potentials, it leaves F_{ab}, and hence E and B, unchanged, and these are the strictly measurable quantities.

In solving particular problems, it is often convenient to reduce the gauge freedom by imposing a constraint on ϕ_a, called a **gauge condition**, which in turn simplifies the problem. For example, an important gauge for discussing electromagnetic radiation is provided by the **Lorentz gauge**

$$\eta^{ab}\phi_{a,b} = 0. \qquad (12.31)$$

Applying this constraint to (12.30), we find that the scalar field is no longer arbitrary but must be a solution of the **wave equation**

$$\Box\psi \equiv \eta^{ab}\psi_{,ab} = 0 \qquad (12.32)$$

where $\Box$ is the d'Alembertian operator

$$\Box \equiv \partial_0^2 - \partial_1^2 - \partial_2^2 - \partial_3^2.$$

The definition (12.29) results in the internal equations (12.23) being **automatically** satisfied, that is, they become identities (exercise). The source equations (12.22) become, in terms of the 4-potential,

$$\partial_b[\eta^{ac}\eta^{bd}(\partial_d\phi_c - \partial_c\phi_d)] = j^a. \qquad (12.33)$$

In the Lorentz gauge, this reduces to (exercise)

$$\Box\phi^a = j^a. \qquad (12.34)$$

In source-free regions, j^a vanishes, and this becomes

$$\Box\phi^a = 0, \qquad (12.35)$$

from which it follows that ϕ^a and F^{ab}, and therefore E and B, all satisfy wave equations.

So far, we have restricted our attention to special relativity in Minkowski coordinates. To obtain the covariant formulation, we simply replace ordinary derivatives by covariant derivatives. However, it is not necessary in equations (12.24) and (12.29) because (exercise)

$$\nabla_{[a}F_{bc]} = \partial_{[a}F_{bc]} \qquad (12.36)$$

and

$$\nabla_{[b}\phi_{a]} = \partial_{[b}\phi_{a]}. \qquad (12.37)$$

The **covariant formulation** of Maxwell's equations **in vacuo** in special relativity is

$$\nabla_b F^{ab} = j^a \qquad (12.38)$$

$$\partial_{[a}F_{bc]} = 0 \qquad (12.39)$$

subject to

$$\nabla_a j^a = 0. \qquad (12.40)$$

In terms of the 4-potential, we still have

$$F_{ab} = \partial_b\phi_a - \partial_a\phi_b. \qquad (12.41)$$

Using the principle of minimal gravitational coupling, we adopt equations (12.38) and (12.39) in general relativity, where, however, the metric is no longer flat but is a solution of the full field equations $G^{ab} = \kappa T^{ab}$ and T^{ab} is the energy–momentum tensor arising from the electromagnetic field — which we now seek.

12.6 The Maxwell energy–momentum tensor

We shall construct the energy–momentum tensor for the electromagnetic field from a variational approach. For simplicity, we shall work **in vacuo** in special relativity in Minkowski coordinates and restrict attention to a source-free region, i.e. a region where j^a vanishes. Consider the Lagrangian for the electromagnetic field defined by

$$\mathscr{L}_{\mathrm{E}}(\phi_a, F_{ab}) = \frac{1}{4\pi}[-\tfrac{1}{2}F_{ab}F^{ab} + (\phi_{a,b} - \phi_{b,a})F^{ab}]. \tag{12.42}$$

Then

$$\frac{\delta \mathscr{L}_{\mathrm{E}}}{\delta \phi_a} = \frac{\partial \mathscr{L}_{\mathrm{E}}}{\partial \phi_a} - \left(\frac{\partial \mathscr{L}_{\mathrm{E}}}{\partial \phi_{a,b}}\right)_{,b}$$

$$= 0 - \frac{1}{4\pi}(F^{ab} - F^{ba})_{,b}$$

and the field equations corresponding to a variation with respect to ϕ_a become

$$(F^{ab} - F^{ba})_{,b} = 0. \tag{12.43}$$

Similarly,

$$\frac{\delta \mathscr{L}_{\mathrm{E}}}{\delta F_{ab}} = \frac{\partial \mathscr{L}_{\mathrm{E}}}{\partial F_{ab}}$$

$$= \frac{\partial}{\partial F_{ab}} \frac{1}{4\pi}[-\tfrac{1}{2}\eta^{ce}\eta^{df}F_{cd}F_{ef} + \eta^{ce}\eta^{df}(\phi_{c,d} - \phi_{d,c})F_{ef}]$$

$$= \frac{1}{4\pi}[-\tfrac{1}{2}\eta^{ae}\eta^{bf}F_{ef} - \tfrac{1}{2}\eta^{ca}\eta^{db}F_{cd} + \eta^{ca}\eta^{db}(\phi_{c,d} - \phi_{d,c})]$$

$$= \frac{\eta^{ac}\eta^{bd}}{4\pi}[-F_{cd} + (\phi_{c,d} - \phi_{d,c})]$$

and the field equations corresponding to a variation with respect to F_{ab} become

$$F_{ab} = \phi_{a,b} - \phi_{b,a}. \tag{12.44}$$

This last equation defines F_{ab} in terms of the 4-potential and reveals that F_{ab} is anti-symmetric. The definition also means that the internal equations are satisfied automatically and (12.43) reduces to

$$F^{ab}{}_{,b} = 0,$$

namely, the source equations (in source-free regions). The result (12.44) also allows us to re-express the Lagrangian as

$$\mathscr{L}_{\mathrm{E}} = \frac{1}{8\pi}\eta^{ac}\eta^{bd}F_{ab}F_{cd}. \tag{12.45}$$

We now make the transition to the full theory and assume that

$$\mathscr{L}_E = \frac{(-g)^{\frac{1}{2}}}{8\pi} \, g^{ac} g^{bd} F_{ab} F_{cd}, \qquad (12.46)$$

together with the definition (12.44) of F_{ab} in terms of ϕ_a. The factor $(-g)^{\frac{1}{2}}$ is included to ensure that $\mathscr{L}_E$ is a scalar density (note that it reduces to 1 in special relativity in Minkowski coordinates). Then we find (exercise)

$$\frac{\partial \mathscr{L}_E}{\partial g^{ab}} = -\frac{(-g)^{\frac{1}{2}}}{4\pi} (-g^{cd} F_{ac} F_{bd} + \tfrac{1}{4} g_{ab} F_{cd} F^{cd}). \qquad (12.47)$$

The analogue of (11.51) for the contravariant metric is

$$\frac{\delta \mathscr{L}_E}{\delta g^{ab}} = -(-g)^{\frac{1}{2}} T_{ab}. \qquad (12.48)$$

These last two equations lead to the definition of the **Maxwell energy–momentum tensor** T_{ab} in source-free regions

$$T_{ab} = \frac{1}{4\pi} (-g^{cd} F_{ac} F_{bd} + \tfrac{1}{4} g_{ab} F_{cd} F^{cd}). \qquad (12.49)$$

Then, in relativistic units, $\kappa = 8\pi$, and the full field equations in source-free regions are called the **Einstein–Maxwell equations** and become

$$G_{ab} = -2g^{cd} F_{ac} F_{bd} + \tfrac{1}{2} g_{ab} F_{cd} F^{cd}. \qquad (12.50)$$

Let us look at some of the components of T_{ab} in special relativity in Minkowski coordinates. In particular, we find that the **energy density** of the electromagnetic field is given by

$$T_{00} = \frac{1}{8\pi} (\boldsymbol{E}^2 + \boldsymbol{B}^2), \qquad (12.51)$$

which agrees with the usual expression for energy density in electrodynamics. Again, the **momentum density** is

$$(T_{01}, T_{02}, T_{03}) = -\frac{1}{4\pi} \boldsymbol{E} \times \boldsymbol{B}, \qquad (12.52)$$

where the vector $\boldsymbol{E} \times \boldsymbol{B}$ is the **Poynting vector** of electrodynamics and represents the momentum density of the electromagnetic field. In addition it is straightforward to verify that Maxwell's equations imply that T^{ab} is divergenceless, i.e.

$$\partial_b T^{ab} = 0. \qquad (12.53)$$

12.7 Other energy–momentum tensors

We have met two methods for obtaining energy–momentum tensors. The first is an **ad hoc** method which constructs likely looking tensors out of the matter and energy fields present and investigates the conservation equations (12.7) in the non-relativistic limit. The second method proceeds from a variational principle formulation and investigates the field equations arising from a proposed Lagrangian. We can construct energy–momentum tensors for other fields or combination of fields using either approach or a combination of them. In particular, we can combine non-interacting fields by superimposing them. For interacting fields, we have to take the interactions into account.

We illustrate this with one example of each procedure. The energy–momentum tensor for a field of charged matter of proper mass density ρ_0 and 4-velocity u^a is (see (12.1) and (12.49))

$$T^{ab} = \rho_0 u^a u^b - F^{ac} F^c{}_b + \tfrac{1}{4} g^{ab} F_{cd} F^{cd}. \tag{12.54}$$

The conservation equations then express the conservation of energy and the equations of motion for the field. The Lagrangian for an elementary particle described by a scalar field $\phi(x)$, for example the π^0-meson, is given by

$$\mathscr{L}_s = (-g)^{\frac{1}{2}} (g^{ab} \phi_{;a} \phi_{;b} + m_0{}^2 \phi^2), \tag{12.55}$$

where m_0 is the rest mass of the particle. The energy–momentum tensor is defined by (12.48) and again the conservation equations express the conservation of energy and the equations of motion of the field.

12.8 The dominant energy condition

In general, the components of any tensor in a particular coordinate system do not have an invariant meaning. However, if we choose an invariantly defined frame and look at the **frame components** of the tensor, then these will have physical significance. In the case of the energy–momentum tensor T_{ab}, we choose a frame at a point by looking for solutions of the **eigenvalue** equation

$$T_a{}^b u^a = \lambda u^b,$$

where u^a is the eigenvector corresponding to the eigenvalue λ. This has characteristic equation

$$|T_a{}^b - \lambda \delta_a{}^b| = 0.$$

If this equation has real non-zero roots, then the corresponding eigenvectors can be normalized to form a frame $e_i{}^a$ of one timelike and three spacelike vectors. The frame components of T_{ab} are

$$T_{ij} = T_{ab} e_i{}^a e_j{}^b = \operatorname{diag}(\mu, p_1, p_2, p_3),$$

since the matrix is diagonal with the eigenvalues as elements. The eigenvalue μ is called the **energy density** and $u^a = e_0{}^a$ is the **4-velocity** of the medium. The eigenvalues p_α ($\alpha = 1, 2, 3$) are called the **principal stresses**, and the corresponding eigenvectors $e_\alpha{}^a$ the **principal axes of stress**. An energy–momentum tensor will only represent a physically realistic matter field if the energy density is non-negative and dominates any stresses present. More precisely, all known matter fields satisfy the **dominant energy condition** of Hawking and Ellis:

$$\mu \geqslant 0, \qquad -\mu \leqslant p_\alpha \leqslant \mu. \tag{12.56}$$

The latter condition can be shown to be equivalent to requiring that the local speed of sound is not greater than the local speed of light.

If, in particular, the three principal stresses are positive and equal, to p say, then setting $\mu = \rho_0$ the energy–momentum tensor takes the form of a perfect fluid (12.14). If the three principal stresses vanish, then the energy–momentum tensor takes the form of dust (12.1).

12.9 The Newtonian limit

In this section, we consider more precisely the Newtonian limit of a slowly varying weak gravitational field. We shall work in non-relativistic units. In the **Newtonian limit**, we assume that there exists a privileged coordinate system

$$(x^a) = (x^0, x^1, x^2, x^3) = (x^0, x^\alpha) = (ct, x, y, z)$$

in which the metric g_{ab} differs only slightly from the Minkowski metric η_{ab}. Moreover, we assume that the field is produced by bodies whose velocities are small compared with the velocity of light. If v is a typical velocity of the bodies, then we let ε denote a small dimensionless parameter of order v/c and our basic assumption is

$$g_{ab} = \eta_{ab} + \varepsilon h_{ab} + O(\varepsilon^2), \tag{12.57}$$

where throughout we shall work to lowest order in ε. In time δt, a body moves a distance δx^α with velocity v, i.e.

$$\delta x^\alpha \sim \text{velocity} \times \text{time} \sim v\delta t \sim (v/c)c\delta t \sim \varepsilon\delta x^0,$$

and so

$$\varepsilon/\delta x^\alpha \sim 1/\delta x^0.$$

Then, for any function f, we assume the **slow-motion approximation**

$$\varepsilon \frac{\partial f}{\partial x^\alpha} \sim \frac{\partial f}{\partial x^0}, \tag{12.58}$$

that is, derivatives with respect to x^0 are of order ε times the spatial derivatives. The conditions (12.57) and (12.58) are the starting assumptions for obtaining the Newtonian limit.

We consider the motion of a free test particle moving with a speed of the order of v on a world-line $x^a = x^a(\tau)$ parametrized by the proper time. It travels on a timelike geodesic

$$\frac{d^2 x^a}{d\tau^2} + \Gamma^a_{bc} \frac{dx^b}{d\tau} \frac{dx^c}{d\tau} = 0. \tag{12.59}$$

By definition,

$$c^2 d\tau^2 = ds^2$$
$$= c^2 dt^2 - dx^2 - dy^2 - dz^2$$
$$= dt^2(c^2 - v^2)$$
$$= c^2 dt^2(1 - \varepsilon^2),$$

and so, taking square roots,

$$\frac{dt}{d\tau} = 1 + O(\varepsilon). \tag{12.60}$$

Hence, working to lowest order in ε, we can replace τ by t in (12.59). Moreover, from our slow-motion approximation,

$$dx^\alpha \sim \varepsilon c\, dt,$$

so that

$$\frac{dx^\alpha}{c\, dt} = O(\varepsilon). \tag{12.61}$$

In addition

$$\Gamma^a_{bc} = \tfrac{1}{2} g^{ad}(\partial_c g_{bd} + \partial_b g_{cd} - \partial_d g_{bc})$$
$$= \tfrac{1}{2} \eta^{ad}\varepsilon(\partial_c h_{bd} + \partial_b h_{cd} - \partial_d h_{bc}) + O(\varepsilon^2), \tag{12.62}$$

so that

$$\Gamma^a_{bc} = O(\varepsilon). \tag{12.63}$$

Since we are interested in the Newtonian limit, we restrict our attention to the spatial part of (12.59), i.e. when $a = \alpha$, and we obtain, by using (12.60) and dividing by c^2,

$$0 = \frac{1}{c^2}\frac{d^2 x^\alpha}{dt^2} + \frac{1}{c^2}\Gamma^\alpha_{bc}\frac{dx^b}{dt}\frac{dx^c}{dt}[1 + O(\varepsilon)]$$

$$= \frac{1}{c^2}\frac{d^2 x^\alpha}{dt^2} + \Gamma^\alpha_{00} + 2\Gamma^\alpha_{0\beta}\left(\frac{dx^\beta}{c\, dt}\right) + \Gamma^\alpha_{\beta\gamma}\left(\frac{dx^\beta}{c\, dt}\right)\left(\frac{dx^\gamma}{c\, dt}\right) + O(\varepsilon^2).$$

From (12.61) and (12.63), the third and fourth terms in this equation are $O(\varepsilon^2)$ and $O(\varepsilon^3)$, respectively. From (12.62), the second term is

$$\Gamma^\alpha_{00} = -\tfrac{1}{2}\varepsilon\left(2\frac{\partial h_{0\alpha}}{\partial x^0} - \frac{\partial h_{00}}{\partial x^\alpha}\right)$$

$$= \tfrac{1}{2}\varepsilon\frac{\partial h_{00}}{\partial x^\alpha} + O(\varepsilon^2)$$

by the slow motion approximation (12.58). So the spatial part of the geodesic equation can be written

$$\frac{d^2 x^\alpha}{dt^2} = -\tfrac{1}{2}c^2\frac{\partial g_{00}}{\partial x^\alpha}[1 + O(\varepsilon)] \tag{12.64}$$

using (12.57). We compare this with the corresponding Newtonian equation (10.4), namely,

$$\frac{d^2 x^\alpha}{dt^2} = -\frac{\partial \phi}{\partial x^\alpha},$$

where ϕ is the Newtonian gravitational potential. Noting that, at large distances from the sources of the field, $\phi \to 0$ and $g_{00} \to 1$, we conclude

$$g_{00} = 1 + \frac{2\phi}{c^2} + O(v/c). \tag{12.65}$$

This is called the **weak-field limit**.

Let us consider the effect of an infinitesimal coordinate transformation

$$x^a \rightarrow x'^a = x^a + \varepsilon X^a(x)$$

which is consistent with the two assumptions (12.57) and (12.58). Then, as in Exercise 11.1, we find

$$g'_{ab} = g_{ab} - \varepsilon(\partial_a X_b + \partial_b X_a) + O(\varepsilon^2), \qquad (12.66)$$

where

$$X_a = \eta_{ab} X^b.$$

To preserve (12.58), we require

$$\frac{\partial X_a}{\partial x^0} \sim \varepsilon \frac{\partial X_a}{\partial x^\alpha},$$

which means from (12.66) that g_{00} is the only component of g_{ab} that does not alter to first order in ε. We have therefore shown that the only component of the metric tensor which is well defined to first order for a slowly varying weak gravitational field is determined to this order by the requirement that the theory should agree with Newtonian theory to this order, and it is given by (12.65). Note that no mention of the field equations has been made in deriving (12.65). It arises purely from assuming geodesic motion and the Newtonian limit as embodied in the equations (12.57) and (12.58).

12.10 The coupling constant

In Chapter 20, we shall see that the assumption (12.57) leads to the full field equations (20.28), which we write

$$\tfrac{1}{2}\varepsilon \, \Box \, (h_{ab} - \tfrac{1}{2}\eta_{ab}\eta^{cd} h_{cd}) = -\kappa T_{ab} + O(\varepsilon^2). \qquad (12.67)$$

Contracting with η^{ab} and applying the slow-motion approximation (12.58), we find (exercise)

$$\tfrac{1}{2}\varepsilon \nabla^2 h_{ab} = \kappa(T_{ab} - \tfrac{1}{2}\eta_{ab}\eta^{cd} T_{cd}) + O(\varepsilon^2). \qquad (12.68)$$

Let us take, as the source of the field, a distribution of dust of **small** proper density ρ_0 moving at low velocity of order v. This assumption means that we neglect terms both of order v/c and $\rho_0 v/c$, and then by (12.6) in relativistic units, the energy–momentum tensor reduces in our privileged coordinate system to

$$T^{ab} = \rho_0 \delta_0^a \delta_0^b, \qquad (12.69)$$

which in turn implies

$$T_{ab} = \rho_0 \delta_a^0 \delta_b^0 \quad \text{and} \quad \eta^{cd} T_{cd} = \rho_0. \qquad (12.70)$$

The zero–zero component of the field equations (12.68) then becomes

$$\varepsilon \nabla^2 h_{00} = \kappa \rho_0 + O(\varepsilon^2). \qquad (12.71)$$

But, by (12.57),

$$g_{00} = 1 + \varepsilon h_{00} + O(\varepsilon^2),$$

so that

$$\nabla^2 g_{00} = \varepsilon \nabla^2 h_{00} + O(\varepsilon^2),$$

and, by (12.65),

$$\nabla^2 g_{00} = \nabla^2 \left(\frac{2\phi}{c^2}\right) + O(\varepsilon).$$

Substituting these results in (12.71), we get

$$\nabla^2 \phi = \tfrac{1}{2}c^2 \kappa \rho_0 + O(\varepsilon).$$

Comparing this with Poisson's equation (4.5), namely,

$$\nabla^2 \phi = 4\pi G \rho_0,$$

we obtain the result (10.44), namely,

$$\kappa = 8\pi G/c^2.$$

Thus, as promised, we have used the correspondence principle with Newtonian theory to obtain the coupling constant κ appearing in the full field equations (10.43).

Exercises

12.1 (§**12.2**) Establish (12.6) from (12.1). Show that (12.7) leads to (12.8) and (12.9).

12.2 (§**12.3**) Show that the conservation equations for a perfect fluid lead to the equation of continuity and the equation of motion.

12.3 (§**12.4**)
(i) Show that Maxwell's equations can be written in the form (12.22) and (12.23), given the definitions (12.20) and (12.21).
(ii) Show that the internal equations can be written in the form (12.24).
(iii) Show that the continuity equation can be written in the form (12.25). Show directly from (12.22) that this equation is an identity.

12.4 (§**12.4**) Find the transformation properties of E, B, ρ, and j under a boost in the x-direction. [Hint: consider F^{ab} and j^a.]

12.5 (§**12.5**)
(i) Show that (12.29) is equivalent to (12.26) and (12.27).
(ii) Show that F_{ab} is invariant under a gauge transformation.

(iii) Show that if F_{ab} is defined in terms of a 4-potential then the internal equations are automatically satisfied.

12.6 (§**12.5**) Show that, in an appropriate gauge, Maxwell's equations reduce to $\Box \phi^a = j^a$ in regions where the source 4-vector is non-zero. What remaining gauge freedom is left? Deduce that E and B satisfy the wave equation in source-free regions.

12.7 (§**12.5**) Check (12.36) and (12.37).

12.8 (§**12.6**)
(i) Establish (12.47) and (12.49).
(ii) Confirm (12.51) and (12.52).
(iii) Investigate the conservation equations (12.53).

12.9 (§**12.7**) Investigate the conservation equations for the energy–momentum tensor arising from (12.55).

12.10 (§**12.9**) Write out the argument fully which deduces (12.65) from (12.64).

12.11 (§**12.9**) Check (12.66). Deduce that g_{00} is the only component not to alter to order ε.

12.12 (§**12.10**) Derive (12.68) from (12.67) and deduce (12.71).

The structure of the field equations

13

13.1 Interpretation of the field equations

Before attempting to solve the field equations we shall consider some of their important physical and mathematical properties in this chapter. The full field equations (in relativistic units) are

$$G_{ab} = 8\pi T_{ab}. \tag{13.1}$$

They can be viewed in three different ways.

1. The field equations are differential equations for determining the metric tensor g_{ab} from a **given energy–momentum tensor T_{ab}**. Here, we are reading the equations from right to left. This is a Machian way of viewing the equations since one specifies a matter distribution and then solves the equations to ascertain the resulting geometry. It is also a natural way of looking at the Einstein–Maxwell equations, namely, what geometry corresponds to a given Maxwell tensor? The most important case of the equations is when $T_{ab} = 0$, in which case we are concerned with finding **vacuum** solutions.

2. The field equations are equations from which the energy–momentum tensor can be read off corresponding to a **given metric tensor g_{ab}**. Here, we are reading the equations from left to right. It was originally thought that this would be a productive way of determining energy–momentum tensors. We simply choose arbitrarily ten functions of the coordinates, namely, the symmetric g_{ab}, and then we can compute G_{ab} and read off T_{ab} from (13.1). However, this rarely turns out to be very useful in practice because the resulting T_{ab} are usually physically unrealistic and violate the dominant energy conditions. In particular, it frequently turns out that the energy density goes negative in some region, which we reject as unphysical because the positive character of energy density dominates gravitation theory.

3. The field equations consist of **ten equations connecting twenty quantities**, namely, the ten components of g_{ab} and the ten components of T_{ab}. Hence, from this point of view, the field equations are to be viewed as **constraints** on the simultaneous choice of g_{ab} and T_{ab}. This approach is used when one can partly specify the geometry and the energy–momentum tensor from physical considerations and then the equations are used to try and determine both quantities completely.

13.2. Determinacy, non-linearity, and differentiability

Let us consider solving the vacuum field equations

$$G_{ab} = 0 \tag{13.2}$$

for g_{ab}. Then, at first sight, the problem seems well posed: there are ten equations for the ten unknowns g_{ab}. However, the equations are not independent but are connected by four differential constraints through the contracted Bianchi identities

$$\nabla_b G^{ab} \equiv 0. \tag{13.3}$$

So we seem to have a problem of **under-determinacy**, since there are fewer equations than unknowns. However, we cannot expect complete determinacy for any set g_{ab}, since they can be transformed with fourfold freedom by a coordinate transformation

$$x^a \to x'^a = x'^a(x) \quad (a = 0, 1, 2, 3).$$

We can in fact use this coordinate freedom to impose four conditions on the g_{ab}. These are known as **coordinate conditions** or **gauge conditions**. For example, we could introduce **Gaussian** or **normal coordinates** in which

$$g_{00} \overset{*}{=} 1, \qquad g_{0\alpha} \overset{*}{=} 0. \tag{13.4}$$

Then the remaining six unknowns $g_{\alpha\beta}$ can be determined by the six independent equations in (13.2). However, there is rather more to the story, but we postpone its consideration until §13.5. Similar remarks apply to the full theory.

The field equations are very difficult to handle because they are **non-linear**. They do not therefore possess a principle of superposition, that is to say, if you have two solutions of the field equations then you cannot add them together to obtain a third. Put another way, it means that you cannot analyse a complicated physical problem by breaking it up into simpler constituent parts. The non-linearity reveals itself physically in the following way: the gravitational field produced by some source contains energy and hence, by special relativity, mass, and this mass in turn is itself a source of a gravitational field; that is to say, the gravitational field is coupled to itself. This non-linearity means that the equations are very difficult to solve in general. Indeed, originally Einstein anticipated that one would never be able to find an exact solution of them. It came as something of a surprise when K. Schwarzschild found an exact solution in less than a year from the publication of the theory in 1915. However, Schwarzschild's solution arises by making a symmetry assumption, indeed the simplest assumption of all, namely, spherical symmetry. Today there are a large number of solutions in existence, probably in excess of four figures (depending on how you count them). Nearly all of them have been obtained by imposing symmetry conditions. It is known that non-linear partial differential equations admit large classes of solutions, many of which are unphysical. It could well be that a large number of the exact solutions are, because of the symmetry assumptions, also unphysical and not in any sense 'close' to a physically meaningful solution, but rather pathologies thrown up by the particular set of partial differential equations. This is still largely an open question.

Ideally, one wants to know what the theory says about physically import-ant situations. In cases where symmetry is absent, or where the symmetry conditions are not strong enough to determine a solution, then recourse has to be made to approximation methods. We met such a method in the Newtonian limit of the last chapter. These approximation methods are based on the **weakness** of the gravitational fields which are most often encountered in nature, or on asymptotic methods applied to isolated sources, so that again the fields are weak a long way from the sources. The weakness means, from a mathematical viewpoint, that the linear terms in certain equations are more important than the rest. We shall meet a linearized form of the field equations in Chapter 20.

There are important mathematical questions concerning the differenti-ability of the solution. However, we shall not take them into account since we will assume that all our fields are smooth or C^∞, so that they can be differentiated indefinitely. This condition can be weakened considerably, for example if we assume that the metric is C^2, which means that it can be differentiated twice, then this ensures that the field equations can be constructed. There are other conditions affecting the differentiability which are connected with surfaces of discontinuities that arise in the theory, for example the surface of a material body. One important set of conditions (analogous to the continuity conditions of potential theory) are the **Lichnero-wicz conditions**: second and higher derivatives of g_{ab} need not be continuous across a surface of discontinuity S, but g_{ab} and $g_{ab,c}$ must be continuous across S.

13.3 The cosmological term

Einstein was rather sceptical about the full field equations (13.1) and regarded the vacuum field equations (13.2) as more fundamental. However, Einstein considered that even these equations were deficient in that they violated Mach's principle in the form M2, since they admit Minkowski space-time as a solution. This means that a test body in an otherwise empty universe would possess inertial properties (as all bodies do in special relativity) even though there is no matter to produce the inertia. As we pointed out before, a set of partial differential equations possesses large classes of solutions many of which are unphysical. In order to decide which solutions are realized in nature, one must also prescribe **boundary conditions**. A natural requirement would be to take space-time to be **asymptotically flat** so that the Riemann tensor vanishes at spatial infinity. However, this requirement does not preclude a flat space solution of the vacuum field equations.

Einstein, realizing the need for prescribing appropriate boundary condi-tions, adopted a different approach. Cosmology, that is, the modelling of the universe, had not really emerged as a separate science prior to general relativity. In as much as there was some generally accepted model of the universe in existence then, it was rather an imprecise one. It suggested that, overall, the universe is **static** (i.e. not undergoing any large-scale motion) and **homogeneous** (i.e. filled uniformly with matter). There are two possible ideas about the spatial extent of the universe, either it is **open** (or **infinite**), in which case it goes on forever in spatial directions, or it is **closed** (**compact** or **finite**), in which case it is bounded in spatial directions. Einstein therefore tried to incorporate a simple model of the universe into the theory and then use this model to prescribe boundary conditions. In particular, he tried to find a static closed solution of the field equations corresponding to a universe uniformly filled with matter. In so doing, he found he was forced to modify the field equations by introducing an extra term, the **cosmological term** Λg_{ab}, where Λ is a constant called the **cosmological constant**, so that they become (with our sign conventions)

$$G_{ab} - \Lambda g_{ab} = 8\pi T_{ab}. \tag{13.5}$$

Since

$$\nabla_b g^{ab} = 0,$$

we see that (13.5) is consistent with the requirement

$$\nabla_b T^{ab} = 0. \tag{13.6}$$

Using the results of §11.3, the corresponding Lagrangian becomes

$$\mathscr{L} = (-g)^{\frac{1}{2}}(R - 2\Lambda) + \mathscr{L}_{M} \tag{13.7}$$

Indeed, if, quite generally, we demand that the gravitational field equations should

(1) be generally covariant,

(2) be of second differential order in g_{ab},

(3) involve the energy–momentum tensor T_{ab} linearly,

then it can be shown that the only equation which meets all of these requirements is

$$R_{ab} + \mu R g_{ab} - \Lambda g_{ab} = \kappa T_{ab}, \tag{13.8}$$

where μ, Λ, and κ are constants. The demand that T_{ab} satisfies the conservation equations (13.6) then leads to $\mu = -\frac{1}{2}$. In fact, it was in the same year as Einstein proposed his equations that the great mathematician Hilbert derived them independently from a variational principle. Of course, they lacked the physical meaningfulness which Einstein had bestowed on them, especially through their reliance on the principle of equivalence.

The full field equations with the cosmological term are Machian in the sense that they no longer admit flat space as a solution. However, shortly after Einstein obtained the static cosmological solution, it was discovered that the universe is not in fact static, but rather is undergoing large-scale expansion, as evidenced by the galactic red shift. Einstein therefore discarded the static solution. At the same time non-static closed solutions of the field equations **without** the cosmological term, corresponding to an expanding distribution of matter, were found. Worse still, from the Machian viewpoint, de Sitter discovered a vacuum solution of the field equations with the cosmological term. These discoveries led Einstein to reject the cosmological term. He did so with some vehemence, describing his original decision to include it as '... the biggest mistake I ever made'. However, despite the fact that its inclusion does not make the theory any more Machian, there is no a priori reason to leave it out. The constant Λ is assumed to be 'very small' in some sense and only of significance on a cosmological scale (or, somewhat bizarrely, on a quantum scale). Most treatments of cosmology include the term, but it is usually omitted for considerations connected with terrestrial or solar system phenomena and, indeed, we shall neglect it until we come to relativistic cosmology. It is worth noting that it is possible to incorporate a number of **ad hoc** assumptions into Newtonian theory and obtain a cosmological theory which has much in common with relativistic cosmology (see §22.3). In the Newtonian model, if $\Lambda > 0$, then all matter experiences a 'cosmic repulsion', which tends to disperse the matter to spatial infinity. Conversely, $\Lambda < 0$ corresponds to a cosmic attraction. Since all matter experiences the force, it provides, in some sense, a realization of a long-range Machian-type interaction.

13.4 The conservation equations

We have suggested an axiomatic formulation of general relativity which replaces $R^a{}_{bcd} = 0$ by $G_{ab} = 8\pi T_{ab}$ in Axiom I(iii) of §8.5. However, it turns out that, rather surprisingly, the geodesic Axioms II(ii) and II(iii) need not be stated separately in general relativity because it can be shown that they must hold automatically **by virtue of the field equations** themselves. That this is possible can be made plausible by considering more carefully the motion of a test particle or photon in a gravitational field. Strictly speaking, the test particle or photon is itself part of the energy and matter present and so should be contained in the energy–momentum tensor. This tensor, in turn, being the source term in the field equations, determines the geometry of space-time and in particular its geodesic structure. In this sense, the motion of a test particle should somehow be contained in the field equations. In fact, it is coded into the Bianchi identities since they lead to the requirement that

$$\nabla_b T^{ab} = 0, \tag{13.9}$$

namely, the conservation equations. It is possible to show that these equations specify unique equations of motion for a point particle in a gravitational field and that the ensuing trajectory of that particle is a geodesic of the corresponding metric. The original demonstration of this result was started by Einstein and Grommer, and developed further by Einstein with contributions from Infeld, Plebanski, and Fock. It rests on treating test particles as singularities in the field and, as a consequence, relies on special mathematical apparatus which they had to construct to cope adequately with these singularities. The resulting work is both very complicated and voluminous and we will make no attempt to describe it. Indeed, not all of the work was fully published. However, the results have been confirmed subsequently by several workers using more powerful mathematical machinery.

There is one neat little calculation which is very suggestive of what happens in essence in the general case. It consists of investigating the equations for a distribution of dust,

$$T^{ab} = \rho_0 u^a u^b.$$

Then the conservation equations (13.9) require

$$\nabla_b [\rho_0 u^a u^b] = 0.$$

The trick is to think of the term in square brackets as being the product $[(\rho_0 u^b) u^a]$ and apply the Leibniz rule to this product:

$$u^a \nabla_b (\rho_0 u^b) + \rho_0 u^b (\nabla_b u^a) = 0. \tag{13.10}$$

We next contract this equation with u_a and use the result

$$u_a u^a = 1 \quad \Rightarrow \quad u_a(\nabla_b u^a) = 0,$$

which makes the second term vanish, leaving

$$\nabla_b(\rho_0 u^b) = 0.$$

Substituting this result back in (13.10) and dividing by $\rho_0 \neq 0$, we get

$$u^b \nabla_b u^a = 0,$$

which is the condition for u^a to be tangent to a geodesic. In other words, the conservation equations necessitate geodesic motion for the dust particles.

13.5 The Cauchy problem

In this section, we look in some detail at the following mathematical problem.

> Given the metric tensor g_{ab} and its first derivatives at one time x^0, then construct the metric which corresponds to a vacuum space-time for all future time.

This is the problem of finding the causal development of a physical system from initial data and is a fundamental problem in the theory of partial differential equations. It is known as the **Cauchy problem** or **initial value problem**, or IVP for short.

We start with a three-dimensional spacelike hypersurface S in the manifold, which we can take without loss of generality to be given by $x^0 = 0$. We specify g_{ab} and its first derivatives $g_{ab,c}$ on S (Fig. 13.1). However, if we know g_{ab} everywhere on S, then we know its spacelike derivatives $g_{ab,\alpha}$ everywhere in S. Hence, it is sufficient to specify the following **initial data** on S:

$$g_{ab}, \quad g_{ab,0},$$

that is, the metric potentials and their time derivatives. Our problem is then to use the second-order vacuum field equations, which we take in the form $R_{ab} = 0$, to try and solve for the second time derivatives $g_{ab,00}$. Let us suppose that we have found some equations for determining $g_{ab,00}$. Then, by repeatedly differentiating these equations with respect to time, we can get all higher time derivatives of g_{ab}. It follows that, if we assume that g_{ab} is an **analytic function** of x^0, we can develop it in a power series in x^0. More precisely, if P and Q are the points $(0, x_0^\alpha)$ and (x^0, x_0^α), so that Q lies on the

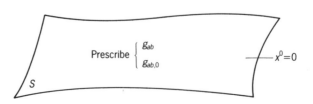

Fig. 13.1 The initial data for the Cauchy IVP.

Prescribe $\begin{cases} g_{ab} \\ g_{ab,0} \end{cases}$

$x^0 = 0$

S

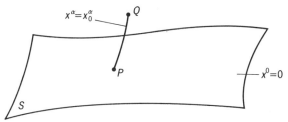

Fig. 13.2 Determining the metric at a later time x^0.

x^0-curve passing through P (Fig. 13.2), then, by Taylor's theorem,

$$g_{ab}(Q) = g_{ab}(P) + g_{ab,0}(P)x^0 + \sum_{n=2}^{\infty} \frac{1}{n!} \partial_0^n g_{ab}|_P (x^0)^n. \qquad (13.11)$$

A straightforward calculation (exercise) reveals that the field equations can be written in the following form:

$$R_{00} = -\tfrac{1}{2} g^{\alpha\beta} g_{\alpha\beta,00} + M_{00} = 0, \qquad (13.12)$$

$$R_{0\alpha} = \tfrac{1}{2} g^{0\beta} g_{\alpha\beta,00} + M_{0\alpha} = 0, \qquad (13.13)$$

$$R_{\alpha\beta} = -\tfrac{1}{2} g^{00} g_{\alpha\beta,00} + M_{\alpha\beta} = 0, \qquad (13.14)$$

where the terms involving M can be expressed **solely in terms of the initial data on S**. This gives rise to two problems of determination:

1. The system (13.12)–(13.14) does not contain $g_{0a,00}$; hence we have a problem of **under-determination**.

2. The system (13.12)–(13.14) represents ten equations in the six unknowns $g_{\alpha\beta,00}$; hence we have a problem of **over-determination**. This means that there must be compatibility requirements for the initial data on S.

We have met Problem 1 before, and it is not unexpected since it relates to the fourfold freedom of coordinate transformations. Let us exploit this coordinate freedom and carry out a coordinate transformation which leaves g_{ab} and $g_{ab,0}$ unchanged on S but which makes

$$g_{0a,00} = 0 \quad \text{on} \quad S. \qquad (13.15)$$

Consider the transformation

$$x^a \to x'^a = x^a + \tfrac{1}{6}(x^0)^3 C^a(x). \qquad (13.16)$$

Then the hypersurface $x^0 = 0$ gets mapped to $x'^0 = 0$ (check). Moreover, **on S**,

$$\left(\frac{\partial x'^a}{\partial x^b}\right) = \delta_b^a, \quad \left(\frac{\partial x'^a}{\partial x^b}\right)_{,c} = 0, \quad \left(\frac{\partial x'^a}{\partial x^0}\right)_{,00} = C^a, \quad \left(\frac{\partial x'^a}{\partial x^\alpha}\right)_{,00} = 0. \quad (13.17)$$

Using

$$g_{ab} = g'_{cd} \frac{\partial x'^c}{\partial x^a} \frac{\partial x'^d}{\partial x^b},$$

we find on S (exercise)

$$\left.\begin{aligned}
g_{ab} &= g'_{ab}, \\
g_{ab,c} &= g'_{ab,c}, \\
g_{ab,\alpha\alpha} &= g'_{ab,\alpha\alpha}, \\
g_{00,00} &= g'_{00,00} + 2g'_{0a}C^a = g'_{00,00} + 2g_{0a}C^a, \\
g_{0\alpha,00} &= g'_{0\alpha,00} + g'_{\alpha a}C^a = g'_{0\alpha,00} + g_{\alpha a}C^a, \\
g_{\alpha\beta,00} &= g'_{\alpha\beta,00}.
\end{aligned}\right\} \tag{13.18}$$

Then choosing C^a such that

$$g_{00,00} = 2g_{0a}C^a \quad \text{and} \quad g_{0\alpha,00} = g_{\alpha a}C^a,$$

which is always possible because $\det(g_{ab}) \neq 0$ and so these equations can be viewed as four independent equations for four unknowns, our result (13.15) follows. The conditions (13.15) are known as **normalization conditions** and overcome Problem 1.

Turning to Problem 2, then, as long as $g^{00} \neq 0$, we can consider the equations (13.14) as serving to determine the six unknowns $g_{\alpha\beta,00}$. We call these the **evolution**, **dynamical**, or **main** equations. Then the remaining four equations, namely,

$$R_{00} = R_{0\alpha} = 0,$$

will act as **constraints** on the initial data. To see this more clearly, we get, from (13.12) and (13.14),

$$g^{00}R_{00} - g^{\alpha\beta}R_{\alpha\beta} = g^{00}M_{00} - g^{\alpha\beta}M_{\alpha\beta} = 0, \tag{13.19}$$

and, from (13.13) and (13.14),

$$g^{00}R_{0\alpha} + g^{0\beta}R_{\alpha\beta} = g^{00}M_{0\alpha} + g^{0\beta}M_{\alpha\beta} = 0. \tag{13.20}$$

It should be immediate from these last two equations that if the evolution equations $R_{\alpha\beta} = 0$ are satisfied then the remaining equations only involve the initial data. It proves more convenient to write these constraints in terms of the mixed Einstein tensor $G_a{}^b$. A simple calculation reveals (exercise)

$$G_0{}^0 = \tfrac{1}{2}(g^{00}M_{00} - g^{\alpha\beta}M_{\alpha\beta}), \tag{13.21}$$

$$G_\alpha{}^0 = g^{00}M_{0\alpha} + g^{0\beta}M_{\alpha\beta}, \tag{13.22}$$

which are equivalent to (13.19) and (13.20). Then, following Lichnerowicz, we may write the vacuum field equations in the **normal form**

$$R_{\alpha\beta} = 0, \quad G_a{}^0 = 0,$$

where the first six equations are **evolution** equations for $g_{\alpha\beta,00}$ and the last four equations are **constraints equations** which the initial data must satisfy on S. This resolves Problem 2.

We now prove a remarkable result

If the constraint equations are satisfied on S, i.e $[G_a{}^0]_S = 0$, then they are satisfied for all time, i.e. $G_a{}^0 = 0$, by virtue of the contracted Bianchi identities.

To show this, we assume that the evolution equations hold and then we can write the mixed Einstein tensor in terms of the Ricci components R_{00} and $R_{0\alpha}$, i.e.

$$\left.\begin{array}{l} G_0{}^0 = \tfrac{1}{2}g^{00}R_{00}, \\[4pt] G_\alpha{}^0 = g^{00}R_{0\alpha}, \\[4pt] G_0{}^\alpha = g^{0\alpha}R_{00} + g^{\alpha\beta}R_{0\beta}, \\[4pt] G_\alpha{}^\beta = g^{0\beta}R_{0\alpha} - \tfrac{1}{2}\delta_\alpha^\beta(g^{00}R_{00} + 2g^{0\gamma}R_{0\gamma}). \end{array}\right\} \tag{13.23}$$

By (13.19)–(13.22), the equations $G_\alpha{}^0 = 0$ are equivalent to $R_{0a} = 0$ if the evolution equations hold. It follows that $G_a{}^\alpha$ depends linearly on $G_a{}^0$ with coefficients involving the metric tensor. It is then straightforward to show (exercise) that the identities

$$\nabla_b G_a{}^b = \nabla_0 G_a{}^0 + \nabla_\alpha G_a{}^\alpha \equiv 0 \tag{13.24}$$

can be written in the form

$$G_a{}^0{}_{,0} = C^{b\alpha}{}_a G_b{}^0{}_{,\alpha} + D^b{}_a G_b{}^0, \tag{13.25}$$

where $C^{b\alpha}{}_a$ and $D^b{}_a$ depend only on the metric tensor and its first derivatives. Then (13.25) form a system of four first-order partial differential equations for the time derivatives of $G_a{}^0$. This system is in so-called **normal form** and therefore possesses a **unique** solution. Indeed, since $[G_a{}^0]_S = 0$, it follows that $G_a{}^0 \equiv 0$ everywhere, which is the promised result.

Let us summarize our results. We prescribe initial data g_{ab} and $g_{ab,0}$ on S **subject to the constraints** $[G_a{}^0]_S = 0$. We next prescribe the four components g_{0a} quite arbitrarily in space and time apart from the requirement that they match the initial data on S and satisfy the normalization conditions $[g_{0a,00}]_S = 0$. Then, assuming $g^{00} \neq 0$, we find that the evolution equations

$$g_{\alpha\beta,00} = 2(g^{00})^{-1}M_{\alpha\beta}$$

determine $g_{\alpha\beta,00}$ on S. By repeated differentiation of this equation, we can find all higher time derivatives of $g_{\alpha\beta}$ on S, and so we can develop $g_{\alpha\beta}$ in a Taylor series in x^0. This determines $g_{\alpha\beta}$ everywhere, and, together with the prescribed g_{0a}, we have determined a vacuum metric g_{ab}.

This procedure relies on the assumption that our solution is analytic in x^0. This assumption is unnatural because Einstein's equations are of hyperbolic type and do not require analytic solutions. The proof of existence and uniqueness of the Einstein equations for finite development in time under certain simple differentiability hypotheses has been given by Choquet-Bruhat. The questions of existence, uniqueness, and stability (i.e. do 'small' variations of the initial data result in 'small' variations in the solution?), and the extent to which solutions can be developed in general relativity, are deep and complex questions, and are the topics of current research.

13.6 The hole problem

We have, in fact, been somewhat imprecise in setting up the Cauchy problem and in so doing we have covered up something which had originally caused Einstein considerable difficulty. We defined the Cauchy problem as starting with a manifold with no metric on it (a so-called 'bare' manifold), prescribing initial data on a hypersurface in the manifold, and then using the field equations to generate a unique solution for the metric g. However, as we know from the principle of general covariance, we may then apply a coordinate transformation to g and so obtain another solution $\bar{g}$, say. How are the solutions g and $\bar{g}$ related physically?

This question had troubled Einstein and was one of the reasons why, even though the principle of general covariance was formulated in 1907, another eight years were to elapse before the field equations were finally obtained. Einstein raised the question in the form of the 'hole problem'. Suppose that the matter distribution is known everywhere outside of some hole H in the manifold. Then the field equations together with the boundary conditions will enable the metric g to be determined inside H and, in particular, at some point P, say. Now carry out a coordinate transformation which leaves everything outside H fixed, but which (from the active viewpoint) moves points around inside H, for example moving P to P', say (Fig. 13.3). Next, determine afresh the metric $\bar{g}$ in H. Is $\bar{g}$ the same as g? The answer is that, although $\bar{g}$ will in general be functionally different from g (i.e. the components of $\bar{g}$ will involve different functions of its coordinates compared with g), it will still represent the **same** physical solution. How can this be so if the points inside H have moved? The nub of the argument is that the point P in the bare manifold is not distinguished from any other point. It does not become a point with physical meaning (that is, an **event**) until a metric is determined in H. As John Stachel puts it so succinctly, 'no metric, no nothing'. Thus, a physical solution, that is, a **space-time**, consists of a manifold together with a metric. Two space-times are physically equivalent, in other words, give rise to the same gravitational field, if the two metrics can be transformed into each other. Mathematically, we should regard physical solutions as equivalence classes of space-times possessing metrics which are related by coordinate transformations.

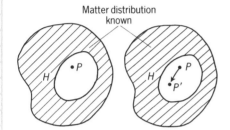

Matter distribution known

Fig. 13.3 The hole problem.

13.7 The equivalence problem

The question which then arises is, Given two metrics, g and $\bar{g}$, are they in fact the same, that is, does there exist a coordinate transformation transforming one into the other? This is a classic problem in differential geometry, known as the **equivalence problem**, and its classic solution by E. Cartan involves computation and comparison of the 10th covariant derivatives of the Riemann tensors of g and $\bar{g}$.

As one discovers in working out the Riemann tensor, even for something as simple as the Schwarzschild solution (see Exercise 6.31), it is a non-trivial task. It is all too easy to make slips in a long hand calculation. In fact, this task of undertaking large amounts of algebraic calculation has been made much more tractable and less error-prone with the advent of general purpose computer algebra systems, the best known of which include REDUCE, MACSYMA, and MAPLE. The system most used in general relativity (for which it was specifically designed) is the system SHEEP, together with its extensions CLASSI (for classifying metrics) and STENSOR (for symbolic tensor manipulation). These systems make possible computations which would have been impossible to contemplate undertaking by hand. Even so, they are not capable currently of computing anything like 10th covariant derivatives of Riemann tensors and so appear to be of little use in the equivalence problem.

The situation has been improved profoundly by the work of A. Karlhede. We will not pursue the details, but in broad outline the Karlhede approach is to classify a geometry by introducing a frame or tetrad, which is defined in stages, such that the Riemann tensor and its covariant derivatives take on a simple or rather **canonical** form at each stage. This is a well-defined procedure leading to a set of invariant quantities characterizing a given geometry. With this approach, the worst case theoretically involves computing the 7th covariant derivative. However, experience in using the algorithm suggests that one may never need go beyond the second derivative and often the first derivative is enough. This makes computer calculation a viable proposition. Thus, given two metrics, one first computes their invariant classification. If the two sets are different, then so are the metrics. If they are the same, then there may be a transformation relating them. The problem is then reduced to solving a set of four algebraic equations to determine the transformation. In general this is non-algorithmic, but in practice it is often manageable.

As a direct consequence of this advance, a project is currently underway to construct a computer database of exact solutions of the Einstein equations, fully classified and documented. The long-term goal is to put all known solutions in the database. Then any possibly new solutions which are discovered can be checked against the database, which can be updated accordingly. Not only would this provide a valuable resource to the research community, but it would also prevent the reporting of already-known solutions. Examples of this are well known: indeed, the Schwarzschild solution itself has apparently been 'discovered' in the literature on some 20 different occasions!

Exercises

13.1 (§**13.3**) Show that the Lagrangian (13.7) gives rise to the full field equations with cosmological term (13.5).

13.2 (§**13.3**) Show that if (13.8) is to be consistent with (13.6) then $\mu = -\frac{1}{2}$.

13.3 (§**13.3**) Show that the trace of the Maxwell energy–momentum tensor is zero. If $\Lambda = 0$, then what value of μ ensures that both sides of (13.8) are trace-free? Hence, propose an alternative Einstein–Maxwell theory.

13.4 (§**13.3**) Show that flat space is not a solution of (13.5).

13.5 (§**13.4**)
(i) Show that the conservation equations for a perfect fluid lead to

$$(\rho_0 + p)u^a \nabla_a u^b + (u^a u^b - g^{ab})\nabla_a p = 0.$$

(ii) We suppose that $\rho_0 = \rho_0(p)$ and define the following quantities:

$$f = \exp\left(\int \frac{dp}{p + \rho_0(p)}\right),$$

$$C_a = f u_a,$$

$$\Omega_{ab} = \frac{1}{2}(\nabla_b C_a - \nabla_a C_b).$$

Deduce that $C^a \Omega_{ab} = 0$.

13.6 (§**13.5**) If g_{ab} is known everywhere on S, then establish that $g_{ab, \alpha}$ is known everywhere on S.

13.7 (§**13.5**) Establish the equations (13.12), (13.13), and (13.14).

13.8 (§**13.5**) Check the results (13.17) and (13.18).

13.9 (§**13.5**) Derive (13.21) and (13.22). [Hint: use the device of breaking up all Latin indices into their zero and Greek constituents, e.g. $g^{0a}R_{0a} = g^{00}R_{00} + g^{0\alpha}R_{0\alpha}$, etc.]

13.10 (§**13.5**)
(i) Establish (13.23).
(ii) Confirm (13.25).
(iii) Assume that $G_a{}^0$ is an analytic function of x^0 and use the result (13.25) to develop it in a formal power series in x^0. Show that if $[G_a{}^0]_S = 0$ then $G_a{}^0 \equiv 0$.

14 The Schwarzschild solution

14.1 Stationary solutions

We now turn our attention to solving the vacuum field equations in the simplest case, namely, that of spherical symmetry. As a preliminary, in the next two sections we make clear the distinction between stationary and static solutions. In simple terms, a solution is stationary if it is time-independent. This does not mean that the solution is in no way evolutionary, but simply that the time does not enter into it explicitly. On the other hand, the stronger requirement that a solution is static means that it cannot be evolutionary. In such a case, nothing would change if at any time we ran time backwards, i.e. static means time-symmetric about any origin of time. Think of the motion of a gas in a pipe (Fig. 14.1). If it is being pumped by some time-dependent device, then the motion will be non-stationary. If the gas travels with constant velocity at each point in the pipe, then the motion is stationary. If the gas velocity is zero everywhere, then the system is static.

A metric will be stationary if there exists a special coordinate system in which the metric is visibly time-independent, i.e.

$$\frac{\partial g_{ab}}{\partial x^0} \overset{*}{=} 0, \tag{14.1}$$

where x^0 is a timelike coordinate. Of course, in an arbitrary coordinate system the metric will probably depend explicitly on all the coordinates; so we need to make the statement (14.1) coordinate-independent. If we define a vector field

$$X^a \overset{*}{=} \delta_0^a \tag{14.2}$$

in the special coordinate system, then

$$L_X g_{ab} = X^c g_{ab,c} + g_{ac} X^c{}_{,b} + g_{bc} X^c{}_{,a}$$
$$\overset{*}{=} \delta_0^c g_{ab,c} = g_{ab,0} = 0$$

by (14.1). $L_X g_{ab}$ is a tensor, so if it vanishes in one coordinate system it vanishes in all coordinate systems. Hence, it follows that X^a is a **Killing vector field**. Conversely, given a **timelike** Killing vector field X^a, then there always exists a coordinate system which is **adapted** to the Killing vector field, that is, in which (14.2) holds, and then

$$0 = L_X g_{ab} \overset{*}{=} g_{ab,0},$$

and so the metric is stationary. We have therefore established the coordinate-

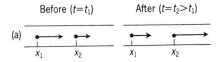

Before ($t = t_1$)   After ($t = t_2 > t_1$)

(a)

x_1 x_2 x_1 x_2

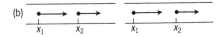

(b)

x_1 x_2 x_1 x_2

(c)

x_1 x_2 x_1 x_2

Fig. 14.1 Two gas particles in a pipe in (a) non-stationary, (b) stationary, and (c) static flow.

independent definition

> A space-time is said to be **stationary** if and only if it admits a timelike
> Killing vector field.

14.2 Hypersurface-orthogonal vector fields

In order to discuss static solutions in a coordinate-independent way, we need
to introduce the concept of a hypersurface-orthogonal vector field, which we
do in this section. We start with the equation of a **family** of hypersurfaces
given by

$$f(x^a) = \mu, \tag{14.3}$$

where different members of the family correspond to different values of μ
(Fig. 14.2). Consider two neighbouring points P and Q with coordinates (x^a)
and $(x^a + dx^a)$, respectively, lying in one of the hypersurfaces, S say. Since
$(x^a + dx^a)$ lies in S, we also have, by (14.3),

$$\mu = f(x^a + dx^a) = f(x^a) + \frac{\partial f}{\partial x^a}\, dx^a$$

to first order. Subtracting (14.3) from this equation, we find

$$\frac{\partial f}{\partial x^a}\, dx^a = 0 \tag{14.4}$$

at P. If we define the **covariant vector field** n_a to the family of hypersurfaces
by

Fig. 14.2 A family of hypersurfaces labelled by μ.

$$n_a \equiv \frac{\partial f}{\partial x^a}, \tag{14.5}$$

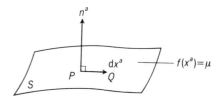

Fig. 14.3 The normal vector field n^a at a point P.

then (14.4) becomes

$$n_a dx^a = g_{ab} n^a dx^b = 0$$

at P, which tells us that n^a is orthogonal to the infinitesimal contravariant
vector field dx^a. Since dx^a lies in S by construction, it follows that n^a is
orthogonal to S and is therefore known as the **normal vector field** to S at P
(Fig. 14.3). Any other vector field X^a is said to be **hypersurface-orthogonal** if
it is everywhere orthogonal to the family of hypersurfaces, in which case it
must be proportional to n^a everywhere, i.e.

$$X^a = \lambda(x) n^a \tag{14.6}$$

for some proportionality factor λ, which in general will vary from point to
point. Then the orbits of X^a are orthogonal to the family of hypersurfaces
(Fig. 14.4). From (14.6) and (14.5), the hypersurface-orthogonal condition can

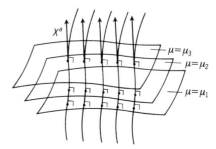

Fig. 14.4 A hypersurface-orthogonal vector field X^a.

also be written

$$X_a = \lambda f_{,a}, \tag{14.7}$$

and so

$$X_a \partial_b X_c = \lambda f_{,a} \lambda_{,b} f_{,c} + \lambda^2 f_{,a} f_{,cb}.$$

Taking the totally anti-symmetric part of this equation and noting that the first term on the right is symmetric in a and c and the second term is symmetric in b and c, so that their totally anti-symmetric parts vanish, then

$$X_{[a} \partial_b X_{c]} = 0. \tag{14.8}$$

This equation is unchanged if we replace the ordinary derivative by a covariant derivative (exercise), namely,

$$X_{[a} \nabla_b X_{c]} = 0. \tag{14.9}$$

We have shown that any hypersurface-orthogonal vector field satisfies (14.9). We shall now establish a partial converse, namely, any non-null Killing vector field satisfying (14.9) is necessarily hypersurface-orthogonal. Since X^a is a Killing vector, it satisfies (7.52), namely,

$$L_X g_{ab} = \nabla_b X_a + \nabla_a X_b = 0.$$

It follows that interchanging indices on the covariant derivative of X_a introduces a minus sign:

$$\nabla_a X_b = -\nabla_b X_a. \tag{14.10}$$

Using this, the six terms in (14.9) reduce to three terms:

$$X_a \nabla_b X_c + X_c \nabla_a X_b + X_b \nabla_c X_a = 0.$$

Contracting with X^c and writing $X^2 = X^a X_a$, we get

$$X_a X^c \nabla_b X_c + X^2 \nabla_a X_b + X_b X^c \nabla_c X_a = 0,$$

or, using (14.10),

$$X_a X^c \nabla_b X_c + X^2 \nabla_a X_b - X_b X^c \nabla_a X_c = 0.$$

Interchanging the raised and lowered index for the dummy index c (why can we do this?) and using (14.10) on the middle term, this becomes

$$X_a X_c \nabla_b X^c - X^2 \nabla_b X_a - X_b X_c \nabla_a X^c = 0.$$

Adding these last two equations, we get

$$X_a \nabla_b X^2 - X_b \nabla_a X^2 + X^2 (\nabla_a X_b - \nabla_b X_a) = 0,$$

or, since X^2 is a scalar field and the terms in the parentheses involving the connection vanish (see (12.37)),

$$X_a \partial_b X^2 - X_b \partial_a X^2 + X^2 (\partial_a X_b - \partial_b X_a) = 0.$$

We write this in the form

$$X^2 \partial_a X_b - X_b \partial_a X^2 = X^2 \partial_b X_a - X_a \partial_b X^2,$$

or equivalently, dividing by X^4,

$$\partial_a\left(\frac{X_b}{X^2}\right) = \partial_b\left(\frac{X_a}{X^2}\right), \tag{14.11}$$

since X^a is non-null by assumption and so $X^2 \neq 0$. This last equation requires that the term in parentheses be a gradient of some scalar field, f say, i.e.

$$\frac{X_a}{X^2} = f_{,a}, \tag{14.12}$$

and so finally

$$X_a = X^2 f_{,a}. \tag{14.13}$$

This is the hypersurface-orthogonal condition (14.7) with $\lambda = X^2$.

14.3 Static solutions

If a solution is stationary, then, in an adapted coordinate system, the metric will be time-independent but the line element will still in general contain cross terms in $dx^0\, dx^\alpha$. If, in addition, the metric is static, we would expect these cross terms to be absent for the following reason. Consider the interval between two events (x^0, x^1, x^2, x^3) and $(x^0 + dx^0, x^1 + dx^1, x^2, x^3)$ in our special coordinate system. Then

$$ds^2 \overset{*}{=} g_{00}(dx^0)^2 + 2g_{01}dx^0dx^1 + g_{11}(dx^1)^2, \tag{14.14}$$

where all the g_{ab} depend on x^α only (why?). Under a time reversal

$$x^0 \to x'^0 = -x^0, \tag{14.15}$$

the g_{ab} remain unchanged, but ds^2 becomes

$$ds^2 \overset{*}{=} g_{00}(dx^0)^2 - 2g_{01}dx^0dx^1 + g_{11}(dx^1)^2. \tag{14.16}$$

The assumption that the solution is static means that ds^2 is invariant under a time reversal about any origin of time, and so, equating (14.14) and (14.16), we find that g_{01} vanishes. Similarly g_{02} and g_{03} must vanish, and so we have shown that there are no cross terms $dx^0\, dx^\alpha$ in the line element in the special coordinate system.

Let us investigate the hypersurface-orthogonal condition (14.13) in a stationary space-time in a coordinate system adapted to the timelike Killing vector field, that is, $X^a \overset{*}{=} \delta_0^a$. Then

$$X_a = g_{ab}X^b \overset{*}{=} g_{ab}\delta_0^b = g_{0a}$$

and

$$X^2 = X_aX^a \overset{*}{=} g_{0a}\delta_0^a = g_{00}.$$

So (14.13) gives

$$g_{0a} \overset{*}{=} g_{00}f_{,a} \tag{14.17}$$

for some scalar field f. When $a = 0$, this produces $f_{,0} \overset{*}{=} 1$, and so integration gives

$$f \overset{*}{=} x^0 + h(x^\alpha),$$

where h is an arbitrary function of the spacelike coordinates only. Consider the coordinate transformation defined by

$$x^0 \to x'^0 = x^0 + h(x^\alpha), \qquad x^\alpha \to x'^\alpha = x^\alpha. \tag{14.18}$$

Then we find, in the new coordinate system (exercise),

$$X'^a \overset{*}{=} \delta_0^a, \tag{14.19}$$

$$g'_{ab,0} \overset{*}{=} 0, \tag{14.20}$$

$$g'_{00} \overset{*}{=} g_{00}, \tag{14.21}$$

$$g'_{0\alpha} \overset{*}{=} 0. \tag{14.22}$$

The last equation reveals that there are no cross terms in $\mathrm{d}x^0\,\mathrm{d}x^\alpha$ and so the solution is static. We therefore have established the following definition.

> A space-time is said to be **static** if and only if it admits a hypersurface-orthogonal timelike Killing vector field.

Moreover, we have established the following important result.

> In a static space-time, there exists a coordinate system adapted to the timelike Killing vector field in which the metric is time-independent and no cross terms appear in the line element involving the time, i.e. $g_{0\alpha} \overset{*}{=} 0$.

It can be shown (exercise) that there still exists the coordinate freedom

$$x^0 \to x'^0 = A x^0 + B, \qquad x^\alpha \to x'^\alpha = h'^\alpha(x^\beta), \tag{14.23}$$

where A and B are constants and the functions h'^α are arbitrary. If the boundary conditions require $g_{00} \to 1$ at spatial infinity, then this requires $A = \pm 1$. Neglecting time reversal, then this fixes A to be 1, and so we have defined a time coordinate, called **world time**, which is defined to within an unimportant additive constant. Thus, in a static space-time, we have regained the old Newtonian idea of an absolute time in the sense that the manifold can be sliced-up in a well-defined way into hypersurfaces $t = $ constant (Fig. 14.5). Then there exist a privileged class of observers who measure world time and hence can agree on events being simultaneous. The corresponding coordinates are Gaussian since $g_{0a} \overset{*}{=} \delta_a^0$.

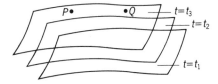

Fig. 14.5 Two 'simultaneous' events in world time.

14.4 Spherically symmetric solutions

Spherical symmetry can be defined rigorously in terms of Killing vector fields as follows.

> A space-time is said to be **spherically symmetric** if and only if it admits three linearly independent spacelike Killing vector fields X^α whose orbits are closed (i.e. topological circles) and which satisfy
>
> $$[X^1, X^2] = X^3, \quad [X^2, X^3] = X^1, \quad [X^3, X^1] = X^2.$$

Then (see Exercise 8.5) there exists a coordinate system in which the Killing vectors take on a standard form as expressed in the following result.

> In a spherically symmetric space-time, there exists a coordinate system (x^a) (called Cartesian) in which the Killing fields X^a are of the form
> $$X^0 \stackrel{*}{=} 0,$$
> $$X^\alpha \stackrel{*}{=} \omega^\alpha{}_\beta x^\beta, \qquad \omega_{\alpha\beta} = -\omega_{\beta\alpha}.$$

The quantity $\omega_{\alpha\beta}$ depends on three parameters which specify three spacelike rotations. These results then lead to a canonical form for the line element. The calculation is rather detailed, so we shall proceed in a different manner and present a heuristic argument for determining the form of the line element.

Intuitively, spherical symmetry means that there exists a privileged point, called the origin O, such that the system is invariant under spatial rotations about O. Then, if we fix the time and consider a point P a distance a from O, the spatial rotations will result in P sweeping out a 2-sphere centred on O. We can then introduce an axial coordinate ϕ and an azimuthal coordinate θ on the sphere in the usual way. Dropping a perpendicular from P to the equational plane ($z = 0$) at Q, then ϕ is the angle which OQ makes with the positive x-axis and θ is the angle which OP makes with the positive z-axis (Fig. 14.6). All points on the 2-sphere will be covered by the coordinate ranges

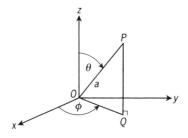

Fig. 14.6 The standard spherical coordinates θ and ϕ.

$$0 \leqslant \theta \leqslant \pi, \tag{14.24}$$

$$-\pi < \phi \leqslant \pi. \tag{14.25}$$

Moreover, the line element of the 2-sphere is (exercise)

$$ds^2 = a^2(d\theta^2 + \sin^2\theta\,d\phi^2). \tag{14.26}$$

It is then natural to assume that in four dimensions we can augment θ and ϕ with an arbitrary timelike coordinate t and some radial-type parameter r, so that the line element reduces to the form (14.26) on a 2-sphere t = constant, r = constant. Spherical symmetry requires that the line element does not vary when θ and ϕ are varied, so that θ and ϕ only occur in the line element in the form $(d\theta^2 + \sin^2\theta\,d\phi^2)$. Moreover, using an analogous argument to the one we used at the beginning of §14.3, there can be no cross terms in $d\theta$ or $d\phi$ (exercise) because the metric must be invariant separately under the reflections

$$\theta \to \theta' = \pi - \theta \tag{14.27}$$

and

$$\phi \to \phi' = -\phi. \tag{14.28}$$

Our starting ansatz, then, is that there exists a special coordinate system

$$(x^a) = (x^0, x^1, x^2, x^3) = (t, r, \theta, \phi)$$

in which the line element has the form

$$ds^2 = A\,dt^2 - 2B\,dt\,dr - C\,dr^2 - D(d\theta^2 + \sin^2\theta\,d\phi^2), \tag{14.29}$$

where A, B, C, and D are as yet undetermined functions of t and r, i.e.

$$A = A(t, r), \qquad B = B(t, r), \qquad C = C(t, r), \qquad D = D(t, r).$$

If we introduce a new radial coordinate by the transformation

$$r \rightarrow r' = D^{\frac{1}{4}},$$

then (14.29) becomes

$$ds^2 = A'(t, r')\,dt^2 - 2B'(t, r')\,dt\,dr' - C'(t, r')\,dr'^2 - r'^2(d\theta^2 + \sin^2\theta\,d\phi^2).$$
(14.30)

Consider the differential

$$A'(t, r')\,dt - B'(t, r')\,dr'.$$

The theory of ordinary differential equations tells us that we can always multiply this by an integrating factor, $I = I(t, r')$ say, which makes it a perfect differential. We use this result to define a new time coordinate t' by requiring

$$dt' = I(t, r')[A'(t, r')\,dt - B'(t, r')\,dr'].$$

Squaring, we obtain

$$dt'^2 = I^2(A'^2\,dt^2 - 2A'B'\,dt\,dr' + B'^2\,dr'^2),$$

and so

$$A'\,dt^2 - 2B'\,dt\,dr' = A'^{-1}I^{-2}\,dt'^2 - A'^{-1}B'^2\,dr'^2,$$

and the line element (14.30) becomes

$$ds^2 = A'^{-1}I^{-2}\,dt'^2 - (C' - A'^{-1}B'^2)\,dr'^2 - r'^2(d\theta^2 + \sin^2\theta\,d\phi^2).$$

Defining two new functions v and λ by

$$A'^{-1}I^{-2} = e^v$$
(14.31)

and

$$C' - A'^{-1}B'^2 = e^\lambda$$
(14.32)

and dropping the primes, we finally obtain the form

$$ds^2 = e^v\,dt^2 - e^\lambda\,dr^2 - r^2(d\theta^2 + \sin^2\theta\,d\phi^2),$$
(14.33)

where

$$v = v(t, r), \qquad \lambda = \lambda(t, r).$$

The definitions of v and λ in (14.31) and (14.32) are given in terms of exponentials, which, since they are always positive, guarantees that the signature of the metric is -2. In fact, there are rigorous arguments which confirm that the most general spherically symmetric line element in four dimensions (with signature -2) can be written in the canonical form (14.33).

14.5 The Schwarzschild solution

We now use Einstein's vacuum field equations to determine the unknown functions v and λ in (14.33). The covariant metric is

$$g_{ab} = \text{diag}(e^v, -e^\lambda, -r^2, -r^2\sin^2\theta)$$
(14.34)

and, since the metric is diagonal, its contravariant form is

$$g^{ab} = \text{diag}(e^{-v}, -e^{-\lambda}, -r^{-2}, -r^{-2}\sin^{-2}\theta).$$
(14.35)

If we denote derivatives with respect to t and r by dot and prime, respectively, then, by Exercise 6.31(v), the non-vanishing components of the mixed Einstein tensor are

$$G_0{}^0 = e^{-\lambda}\left(\frac{\lambda'}{r} - \frac{1}{r^2}\right) + \frac{1}{r^2}, \tag{14.36}$$

$$G_0{}^1 = -e^{-\lambda}r^{-1}\dot{\lambda} = -e^{\lambda-\nu}G_1{}^0, \tag{14.37}$$

$$G_1{}^1 = -e^{-\lambda}\left(\frac{\nu'}{r} + \frac{1}{r^2}\right) + \frac{1}{r^2}, \tag{14.38}$$

$$G_2{}^2 = G_3{}^3 = \frac{1}{2}e^{-\lambda}\left(\frac{\nu'\lambda'}{2} + \frac{\lambda'}{r} - \frac{\nu'}{r} - \frac{\nu'^2}{2} - \nu''\right) + \frac{1}{2}e^{-\nu}\left(\ddot{\lambda} + \frac{\dot{\lambda}^2}{2} - \frac{\dot{\lambda}\dot{\nu}}{2}\right). \tag{14.39}$$

The contracted Bianchi identities reveal that equation (14.39) vanishes **automatically** if the equations (14.36), (14.37), and (14.38) all vanish (exercise). Hence, there are three independent equations to solve, namely,

$$e^{-\lambda}\left(\frac{\lambda'}{r} - \frac{1}{r^2}\right) + \frac{1}{r^2} = 0, \tag{14.40}$$

$$e^{-\lambda}\left(\frac{\nu'}{r} + \frac{1}{r^2}\right) - \frac{1}{r^2} = 0, \tag{14.41}$$

$$\dot{\lambda} = 0. \tag{14.42}$$

Adding (14.40) and (14.41), we get

$$\lambda' + \nu' = 0$$

and integration gives

$$\lambda + \nu = h(t), \tag{14.43}$$

where h is an arbitrary function of integration. Here, λ is purely a function of r by (14.42), and so (14.40) is simply an **ordinary** differential equation, which we write

$$e^{-\lambda} - re^{-\lambda}\lambda' = 1,$$

or equivalently

$$(re^{-\lambda})' = 1.$$

Integrating, we get

$$re^{-\lambda} = r + \text{constant}.$$

Choosing the constant of integration to be $-2m$, for later convenience, we then obtain

$$e^{\lambda} = (1 - 2m/r)^{-1}. \tag{14.44}$$

So, at this stage, the metric has been reduced, by (14.43) and (14.44), to

$$g_{ab} = \text{diag}\left[e^{h(t)}(1 - 2m/r), -(1 - 2m/r)^{-1}, -r^2, -r^2\sin^2\theta\right]. \tag{14.45}$$

The final stage is to eliminate $h(t)$. This is done by transforming to a new time coordinate t', i.e. $t \to t'$, where t' is determined by the relation

$$t' = \int_c^t e^{\frac{1}{2}h(u)}\,du \tag{14.46}$$

where c is an arbitrary constant. Then the only component of the metric which changes is (exercise)

$$g'_{00} = (1 - 2m/r).$$

Dropping primes, we have shown that it is always possible to find a coordinate system in which the most general spherically symmetric solution of the vacuum field equations is

$$ds^2 = (1 - 2m/r)\,dt^2 - (1 - 2m/r)^{-1}\,dr^2 - r^2(d\theta^2 + \sin^2\theta\,d\phi^2). \quad (14.47)$$

This is the famous **Schwarzschild line element.**

14.6 Properties of the Schwarzschild solution

It is immediate from (14.47) that $g_{ab,0} \overset{*}{=} 0$, and so the solution is **stationary.** Moreover, the coordinates are **adapted** to the Killing vector field $X^a \overset{*}{=} \delta_0^a$. Since

$$X_a = g_{ab}X^b \overset{*}{=} g_{ab}\delta_0^b = g_{0a} = g_{00}\delta_a^0 = (1 - 2m/r, 0, 0, 0),$$

we see that X^a is **hypersurface-orthogonal**, that is, $X_a = \lambda f_{,a}$, with

$$\lambda = X^2 \overset{*}{=} g_{00} \quad \text{and} \quad f(x^a) \overset{*}{=} t = \text{constant}.$$

Alternatively, we can check (exercise) that

$$X_{[a}\partial_b X_{c]} = 0. \quad (14.48)$$

Thus, the timelike Killing vector field X^a is hypersurface-orthogonal to the family of hypersurfaces $t = \text{constant}$, and hence the solution is **static** and t is a **world time.** Alternatively, it is immediate from (14.47) that the solution is **time-symmetric**, since it is invariant under the time reflection $t \to t' = -t$, and **time translation invariant**, since it is invariant under the transformation $t \to t' = t + \text{constant}$, and so again it is static (see Exercise 14.1). We have thus proved the following somewhat unexpected result.

> **Birkhoff's theorem**: A spherically symmetric vacuum solution is necessarily static.

This is unexpected because in Newtonian theory spherical symmetry has nothing to do with time dependence. This highlights the special character of non-linear partial differential equations and the solutions they admit. In particular, Birkhoff's theorem implies that if a spherically symmetric source like a star changes its shape, but does so always remaining spherically symmetric, then it cannot propagate any disturbances into the surrounding space. Looking ahead, this means that a pulsating spherically symmetric star cannot emit gravitational waves (Fig. 14.7). If a spherically symmetric source is restricted to the region $r \leqslant a$ for some a, then the solution for $r > a$ must be the Schwarzschild solution, or to give it its full name the Schwarzschild **exterior** solution. However, the converse is not true: a source which gives rise to an exterior Schwarzschild solution is **not** necessarily spherically symmetric. Some counter-examples are known. Thus, in general, a source need not inherit the symmetry of its external field.

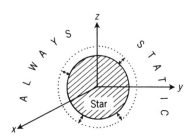

Fig. 14.7 A pulsating spherical star cannot emit gravitational waves.

If we take the limit of (14.47) as $r \to \infty$, then we obtain the flat space metric of special relativity in spherical polar coordinates, namely,

$$ds^2 = dt^2 - dr^2 - r^2(d\theta^2 + \sin^2\theta\, d\phi^2). \qquad (14.49)$$

We have therefore shown that a spherically symmetric vacuum solution is necessarily **asymptotically flat**. Some authors obtain the Schwarzschild solution from the starting assumptions that the solution is spherically symmetric, static, and asymptotically flat. However, as we have seen, there is no need to adopt these last two assumptions a priori, because the field equations force them on you. Let us attempt an interpretation of the constant m appearing in the solution, by considering the Newtonian limit. A point mass M situated at the origin O in Newtonian theory gives rise to a potential $\phi = -GM/r$. Inserting this into the weak-field limit (12.65) gives

$$g_{00} \simeq 1 + 2\phi/c^2 = 1 - 2GM/c^2r,$$

and, comparing this with (14.47), we see that

$$m = GM/c^2 \qquad (14.50)$$

in non-relativistic units. In other words, if we interpret the Schwarzschild solution as due to a point particle situated at the origin, then the constant m is simply the mass of the particle in relativistic units. It is clear from (14.47) that m has the dimensions of length. It is sometimes known as the **geometric mass**. We postpone a discussion of the coordinate ranges and the interpretation of the coordinates until Chapter 16. We end this section by summarizing the properties we have met. The Schwarzschild solution:

(1) is spherically symmetric;

(2) is stationary;

(3) has coordinates adapted to the timelike Killing vector field X^a;

(4) is static $\Leftrightarrow$ is time-symmetric and time translation invariant,
$\qquad\qquad \Leftrightarrow$ has a hypersurface-orthogonal timelike Killing vector field X^a;

(5) is asymptotically flat;

(6) has geometric mass $m = GMc^{-2}$.

14.7 Isotropic coordinates

In this section, we seek an alternative set of coordinates in which the time slices $t = $ constant are as close as we can get them to Euclidean 3-space. More specifically, we attempt to write the line element in the form

$$ds^2 = A(r)\,dt^2 - B(r)\,d\sigma^2,$$

where $d\sigma^2$ is the line element of Euclidean 3-space, namely,

$$d\sigma^2 = dx^2 + dy^2 + dz^2$$

in Cartesian coordinates, or equivalently

$$d\sigma^2 = dr^2 + r^2 d\theta^2 + r^2 \sin^2\theta\, d\phi^2$$

in spherical polar coordinates. In this form, the metric in a slice $t = $ constant is **conformal** to the metric of Euclidean 3-space, and hence, in particular, angles between vectors and ratios of lengths are the same for each metric (see Exercise 6.27).

We consider a transformation in which the coordinates θ, ϕ, and t remain unchanged while

$$r \rightarrow \rho = \rho(r), \tag{14.51}$$

so that ρ is some other radial coordinate, and we attempt to put the solution in the form

$$ds^2 = (1 - 2m/r)\,dt^2 - [\lambda(\rho)]^2[d\rho^2 + \rho^2(d\theta^2 + \sin^2\theta\,d\phi^2)]. \tag{14.52}$$

We could consider how (14.47) transforms under the transformation (14.51), but it is easier to proceed as follows. Comparing (14.52) with (14.47), the coefficients of $d\theta^2 + \sin^2\theta\,d\phi^2$ must be equal, which requires

$$r^2 = \lambda^2 \rho^2. \tag{14.53}$$

Equating the two radial elements produces

$$(1 - 2m/r)^{-1}\,dr^2 = \lambda^2\,d\rho^2. \tag{14.54}$$

Eliminating λ and taking square roots, we find

$$\frac{dr}{(r^2 - 2mr)^{\frac{1}{2}}} = \pm\frac{d\rho}{\rho}. \tag{14.55}$$

This is an ordinary differential equation in which the variables are separated. Since we require $\rho \rightarrow \infty$ as $r \rightarrow \infty$, we take the positive sign, and by integration we find (exercise)

$$r = \rho(1 + \tfrac{1}{2}m/\rho)^2 \tag{14.56}$$

and so, from (14.53),

$$\lambda^2 = (1 + \tfrac{1}{2}m/\rho)^4. \tag{14.57}$$

Using (14.56) to eliminate r, we find that the Schwarzschild solution can be written in the following **isotropic form**:

$$ds^2 = \frac{(1 - \tfrac{1}{2}m/\rho)^2}{(1 + \tfrac{1}{2}m/\rho)^2}\,dt^2 - (1 + \tfrac{1}{2}m/\rho)^4[d\rho^2 + \rho^2(d\theta^2 + \sin^2\theta\,d\phi^2)].$$
$$\tag{14.58}$$

Exercises

14.1 (§**14.1**) A system is time-symmetric if it is invariant under

$$t \rightarrow t' = -t.$$

Give an example of a non-stationary time-symmetric system. Show that if a time-symmetric system is also time translation invariant, i.e. invariant under

$$t \rightarrow t' = t + \text{constant},$$

then the system is static. Deduce that a stationary time-symmetric system is necessarily static.

14.2 (§**14.1**) Show that if g_{ab} is stationary then there exists a privileged coordinate system (t, x^α) in which the Killing vector field X reduces to $X = \partial/\partial t$, with $X(g_{ab}) = 0$. Show that X generates a time translation invariance

$$t \rightarrow t' = t + \text{constant}.$$

14.3 (§**14.2**)
  (i) Take the differential of (14.3) to confirm (14.4).
  (ii) Show that (14.9) is equivalent to (14.8).
  (iii) Check that (14.12) is consistent with (14.11).

14.4 (§**14.3**)
(i) Establish (14.19)–(14.22) under the transformation (14.18).
(ii) Show that there still remains the coordinate freedom (14.23).

14.5 (§**14.4**) Consider a point P on a 2-sphere of radius a centred at the origin. Find the distance P travels under an increase of coordinates
(i) $\theta \to \theta + d\theta$,
(ii) $\phi \to \phi + d\phi$.
Use Pythagoras' theorem to obtain the line element (14.26) for a 2-sphere.

14.6 (§**14.4**) Show that a spherically symmetric line element cannot possess cross terms in $d\theta$ and $d\phi$ because the metric must be invariant under the reflections (14.27) and (14.28). [Hint: assume that all the metric components g_{ab} $(a, b \neq 3)$ and $g_{33} \sin^{-2}\theta$ do not depend on θ or ϕ.]

14.7 (§**14.5**) Show that if (14.36), (14.37), and (14.38) vanish then so does (14.39) by the contracted Bianchi identities.

14.8 (§**14.5**) Show that, under the transformation to a new time coordinate t' given by (14.46), the line element (14.45) is transformed into the form (14.47), where primes have been dropped in (14.47).

14.9 (§**14.6**) Check that (14.48) holds for the Schwarzschild line element where X^a is the timelike Killing vector field.

14.10 (§**14.6**) Find the dimensions of G. [Hint: use (4.4)

and Newton's second law.] Use (14.50) to show that m has the dimensions of a length.

14.11 (§**14.6**) Find the non-zero components of R_{abcd} for the Schwarzschild solution.

14.12 (§**14.7**)
(i) Show that (14.55) taken with the positive sign integrates to give (14.56).
(ii) Use (14.52), (14.56), and (14.57) to derive (14.58).

14.13 (§**14.7**) Consider (14.58) in the weak-field limit $m \ll \rho$ to show that $g_{00} \simeq 1 - 2m/\rho$ and confirm (14.50).

14.14 (§**14.7**) Which of the six properties listed at the end of §14.6 still hold for the isotropic form of the Schwarzschild line element?

14.15 (§**14.7**) Show that the isotropic form of the Schwarzschild solution

$$ds^2 = \frac{(1 - \tfrac{1}{2}m/\rho)^2}{(1 + \tfrac{1}{2}m/\rho)^2} dt^2 - (1 + \tfrac{1}{2}m/\rho)^4 [dx^2 + dy^2 + dz^2],$$

where

$$r = (x^2 + y^2 + z^2)^{\frac{1}{2}} = \rho(1 + \tfrac{1}{2}m/\rho)^2$$

admits the Killing vector fields

$$\frac{\partial}{\partial t}, \quad x\frac{\partial}{\partial y} - y\frac{\partial}{\partial x}, \quad y\frac{\partial}{\partial z} - z\frac{\partial}{\partial y}, \quad z\frac{\partial}{\partial x} - x\frac{\partial}{\partial z}.$$

[Hint: use the symmetry in x, y, and z.] Find all their commutators.

15 Experimental tests of general relativity

15.1 Introduction

In this chapter, we shall consider the experimental status of general relativity. We shall see that the status is very different from that of special relativity, which has been subjected to a welter of different tests. Indeed, there are very few theories whose experimental footing has been so well established. In contrast, there are very few tests of the general theory. The main reason for this small number is that the gravitational fields experienced in our locality are very weak and their effects are not significantly different from the corresponding Newtonian ones. The extent to which the tests we have are actually tests of the particular set of field equations of general relativity as opposed to much weaker statements like the principle of equivalence is also arguable. The first tests of the theory were the three so-called 'classical tests' of general relativity, namely, the precession of the perihelion of Mercury, the bending of light, and the gravitational red shift. These tests have been augmented more recently by a fourth classical test, the delay of a light signal in a gravitational field. Perhaps the most significant test of the theory involves the orbital motion of the binary pulsar PSR 1913 + 16, because of its indirect indication of gravitational radiation. It seems likely that the tests which will prove to be the most conclusive are those which occur on a cosmological scale, in particular, through the possible detection of black holes and gravitational waves. We shall postpone consideration of black holes and gravitational waves until Parts D and E of the book, respectively.

There have probably been at least a score of alternative relativistic theories of gravitation proposed since the advent of special relativity. These are classical alternative theories as opposed to quantum ones. The one which has enjoyed most attention to date is the Brans–Dicke theory, especially in the mid-1960s when there was a reported detection of solar oblateness. We shall consider briefly the story concerning solar oblateness in this chapter. We shall end with a brief chronology of the main experimental or observational events connected with general relativity. Although the experimental tests are few in number, those that are there support general relativity as being the best and simplest classical theory that we have.

15.2 Classical Kepler motion

We first review the classical Kepler problem, namely, the motion of a test particle in the gravitational field of a massive body, before considering its general relativistic counterpart. It starts from the assumption that a particle of mass m moves under the influence of an inverse square law force whose centre of attraction is at the origin O, that is,

$$F = -m\frac{\mu}{r^2}\hat{r}, \tag{15.1}$$

where μ is a constant. Then Newton's second law is

$$m\ddot{r} = -m\frac{\mu}{r^2}\hat{r}. \tag{15.2}$$

The **angular momentum** of m is defined as

$$L = r \times m\dot{r}, \tag{15.3}$$

and so

$$\frac{dL}{dt} = \dot{r} \times m\dot{r} + r \times m\ddot{r}$$

$$= r \times \left(-m\frac{\mu}{r^2}\hat{r}\right)$$

$$= 0,$$

where the cross products of $\dot{r}$ with itself and r with $\hat{r}$ both vanish because the vectors are parallel. Hence, the angular momentum is conserved and

$$L = mh, \tag{15.4}$$

where h is a constant vector. Assuming $h \neq 0$, it follows from (15.3) that r is always perpendicular to h, and so the particle is restricted to move in a plane. If we introduce plane polar coordinates (R, ϕ), then the equation of motion (15.2) becomes

$$(\ddot{R} - R\dot{\phi}^2)\hat{R} + \frac{1}{R}\frac{d}{dt}(R^2\dot{\phi})\hat{\phi} = -\frac{\mu}{R^2}\hat{R}. \tag{15.5}$$

Taking the scalar product with $\hat{\phi}$ throughout and integrating produces

$$R^2\dot{\phi} = h, \tag{15.6}$$

which is conservation of angular momentum again, where h is the magnitude of the angular momentum per unit mass. Taking the scalar product with $\hat{R}$ throughout (15.5) gives

$$\ddot{R} - R\dot{\phi}^2 = -\mu/R^2. \tag{15.7}$$

We are interested in obtaining the equation of the orbit of the particle, which in plane polar coordinates is

$$R = R(\phi). \tag{15.8}$$

If we introduce the new variable $u = R^{-1}$, then this can also be written as

$u = u(\phi)$. Using the function of a function rule, we find

$$\dot{R} = \frac{\mathrm{d}R}{\mathrm{d}t} = \frac{\mathrm{d}}{\mathrm{d}t}\left(\frac{1}{u}\right) = -\frac{1}{u^2}\frac{\mathrm{d}u}{\mathrm{d}\phi}\frac{\mathrm{d}\phi}{\mathrm{d}t} = -\frac{1}{u^2}hu^2\frac{\mathrm{d}u}{\mathrm{d}\phi} = -h\frac{\mathrm{d}u}{\mathrm{d}\phi}$$

by (15.6). Similarly (exercise),

$$\ddot{R} = -h^2u^2\frac{\mathrm{d}^2u}{\mathrm{d}\phi^2}, \tag{15.9}$$

and so (15.7) becomes **Binet's equation**

$$\frac{\mathrm{d}^2u}{\mathrm{d}\phi^2} + u = \frac{\mu}{h^2}. \tag{15.10}$$

Binet's equation is the orbital differential equation for the particle, and has solution (exercise)

$$u = \frac{\mu}{h^2} + C\cos(\phi - \phi_0), \tag{15.11}$$

where C and ϕ_0 are constants. This can be written in terms of R as

$$l/R = 1 + e\cos(\phi - \phi_0), \tag{15.12}$$

where $l = h^2/\mu$ and $e = Ch^2/\mu$. This is the polar equation of a conic section in which l (semi-latus rectum) determines the scale, e (eccentricity) the shape, and ϕ_0 the orientation (relative to the x-axis). In particular, if $0 < e < 1$ then the conic is an ellipse (Fig. 15.1), and the point of nearest approach to the origin is called the **perihelion**.

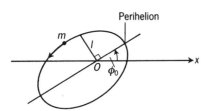

The motion of a test particle in the field of a massive body is called the **one-body problem**. We shall establish the classic result that in Newtonian theory the **two-body problem** of two point masses moving under their mutual gravitational attraction can be reduced to a one-body problem. Consider two masses m_1 and m_2 with position vectors r_1 and r_2, respectively (Fig. 15.2). Define the position vector of m_1 (say) relative to m_2 by

Fig. 15.1 Kepler motion in an ellipse.

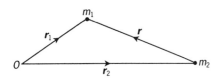

Fig. 15.2 The two-body problem.

$$r = r_1 - r_2.$$

If F_{12} is the force on m_1 due to m_2, and F_{21} the force on m_2 due to m_1, then, by Newton's third law,

$$F_{21} = -F_{12}. \tag{15.13}$$

Using Newton's second law, (15.13), and Newton's universal law of gravitation (4.4), we obtain

$$F_{12} = m_1\ddot{r}_1 = -m_2\ddot{r}_2 = -\frac{Gm_1m_2}{r^2}\hat{r},$$

and so

$$\ddot{r} = \ddot{r}_1 - \ddot{r}_2 = -\frac{Gm_2}{r^2}\hat{r} - \frac{Gm_1}{r^2}\hat{r} = -\frac{G(m_1 + m_2)}{r^2}\hat{r}.$$

We find finally that the equation of motion can be written as

$$F_{12} = m\ddot{r} = -m\frac{\mu}{r^2}\hat{r}, \qquad (15.14)$$

where m, the **reduced mass**, is given by

$$m = m_1 m_2/(m_1 + m_2) \qquad (15.15)$$

and

$$\mu = G(m_1 + m_2). \qquad (15.16)$$

Comparing (15.14) with (15.2), we see that this is the one-body problem we discussed earlier. In the simplest model of planetary motion, we take m_2 to be the mass of the sun and m_1 to be the mass of the planet. Then, suitably interpreted (see Exercise 15.6), the motion of a planet is again a Kepler ellipse.

15.3 Advance of the perihelion of Mercury

We now look at the one-body problem in general relativity. We assume that the central massive body produces a spherically symmetric gravitational field. The appropriate solution in general relativity is then the Schwarzschild solution. Moreover, a test particle moves on a timelike geodesic, and so we begin by studying some of the geodesics of the Schwarzschild solution. The simplest approach is to employ the variational method of §7.6. Letting a dot denote differentiation with respect to proper time τ, we then find, for timelike geodesics,

$$2K = (1 - 2m/r)\dot{t}^2 - (1 - 2m/r)^{-1}\dot{r}^2 - r^2\dot{\theta}^2 - r^2\sin^2\theta\dot{\phi}^2 = 1. \quad (15.17)$$

We next work out the Euler–Lagrange equations. It turns out to be sufficient to restrict attention to the three simplest equations, which are given when $a = 0, 2, 3$ in (7.46), and which are

$$\frac{d}{d\tau}[(1 - 2m/r)\dot{t}] = 0, \qquad (15.18)$$

$$\frac{d}{d\tau}(r^2\dot{\theta}) - r^2\sin\theta\cos\theta\dot{\phi}^2 = 0, \qquad (15.19)$$

$$\frac{d}{d\tau}(r^2\sin^2\theta\dot{\phi}) = 0. \qquad (15.20)$$

This is because we need four differential equations to determine our four unknowns, namely,

$$t = t(\tau), \qquad r = r(\tau), \qquad \theta = \theta(\tau), \qquad \phi = \phi(\tau).$$

However, (15.17) is itself an integral of the motion and so, together with (15.18)–(15.20), provides the four equations needed. We have seen in Newtonian theory that the corresponding motion is confined to a plane. Let us see if planar motion is possible in general relativity. Specifically, let us consider motion in the plane equatorial $\theta = \frac{1}{2}\pi$ (the (x, y) plane). In this plane, $\dot{\theta} = 0$, and hence, by (15.19), it follows that $\ddot{\theta} = 0$. Differentiating (15.19), we can show that all higher derivatives of θ must vanish as well, and hence it follows that planar motion is possible (why?). Then (15.20) can be integrated directly to give

$$r^2 \dot{\phi} = h, \tag{15.21}$$

where h is a constant. This is conservation of angular momentum (compare with (15.6) and note that, in the equatorial plane, the spherical polar coordinate r is the same as the plane polar coordinate R). Similarly, (15.18) gives

$$(1 - 2m/r)\dot{t} = k, \tag{15.22}$$

where k is a constant. Substituting in (15.17), we obtain

$$k^2(1 - 2m/r)^{-1} - (1 - 2m/r)^{-1}\dot{r}^2 - r^2\dot{\phi}^2 = 1. \tag{15.23}$$

We proceed as we did in the classical theory and set $u = r^{-1}$, which leads to

$$\dot{r} = -h\frac{du}{d\phi},$$

then, using (15.21), we find (15.23) becomes

$$\left(\frac{du}{d\phi}\right)^2 + u^2 = \frac{k^2 - 1}{h^2} + \frac{2m}{h^2}u + 2mu^3. \tag{15.24}$$

This is a first-order differential equation for determining the orbit of a test particle, or more precisely the trajectory of the test body projected into a slice $t = $ constant (Fig. 15.3). It can be integrated directly by using elliptic functions. We shall use an approximation method to solve it.

Differentiating (15.24), we obtain the second-order equation

$$\frac{d^2u}{d\phi^2} + u = \frac{m}{h^2} + 3mu^2. \tag{15.25}$$

This is the relativistic version of Binet's equation (15.10) and differs from the Newtonian result by the presence of the last term. For planetary orbits, this

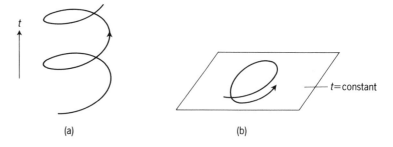

Fig. 15.3 Motion of a test particle (a) in space-time and (b) projected on to $t = $ constant.

(a)

(b)

$t = $constant

last term is small, because the ratio of the two terms on the right-hand side of (15.25) is $3h^2/r^2$, which for Mercury, for example, goes like 10^{-7}. On this assumption, we may solve the equation approximately by a perturbation method. We introduce the parameter

$$\varepsilon = 3m^2/h^2, \tag{15.26}$$

which in non-relativistic units is dimensionless (remember we have set $c = 1$). If we denote differentiation with respect to ϕ by a prime, then (15.25) becomes

$$u'' + u = \frac{m}{h^2} + \varepsilon\left(\frac{h^2 u^2}{m}\right). \tag{15.27}$$

We assume that this has a solution of the form

$$u = u_0 + \varepsilon u_1 + O(\varepsilon^2). \tag{15.28}$$

Substituting in (15.27), we find

$$u_0'' + u_0 - \frac{m}{h^2} + \varepsilon\left(u_1'' + u_1 - \frac{h^2 u_0^2}{m}\right) + O(\varepsilon^2) = 0. \tag{15.29}$$

If we equate the coefficients of different powers of ε to zero, then the zeroth-order solution u_0 is the usual conic section (15.11)

$$u_0 = \frac{m}{h^2}(1 + e\cos\phi),$$

where, for convenience, we have taken $\phi_0 = 0$. The first-order equation is

$$u_1'' + u_1 = \frac{h^2}{m}u_0^2 \tag{15.30}$$

and so substituting for u_0 we get

$$u_1'' + u_1 = \frac{m}{h^2}(1 + e\cos\phi)^2$$

$$= \frac{m}{h^2}(1 + 2e\cos\phi + e^2\cos^2\phi)$$

$$= \frac{m}{h^2}(1 + \tfrac{1}{2}e^2) + \frac{2me}{h^2}\cos\phi + \frac{me^2}{2h^2}\cos 2\phi.$$

If we try a particular solution of the form

$$u_1 = A + B\phi\sin\phi + C\cos 2\phi, \tag{15.31}$$

then we find (exercise)

$$A = \frac{m}{h^2}(1 + \tfrac{1}{2}e^2), \qquad B = \frac{me}{h^2}, \qquad C = -\frac{me^2}{6h^2}. \tag{15.32}$$

Thus, the general solution of (15.27) to first order is

$$u \simeq u_0 + \varepsilon\frac{m}{h^2}[1 + e\phi\sin\phi + e^2(\tfrac{1}{2} - \tfrac{1}{6}\cos 2\phi)]. \tag{15.33}$$

The most important correction to u_0 is the term involving $e\phi\sin\phi$, because after each revolution it gets larger and larger. If we neglect the other corrections, this becomes

$$u \simeq \frac{m}{h^2}[1 + e\cos\phi + \varepsilon e\phi\sin\phi],$$

or

$$u \simeq \frac{m}{h^2}\{1 + e\cos[\phi(1-\varepsilon)]\}, \qquad (15.34)$$

again neglecting terms of order ε^2 (check). Thus, the orbit of the test body is only approximately an ellipse. The orbit is still periodic, but no longer of period 2π; rather it is of period

$$\frac{2\pi}{1-\varepsilon} \simeq 2\pi(1+\varepsilon). \qquad (15.35)$$

In simple intuitive terms, a planet will travel in an ellipse but the axis of the ellipse will rotate, moving on by an amount $2\pi\varepsilon$ between two points of closest approach (Fig. 15.4). This is the famous **precession of the perihelion**. In non-relativistic units, this becomes (exercise)

$$2\pi\varepsilon \simeq \frac{24\pi^3 a^2}{c^2 T^2(1-e^2)}, \qquad (15.36)$$

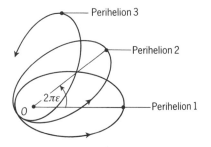

Fig. 15.4 Precession of the perihelion.

where a is the semi-major axis of the ellipse and T is the period of the orbit.

Now, in fact, in Newtonian theory, there is also an advance of the perihelion. This is because the planetary system is not a two-body system but rather an n-body system, and all the other planets produce a perturbation effect on the motion of one particular planet (rather similar in effect to the perturbation in (15.25)). For example, the planet Jupiter produces a measurable perturbation because its mass is relatively large, being about 1% of that of the Sun. Mercury has an orbit with high eccentricity and small period (see (15.36)) and the perihelion position can be accurately determined by observation. Before general relativity, there was a discrepancy between the classical prediction and the observed shift of some 43 seconds of arc per century. Even though this is a very small difference, it is very significant on an astrophysical scale and represents about a hundred times the probable observational error. This discrepancy had worried astronomers since the middle of the last century. In fact, in an attempt to explain the discrepancy, it was suggested that there existed another planet, which was given the name Vulcan, whose orbit was inside the orbit of Mercury. (Indeed, there is a famous incident of its reported 'observation' by a French astronomer.) However, Vulcan does not exist, and general relativity appears to explain the discrepancy, since it gives a theoretical prediction of 43.03 seconds of arc per century. The agreement of the residual perihelion precession with the other planets is not so marked because their observed precessions are very small and some of the observational data involved is not sufficiently accurate. One exception is a measurement in 1971 of the residual precession of the minor planet Icarus, which once again is in good agreement with the predicted values of general relativity (Table 15.1).

Table 15.1 Theoretical and observation values of residual perihelion precession

Planet	GR prediction	Observed
Mercury	43.0	43.1 ± 0.5
Venus	8.6	8.4 ± 4.8
Earth	3.8	5.0 ± 1.2
Icarus	10.3	9.8 ± 0.8

15.4 Bending of light

We next consider the case of the trajectory of a light ray in a spherically symmetric gravitational field. The calculation is essentially the same as that given in the last section, except that a light ray travels on a **null** geodesic and so a dot now denotes differentiation with respect to an affine parameter and the right-hand side of (15.17) is zero. The analogue of (15.25) is easily found to be (exercise)

$$\frac{d^2 u}{d\phi^2} + u = 3mu^2. \tag{15.37}$$

In the limit of **special relativity**, m vanishes and the equation becomes

$$\frac{d^2 u}{d\phi^2} + u = 0, \tag{15.38}$$

the general solution of which can be written in the form

$$u = \frac{1}{D}\sin(\phi - \phi_0), \tag{15.39}$$

where D is a constant. This is the equation of a straight line (exercise) as ϕ goes from ϕ_0 to $\phi_0 + \pi$, where D is the distance of closest approach to the origin. The straight line motion (Fig. 15.5) is the same as is predicted by Newtonian theory.

The equation of a light ray in Schwarzschild space-time (15.37) can again be thought of as a perturbation of the classical equation (15.38). However, this time we treat the dimensionless quantity mu or m/r as small. We therefore seek an approximate solution of the form

$$u = u_0 + 3mu_1, \tag{15.40}$$

where u_0 is the solution (15.39). Again, for convenience, we can take $\phi_0 = 0$. Then, if we neglect terms of order $(mu)^2$, the equation for u_1 becomes

$$u_1'' + u_1 = u_0^2 = \frac{\sin^2 \phi}{D^2}. \tag{15.41}$$

This has $(1 + C\cos\phi + \cos^2\phi)/3D^2$ as solution (exercise), where C is an arbitrary constant of integration, and so the general solution of (15.37) is approximately

$$u \simeq \frac{\sin\phi}{D} + \frac{m(1 + C\cos\phi + \cos^2\phi)}{D^2}. \tag{15.42}$$

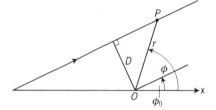

Fig. 15.5 Straight line motion of a light ray in special relativity.

Since m/D is small, this is clearly a perturbation from straight line motion. We are interested in determining the angle of deflection, δ say, for a light ray in the presence of a spherically symmetric source, such as the sun. A long way from the source, $r \to \infty$ and hence $u \to 0$, which requires the right-hand side of (15.42) to vanish. Let us take the values of ϕ for which $r \to \infty$, that is, the angles of the asymptotes, to be $-\varepsilon_1$ and $\pi + \varepsilon_2$, respectively, as shown in

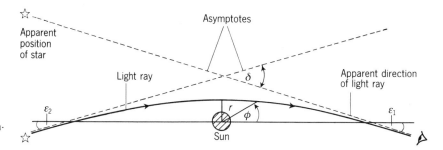

Fig. 15.6 Deflection of light in a gravitational field.

Fig. 15.6. Using the small-angle formulae for ε_1 and ε_2, we get

$$-\frac{\varepsilon_1}{D} + \frac{m}{D^2}(2 + C) = 0, \qquad -\frac{\varepsilon_2}{D} + \frac{m}{D^2}(2 - C) = 0.$$

Adding, we find

$$\delta = \varepsilon_1 + \varepsilon_2 = 4m/D, \qquad (15.43)$$

or, in non-relativistic units,

$$\delta = 4GM/c^2 D. \qquad (15.44)$$

The deflection predicted for a light ray which just grazes the Sun is 1.75 seconds of arc. Attempts have been made to measure this deflection at a time of total eclipse when the light from the Sun is blocked out by the Moon, so that the apparent position of the stars can be recorded. Then, if photographs of a star field in the vicinity of the Sun at a time of total eclipse are compared with photographs of the same region of the sky taken at a time when the sun is not present, they reveal that the stars appear to move out radially because of light deflection (Fig. 15.7).

The first expedition to record a total eclipse was one in 1919 under the leadership of Sir Arthur Eddington. The fact that this took place shortly after the end of World War I (and, moreover, that the expedition was led by an English scientist attempting to confirm a theory of a German scientist) caught the imagination of a war-weary world. When Eddington reported that the observations confirmed Einstein's theory, Einstein became something of a

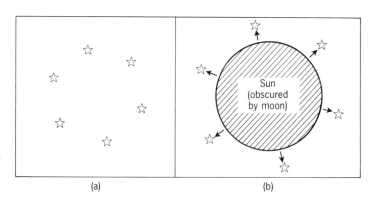

Fig. 15.7 Position of stars in a field (a) when sun is absent and (b) during a total eclipse.

celebrity and the newspapers of the day carried popular articles attempting to explain how we now lived in a curved four-dimensional world. Einstein was so convinced that his theory was right that he reportedly remarked that he would have been sorry for God if the observations had disagreed with the theory. In fact, it is now believed that the observations were not as clear-cut as they then seemed, because of problems associated with the solar corona, systematic errors, and photographic emulsions. In all, there have been some seven attempts to make eclipse measurements. The results have varied markedly from 0.7 to 1.55 times the Einstein prediction. So the best that can be said is that the results are in qualitative agreement with the Einstein prediction, but are uncertain in terms of their quantitative agreement. With the advent of large radio telescopes and the discovery of pointlike sources called **quasars** (quasi-stellar objects), which emit huge amounts of electromagnetic radiation, the deflection can now be measured using interferometric techniques when such a source passes near the Sun. Early measurements range from 1.57 to 1.82 ± 0.2 seconds of arc, but significant improvement on the accuracy of these measurements should be possible.

If one considers a family of curves representing light rays coming in parallel to each other from a distant source, then the presence of a massive object like the sun causes the light rays to converge and produce a caustic line on the axis $\phi = 0$. In this way, a spherically symmetric gravitational field acts as a **gravitational lens** (Fig. 15.8). Moreover, distant point-like sources can produce double images (see Fig. 15.9). There was considerable interest in 1980 when astronomers first reported the identification of what was previously considered two distinct quasars (known as 0957 + 561A, B), separated by 6 seconds of arc. The evidence is that there is a galaxy, roughly a quarter of the way from us to the quasar, which is the principal component of a gravitational lens.

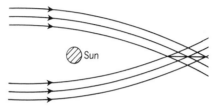

Fig. 15.8 The graviational lens effect of a Schwarzschild field.

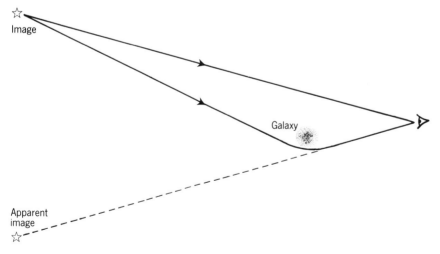

Fig. 15.9 Schematic representation of the double image effect of the gravitational lens.

15.5 Gravitational red shift

The third classical test is the gravitational red shift. At first, it was thought to be a direct test of general relativity since it employed the Schwarzschild solution. However, it is now clear that any relativistic theory of gravitation consistent with the principle of equivalence will predict a red shift. We outline

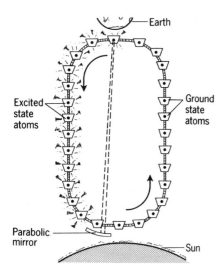

Fig. 15.10 A gravitational *perpetuum mobile*?

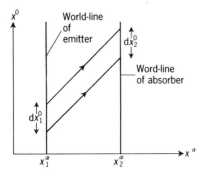

Fig. 15.11 Emission and reception of successive wave crests of a signal.

below a thought experiment which leads directly to the existence of a gravitational red shift. Consider an endless chain running between the Earth and the Sun carrying buckets containing atoms in an excited state on one side and an equal number of atoms in the ground state on the other side (Fig. 15.10). Since the excited atoms possess greater energy, they must have greater mass (using $E = mc^2$). They will be heavier than the ground state atoms and so, by the principle of equivalence, they will fall towards the Sun, whose gravitational field predominates. Suppose we have a device which returns an atom arriving at the Sun to its ground state, collects the emitted energy radiated in a mirror, and reflects it back to the Earth, where it is used to excite an incoming atom in the ground state. Then the rotating chain will run on indefinitely. In this way, we have constructed a **perpetuum mobile**, or perpetual motion machine. Such a device contradicts the principle of conservation of energy, the cornerstone of physics, and so something must be wrong with the argument. It breaks down because the radiation arriving at the Earth is not sufficiently energetic to excite the incoming ground state atom. In other words, it gets downgraded climbing up the gravitational field: the radiation has been shifted to the **red**.

We shall next obtain a quantitative expression for the red shift in the special case of a **static** space-time. The coordinates are taken to be

$$(x^a) = (x^0, x^\alpha),$$

where x^0 is the world time and x^α are spatial coordinates. We consider two observers carrying ideal atomic clocks whose world-lines are $x^\alpha = x_1^\alpha$ and $x^\alpha = x_2^\alpha$, respectively (see Fig. 15.11). Let the first observer possess an atomic system which is sending out radiation to the second observer. We denote the time separation between successive wave crests as measured by the first clock by $d\tau$ in terms of proper time and by dx_1^0 in terms of coordinate time. It follows from the definition of proper time that

$$d\tau^2 = g_{ab}(x_1^\alpha)dx_1^a\,dx_1^b = g_{00}(x_1^\alpha)(dx_1^0)^2, \qquad (15.45)$$

since g_{ab} can only depend on the spatial coordinates. Let the corresponding interval of reception recorded by the second observer be $\alpha\,d\tau$ in proper time and dx_2^0 in coordinate time. Then, similarly,

$$(\alpha\,d\tau)^2 = g_{00}(x_2^\alpha)(dx_2^0)^2. \qquad (15.46)$$

However, the assumption that the space-time is static means that

$$dx_1^0 = dx_2^0, \qquad (15.47)$$

because otherwise there would be a build-up or depletion of wave crests between the two observers, in violation of the static assumption. Dividing (15.45) and (15.46), we find

$$\alpha = \left(\frac{g_{00}(x_2^\alpha)}{g_{00}(x_1^\alpha)}\right)^{\frac{1}{2}}. \qquad (15.48)$$

The factor α records how many times the second clock has ticked between the reception of the two wave crests. It follows that if the atomic system has characteristic frequency ν_0 then the second observer will measure a frequency for the first clock of $\bar{\nu}_0$, where

$$\bar{\nu}_0 = \frac{\nu_0}{\alpha} = \nu_0\left(\frac{g_{00}(x_1^\alpha)}{g_{00}(x_2^\alpha)}\right)^{\frac{1}{2}}. \qquad (15.49)$$

Then, in particular,

$$g_{00}(x_1^{\alpha}) < g_{00}(x_2^{\alpha}) \quad \Rightarrow \quad \bar{v}_0 < v_0, \qquad (15.50)$$

which means that the frequency is shifted to the red. We define the **frequency shift** to be

$$\frac{\Delta v}{v} = \frac{\bar{v}_0 - v_0}{v_0}, \qquad (15.51)$$

which, in the case of the weak-field limit (12.65), namely,

$$g_{00} \simeq 1 + 2\phi/c^2,$$

gives (exercise)

$$\frac{\Delta v}{v} \simeq \frac{\phi_1 - \phi_2}{c^2}. \qquad (15.52)$$

Note that we have obtained this expression **without** recourse to the field equations. In the special case of the Schwarzschild solution, this becomes, in non-relativistic units,

$$\frac{\Delta v}{v} \simeq -\frac{GM}{c^2}\left(\frac{1}{r_1} - \frac{1}{r_2}\right). \qquad (15.53)$$

Then

$$r_1 < r_2 \quad \Rightarrow \quad \Delta v < 0, \qquad (15.54)$$

and so the frequency is shifted to the red.

If we take r_1 to be the observed radius of the Sun and r_2 the radius of the earth's orbit (Fig. 15.12), then (neglecting the Earth's gravitational field)

$$\Delta v/v \simeq -2.12 \times 10^{-6}. \qquad (15.55)$$

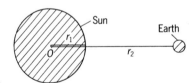

Fig. 15.12 Observation of red shift of atoms near the Sun's edge.

Observations of the Sun's spectra near its edge give results of this order, but there is great difficulty in interpreting the results generally because of lack of knowledge of the detailed structure of the Sun and the solar atmosphere. Similar remarks hold about white dwarfs, which, because of their small radii compared with their masses, have a more pronounced shift.

Since there are difficulties associated with astronomical measurements of the gravitational red shift, there has been interest in the possibility of a terrestrial test. This is a difficult task because the expected shift over a vertical distance of 100 ft, say, is only of the order of 10^{-15}. Fortunately, the discovery of the Mössbauer effect in 1958 gave a method of producing and detecting gamma rays which are monochromatic to 1 part in 10^{12}, and so makes a terrestrial test feasible. Pound and Rebka carried out such a test in 1960. They placed a gamma ray emitter at the bottom of a vertical 72 ft tower with an absorber at the top. Gamma rays emitted at the bottom then suffered a gravitational red shift climbing up the Earth's gravitational field to the top of the tower and were therefore less favourably absorbed. By moving the emitter upwards at a small measured velocity, a compensating Doppler shift was produced which allowed the rays to be resonantly absorbed. The experimental result gave 0.997 ± 0.009 times the predicted shift of 4.92×10^{-15}, that is, an agreement of better than 1%. Other experiments since 1960 have measured the change in the rate of atomic clocks transported on aircraft, rockets, and satellites; these have produced agreement with the theoretical predictions to about the same order of accuracy. One example being the shift experienced by radio signals from the spaceprobe Voyager I in its flight past Saturn in 1980. The accuracy was increased by two more orders of magnitude over the 1960 result in 1976 when a hydrogen maser clock was flown on a Scout rocket to an altitude of some 10 000 km and compared to a similar clock on the ground. It is intriguing to note that the length of the Scout rocket was almost exactly the same as the height of the Jefferson Physical Laboratory tower at Harvard University used for the 1960 experiment.

15.6 Time delay of light

A fourth test which may also be considered a classical test of general relativity was proposed by Shapiro in 1964. The idea is to use radar methods to measure the time travel of a light signal in a gravitational field. Because space-time is curved in the presence of a gravitational field, this travel time is greater than it would be in flat space, and the difference can be tested experimentally.

We begin by considering the path of a light ray in the equatorial plane $\theta = \frac{1}{2}\pi$ in Schwarzschild space-time, where, using (14.47),

$$(1 - 2m/r)\,dt^2 - (1 - 2m/r)^{-1}\,dr^2 - r^2\,d\phi^2 = 0. \qquad (15.56)$$

To find the travel time of a light ray, we need to eliminate ϕ in terms of r and so obtain a differential equation for dt/dr. We could use our solution (15.42) but, since we are only going to work to first order in m/r, it is sufficient to take the straight-line approximation

$$r \sin \phi = D.$$

Differentiating, we get

$$r \cos \phi \, d\phi + dr \sin \phi = 0,$$

so that

$$r\,d\phi = -\tan \phi \, dr$$

and

$$r^2\,d\phi^2 = \tan^2 \phi \, dr^2 = \frac{D^2}{r^2 - D^2}\,dr^2.$$

Substituting in (15.56), we find

$$dt^2 = \left((1 - 2m/r)^{-2} + (1 - 2m/r)^{-1}\,\frac{D^2}{r^2 - D^2} \right) dr^2.$$

Expanding in powers of m/r, we then have, to first order,

$$dt^2 \simeq \left(1 + 4m/r + (1 + 2m/r)\,\frac{D^2}{r^2 - D^2} \right) dr^2$$

$$= \left(\frac{r^2 + 4mr - 2mD^2/r}{r^2 - D^2} \right) dr^2$$

$$= \frac{r^2}{r^2 - D^2}\,(1 + 4m/r - 2mD^2/r^3)\,dr^2.$$

Taking square roots, we get

$$dt \simeq \frac{\pm r}{(r^2 - D^2)^{\frac{1}{2}}} (1 + 4m/r - 2mD^2/r^3)^{\frac{1}{2}}\, dr,$$

which, again to first order, gives

$$dt \simeq \frac{\pm r}{(r^2 - D^2)^{\frac{1}{2}}} (1 + 2m/r - mD^2/r^3)\, dr. \qquad (15.57)$$

We are interested in the travel time for a signal between a planet and the Earth. Integrating, we find the travel time is (exercise)

$$T = [(D_p^2 - D^2)^{\frac{1}{2}} + (D_E^2 - D^2)^{\frac{1}{2}}]$$
$$+ 2m \ln \{[(D_p^2 - D^2)^{\frac{1}{2}} + D_p][(D_E^2 - D^2)^{\frac{1}{2}} + D_E]/D^2\}$$
$$- m[(D_p^2 - D^2)^{\frac{1}{2}}/D_p + (D_E^2 - D^2)^{\frac{1}{2}}/D_E], \qquad (15.58)$$

where D is the closest approach to the Sun, D_p is the planet's orbit radius, and D_E is the Earth's orbit radius (see Fig. 15.13). The first term in square brackets in (15.58) represents the flat space result (as should be clear from the figure and also by setting $m = 0$).

The experimental verification of the delay consists in sending pulsed radar signals from the Earth to Venus and Mercury and timing the echoes as the positions of the Earth and the planet change relative to the Sun. For Venus, the measured delay is about 200 µs, which gives an agreement with the theoretical prediction of better than 5%.

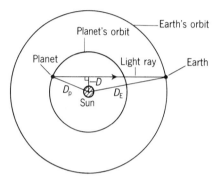

Fig. 15.13 A light ray travelling from a planet to the Earth in the Sun's gravitational field.

15.7 The Eötvös experiment

We have seen that the gravitational red shift is essentially a test of the principle of equivalence. The bending of light is also closely related to the principle of equivalence (see Exercise 9.8). Indeed, Schiff has shown that it is possible to obtain light bending from this principle coupled with Newtonian mechanics and some heuristic arguments about rigid rods (although these latter arguments have resulted in some dispute). Time delay of light signals is intimately connected with light bending. It would therefore seem that only the first classical test of the perihelion advance is a direct test of the theory. We shall consider this result again in the next section.

Since the principle of equivalence is so central to general relativity, we mention briefly here the important Eötvös torsion balance experiment which supports the principle. The experiment grew out of the much earlier work of Newton and Bessel using pendula. The Eötvös experiment consists of two objects of different composition connected by a rod of length l and suspended horizontally by a fine wire (Fig. 15.14). If the gravitational acceleration of the two masses is different, then it can be shown that there will be a torque N on the wire with

$$|N| = \eta l(g \times k) \cdot i, \qquad (15.59)$$

where g is the gravitational acceleration, i and k are unit vectors along the rod and the wire, and η is a limit on the difference in acceleration called the Eötvös ratio. If the apparatus is rotated with angular velocity ω then the torque will be modulated with period $2\pi/\omega$. In the original experiment of

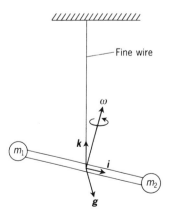

Fig. 15.14 The Eötvös torsion balance.

Baron von Eötvös in the early 1920s, g was the gravitational acceleration due to the Earth and the apparatus was rotated about the direction of the wire. Eötvös found a limit on η of $|\eta| < 5 \times 10^{-9}$.

More recently, the experiment has been repeated and improved by Dicke at Princeton and Braginski at Moscow. In their experiments, g was due to the Sun, and the rotation of the Earth provided the modulation of the torque. The torque was determined by measuring the force required to keep the rod in place in the Princeton experiment and gave a result $|\eta| < 10^{-11}$. In the Moscow experiment, the torque was determined by measuring the torsional motion of the rod and produced $|\eta| < 10^{-12}$, one of the most accurate results in physics.

15.8 Solar oblateness

As we have seen, general relativity cannot be considered a completely Machian theory. In an attempt to produce a relativistic theory of gravitation which better incorporated Mach's principle, Brans and Dicke proposed an alternative theory in 1961. We shall not discuss the details of it here except to say that it is motivated in part by the idea of treating the Newtonian constant G as a function of epoch (time), rather than a constant as in general relativity. The resulting theory has an adjustable parameter in it called ω and if, for suitable boundary conditions, we allow $\omega \to \infty$, then the theory corresponds to general relativity. The theory came into prominence in 1966 when Dicke and Goldenberg made a measurement of the visual oblateness, or flattening, of the Sun's disc and found a difference in the apparent polar and equatorial radii. Any oblateness in the Sun would produce a perturbation in the orbit equation (15.25) in addition to the general relativistic perturbation. The value they reported contributes some 4 seconds of arc per century to Mercury's perihelion shift, and this seriously undermines the remarkable agreement which general relativity has with the observed shift. On the other hand, the Brans–Dicke theory can accommodate this additional source of perihelion shift by setting $\omega \simeq 5$.

This led to considerable controversy in the relativity community and a large number of papers were produced which both supported and opposed solar oblateness. The experiment hinged on the method of placing an occulting disc in front of the Sun and measuring the difference in brightness between the pole and the equator at the limits of the Sun. The main counter-argument was that the difference reported could equally well be interpreted by assuming a sun with negligible oblateness but with a temperature difference between the pole and the equator, which indeed a standard model of the sun would require. The arguments depend critically on the precise model of the sun which is adopted and this is turn rests on complex solar physics theory which is not completely understood.

The controversy died away when Hill and collaborators performed a similar measurement in 1973 and failed to find any visual oblateness. More precisely, their value was at least one-fifth of Dicke's value and produces a perihelion result consistent with general relativity. The disagreement between the two results is unresolved, but mainstream opinion would appear to discount solar oblateness as being of any great significance.

15.9 A chronology of experimental and observational events

We end our considerations of experimental relativity with a brief chronology of the more important experimental and observational events which relate to general relativity.

1919	Eclipse expedition
1922	Eötvös torsion balance experiments Eclipse expedition
1929	Eclipse expedition
1936	Eclipse expedition
1947	Eclipse expedition
1952	Eclipse expedition
1960	Hughs–Drever mass-anisotropy experiments Pound–Rebka gravitational red-shift experiment
1962	Princeton Eötvös experiments
1965	Discovery of 3 K cosmic microwave background radiation
1966	Reported detection of solar oblateness Discovery of pulsars
1968	Planetary radar measurements of time delay First radio deflection measurements
1970	Cygnus XI: first black hole candidate Mariners 6 and 7 time-delay measurements
1972	Moscow Eötvös experiments
1973	Eclipse expedition
1974	Discovery of binary pulsar
1976	Rocket gravitational red-shift experiment Mariner 9 and Viking time-delay results
1978	Measurement of orbit-period decrease in binary pulsar
1979	Scout rocket maser clock red-shift measurements
1980	Discovery of gravitational lens

15.10 Rubber-sheet geometry

We end our considerations of general relativity with the description of a simple model which may help in understanding the theory. The model consists of an open box with a sheet of rubber stretched tightly over it. If a marble is then projected across the sheet, then it will move (approximately) in a straight line with constant velocity. This simulates **flat** space or special relativity, with the marble's path corresponding to the straight line geodesic motion of special relativity (Fig. 15.15). Next, a weight is placed on the centre

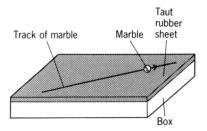

Fig. 15.15 Simulation of straight line geodesic motion in special relativity.

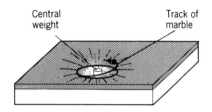

Fig. 15.16 Simulation of precessing elliptical motion in general relativity.

of the sheet, causing the rubber to become curved. If the marble is now projected correctly it will be seen to orbit the central weight. This simulates general relativity, where a central mass curves up space-time in its vicinity in such a way that a particle with suitable initial conditions will orbit the mass. The orbiting marble is performing the 'straightest' motion possible on the curved rubber sheet, or, more precisely, it is travelling on a geodesic of the sheet. Moreover, if the marble is projected carefully, it can be seen to be travelling on an elliptically shaped orbit which, owing to frictional effects between the marble and the rubber sheet, precesses about the central weight in analogy to a planetary orbit (Fig. 15.16).

We can relate this model better to the full theory if we consider an embedding diagram of the Schwarzschild solution in a slice $t = $ constant and in the equatorial plane $\theta = \frac{1}{2}\pi$. The line element then reduces to

$$ds^2 = (1 - 2m/r)^{-1}\, dr^2 + r^2 d\phi^2. \tag{15.60}$$

The curved geometry of this two-dimensonal surface is best understood if it is embedded in the flat geometry of a three-dimensional Euclidean manifold. This is depicted in Fig. 15.17, where the distance between two neighbouring points (r, ϕ) and $(r + dr, \phi + d\phi)$ defined by (15.60) is correctly represented. However, distances measured off the curved surface have no direct physical meaning, nor do points off the curved surface; only the curved 2-surface has meaning. If we fill in the interior of the Schwarzschild solution for $r \leqslant r_0$ $(r_0 > 2m)$, then this represents the gravitational field due to a spherical star and the embedding diagram looks like Fig. 15.18. The surface

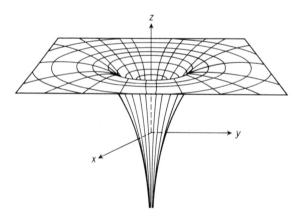

Fig. 15.17 Schwarzschild solution ($t = $ constant, $\theta = \frac{1}{2}\pi$) embedded in Euclidean 3-space.

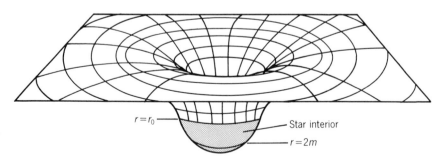

Fig. 15.18 Embedded geometry exterior and interior to a spherical star.

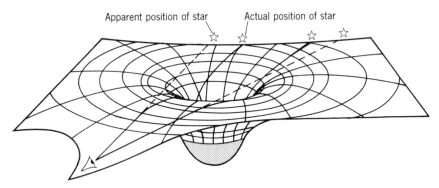

Apparent position of star   Actual position of star

Fig. 15.19 Depiction of light bending in the gravitational field of a star.

depicted in Fig. 15.18 is similar in nature to the curved surface of the rubber sheet in Fig. 15.16. This embedding diagram also helps us to understand the phenomenon of **light bending** (Fig. 15.19). The front cover contains a similar representation of a double image.

Although these diagrams are helpful in providing some insight into the idea of a curved space-time, they need to be used with caution. For example, the actual deflection of light is twice that suggested by Fig. 15.19 because the light travel takes place in **space-time** rather than space. What they do show, however, is how mass curves up space (actually space-time) in its vicinity and how free particles and photons travel in the straightest lines possible, namely, on the geodesics of the curved space. As J. A. Wheeler puts it so succinctly, 'Matter tells space how to curve, and space tells matter how to move.' The model also explains how the influence of the central mass is communicated to free particles and photons. This is very different from the action at a distance theory of Newtonian gravitation, where a central mass communicates its influence on a distant particle in a rather mysterious or at least unexplained way. Moreover, if the central mass changes in any way in Newtonian theory, then its influence is altered at all distant points **instantaneously**. In general relativity any change in the mass of the central source will spread out like a ripple in the rubber-sheet geometry, travelling with the speed of light. This leads to the beginnings of understanding how gravitational waves are generated, which we shall consider further in Part E.

Exercises

15.1 (§**15.2**) Show that (15.3) and (15.4) lead immediately to (15.6) if $h \neq 0$. What is the motion if $h = 0$?

15.2 (§**15.2**) Establish the result (15.9), Binet's equation (15.10), and its solution (15.11) and (15.12).

15.3 (§**15.2**) Establish Kepler's laws of planetary motion for the one-body problem, namely:

K1: Each planet moves about the Sun in an ellipse, with the Sun at one focus.

K2: The radius vector from the Sun to the planet sweeps out equal areas in equal intervals of time.

K3: The squares of the periods τ of any two planets are proportional to the cubes of the semi-major axes a of their respective orbits, i.e. $\tau \sim a^{\frac{3}{2}}$.

15.4 (§**15.2**) Show that the total energy E for the one-body problem can be written in terms of (R, ϕ) as

$$E = \tfrac{1}{2}m(\dot{R}^2 + R^2\dot{\phi}^2) - m\mu/R.$$

Express this in terms of (u, ϕ) and use (15.12) to identify the parameters as

$$l = h^2/\mu, \qquad e = (1 + 2Eh^2/m\mu^2)^{\frac{1}{2}}.$$

15.5 (§**15.2**) Establish (15.14) subject to (15.15) and (15.16) for the two-body problem.

15.6 (§**15.2**) Define the centre of mass R by

$$R = \frac{m_1 r_1 + m_2 r_2}{m_1 + m_2}$$

for the two-body problem and deduce that it moves with constant velocity. Transform to an inertial frame S' in which the centre of mass is at rest and situated at the origin O' of the frame S'. Define position vectors r'_1 and r'_2 of m_1 and m_2 relative to O', and hence describe the motion of m_1 and m_2 relative to O'. How are Kepler's laws modified in the case of the two-body problem? Show that, in particular,

$$\tau \simeq 2\pi a^{\frac{3}{2}} (Gm_{\text{sun}})^{-\frac{1}{2}}.$$

15.7 (§**15.3**) Establish the Euler–Lagrange equations (15.18)–(15.20). Write down the equation corresponding to $a = 1$ and confirm that (15.18)–(15.20) are the three simplest Euler–Lagrange equations.

15.8 (§**15.3**) Derive (15.24) and deduce (15.25) from it. What do the equations become in special relativity?

15.9 (§**15.3**) Show that (15.31) subject to (15.32) is a particular solution of (15.30). Hence establish (15.34).

15.10 (§**15.3**) Establish the result (15.36). [Hint: replace t by ct in (14.47) and use (14.50) and Exercise 15.6.]

15.11 (§**15.4**) Show that the right-hand side of (15.25) may be written in the form

$$\frac{GM}{c^2 r^4} \left(\frac{ds}{d\phi} \right)^2 + 3 \frac{GM}{c^2 r^2}$$

in appropriate units. Hence deduce the limiting case of the orbit equation for a light ray in the equatorial plane.

15.12 (§**15.4**) Show that (15.39) is the general solution of (15.38) and interpret (15.39) geometrically. Hence establish (15.42) as the approximate solution of (15.37).

15.13 (§**15.5**) Show that (15.49) leads to (15.52) in the weak-field limit. Deduce (15.53) for the Schwarzschild solution.

15.14 (§**15.6**) Integrate (15.57). [Hint: use a new variable u, where $r = D \cosh u$.] Deduce (15.58).

D. Black Holes

Non-rotating black holes

16.1 Characterization of coordinates

In this chapter, we are going to make an effort to understand the Schwarz-schild vacuum solution. The solution (14.47) is exhibited in a particular coordinate system. In general, if we wish to write down a solution of the field equations, then we need to do so in some particular coordinate system. But what, if any, is the significance of any particular coordinate system? For example, take the Schwarzschild solution and apply as complicated a co-ordinate transformation as you can imagine, labelling the new coordinates x'^a. Now suppose you had been given this solution and were asked to interpret the solution and identify the coordinates x'^a. The solution will, of course, still satisfy the vacuum field equations, but there is likely to be little or no geometrical significance attached to the coordinates x'^a. For example, one cannot just set $x'^0 = t$, say, and interpret t as a 'time' parameter. As a trivial illustration of this, consider the transformation

$$x'^0 = \theta, \qquad x'^1 = r, \qquad x'^2 = t, \qquad x'^3 = \phi.$$

One thing we can do, however, is establish whether the coordinate hyper-surface

$$x^{(a)} = \text{constant} \tag{16.1}$$

(where the parentheses enclosing the label a mean that it is to be regarded as fixed) is timelike, null, or spacelike at a point. For the normal vector field is

$$n_b = \delta_b^{(a)},$$

or in contravariant form

$$n^c = g^{cb} n_b = g^{cb} \delta_b^{(a)} = g^{c(a)},$$

which has magnitude squared given by

$$n^2 = n^c n_c = g^{c(a)} \delta_c^{(a)} = g^{(a)(a)} \quad \text{(not summed)}.$$

Hence, if the signature is -2, then the hypersurface (16.1) at P is **timelike**, **null**, or **spacelike** depending on whether $g^{(a)(a)}$ is >0, $=0$, or <0. At any point where the coordinate system is regular, the coordinate hypersurfaces may have any character, but the four normal vector fields $n_{(a)}^b$ must be linearly independent. Thus, for example, the hypersurfaces could all be null, timelike, or spacelike, or any combination of the three. We shall be meeting the three most common situations where the four coordinates consist of:

  1 timelike, 3 spacelike
  1 null, 3 spacelike
  2 null, 2 spacelike.

Although a metric may be displayed in any coordinate system, if it possesses symmetries then there will exist preferred coordinates adapted to the symmetries. We have already seen in Chapter 14 that if a solution possesses a Killing vector field then the coordinates may be adapted to the Killing vector field. If a solution possesses more than one Killing vector field, then the coordinates can be adapted to each of them as long as the Killing vector fields commute, that is, their Lie brackets vanish. If they do not commute, then the story is more complicated, but none the less the symmetries can be used to tie down the possible coordinate systems.

With these ideas in mind, let us look at the Schwarzschild solution in the form (14.47) to see if we can characterize the coordinates (t, r, θ, ϕ). First of all, since

$$g^{00} = \left(1 - \frac{2m}{r}\right)^{-1}, \quad g^{11} = -\left(1 - \frac{2m}{r}\right), \quad g^{22} = -\frac{1}{r^2}, \quad g_{33} = -\frac{1}{r^2 \sin^2\theta},$$

$$(16.2)$$

it follows that $x^0 = t$ is timelike and $x^1 = r$ is spacelike as long as $r > 2m$ and both $x^2 = \theta$ and $x^3 = \phi$ are spacelike. Next, since the metric is independent of t and there are no cross terms in dt, it follows that the solution is static and t is the invariantly defined world time of §14.3. The coordinate r is a radial parameter which has the property that the 2-sphere $t = $ constant, $r = $ constant has the standard line element

$$ds^2 = -r^2(d\theta^2 + \sin^2\theta \, d\phi^2),$$

from which it follows that the surface area of the 2-sphere is $4\pi r^2$. This would fail to be the case if we had chosen a different radial parameter, such as the isotropic coordinate ρ in (14.58). Then, finally, θ and ϕ are the usual spherical polar angular coordinates on the 2-spheres, which are invariantly defined by the spherical symmetry. In short, the Schwarzschild coordinates (t, r, θ, ϕ) are canonical coordinates defined invariantly by the symmetries present.

16.2 Singularities

We now turn to another problem associated with coordinates, that is, the fact that in general a coordinate system only covers a portion of the manifold. Thus, for example, the Schwarzschild coordinates do not cover the axis $\theta = 0$, π because the line element becomes degenerate there and the metric ceases to be of rank 4. This degeneracy could be removed by introducing Cartesian coordinates (x, y, z), where, as usual,

$$x = r \sin\theta \cos\phi, \quad y = r \sin\theta \sin\phi, \quad z = r \cos\theta.$$

Such points are called **coordinate singularities** because they reflect deficiencies in the coordinate system used and are therefore **removable**. There are two other values of the coordinates for which the Schwarzschild solution is degenerate, namely, $r = 2m$ and $r = 0$. The value $r = 2m$ is known as the **Schwarzschild radius**. The hypersurface $r = 2m$ again turns out to be a removable coordinate singularity. This is indicated by the Riemann tensor scalar invariant

$$R_{abcd}R^{abcd} = 48m^2 r^{-6},$$

which is finite at $r = 2m$. Since it is a scalar, its value remains the same in all coordinate systems. By the same token, this invariant blows up at the origin $r = 0$. The singularity at the origin is indeed irremovable and is variously called an **intrinsic, curvature, physical, essential**, or **real** singularity. Notice also by (15.48) that, since g_{00} vanishes at the Schwarzschild radius, the surface $r = 2m$ is a surface of infinite **red shift**. We shall pursue this later.

The normal interpretation of the Schwarzschild solution is as a vacuum solution exterior to some spherical body of radius $a > 2m$ (Fig. 16.1). A different metric would describe the body itself for $r < a$, and would then correspond to some distribution of matter resulting in a non-zero energy–momentum tensor. Indeed, Schwarzschild obtained a spherically symmetric static perfect fluid solution known as the **interior** Schwarzschild solution. However, our programme in this chapter will be to investigate the Schwarzschild vacuum solution abstracted away from any source for all values of r. In such a case, it should be clear from (16.2) that $r = 2m$ is a null hypersurface dividing the manifold into two disconnected components:

I. $2m < r < \infty$

II. $0 < r < 2m$.

Inside the region II the coordinates t and r reverse their character, with t now being spacelike and r timelike. It follows that the topology of the Schwarzschild solution is not simply Euclidean.

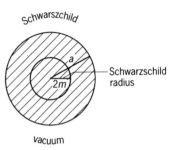

Fig. 16.1 Standard interpretation of the Schwarzschild exterior solution.

16.3 Spatial and space-time diagrams

The main technique we shall use to help interpret the solution is to investigate its local future light cone structure. A local light cone is defined as the locus of points $x_0^a + dx^a$, in the neighbourhood of a point x_0^a, for which

$$g_{ab}dx^a dx^b = 0.$$

The light cone structure puts constraints on the possible histories of an observer, since an observer moves on a timelike world-line whose direction at any point must lie within the future light cone at the point. Various diagrams will help us in trying to understand the nature of the solution.

In a purely **spatial diagram**, we shall be interested in what happens at various points in the manifold at two successive intervals of time, t_1 and t_2 say. At time t_1, a light flash is emitted from each point of interest and the spatial diagram indicates where the wave fronts of these flashes have reached at time t_2. This is illustrated in Fig. 16.2 for Minkowski space-time. In this figure, the light from each point will form a spherical wave front centred on the point. If there are symmetries present, it may be sufficient to consider what happens if we suppress one spatial dimension. For example, Fig. 16.2 becomes Fig. 16.3 in the plane $z = 0$, say, and the spheres now become circles.

In a **space-time diagram**, we are interested in the history of these light flashes. Suppose we take successive 'snapshots' of the wave fronts emanating from some point P at instants t_1, t_2, t_3, etc. (Fig. 16.4). The idea, in a space-time diagram, is to stack these pictures up in time. Since this would involve a four-dimensional picture — and there are enough problems in drawing three-

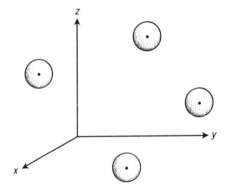

Fig. 16.2 Spatial diagram of Minkowski space–time.

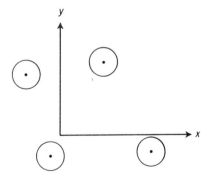

Fig. 16.3 Spatial diagram of Minkowski space time (one spatial dimension suppressed).

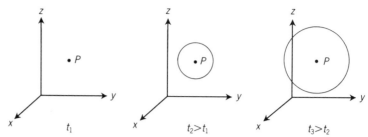

Fig. 16.4 Light flash from a point at three successive times.

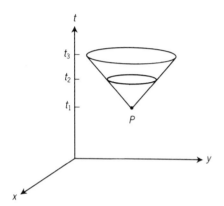

Fig. 16.5 Space-time diagram of light flash (one spatial dimension suppressed).

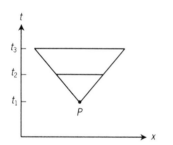

Fig. 16.6 Space-time diagram of light flash (two spatial dimensions suppressed).

dimensional pictures in two dimensions — we suppress one spatial dimension and, as in Chapter 2, we draw the time axis vertically. To be specific, let us restrict attention to the plane $z = 0$ and then the wave fronts will become circles (which will appear as ellipses in the diagram to take some account of perspective) lying on the future light cone through P (Fig. 16.5). In the same way, we can include the past light cone which can be thought of as an imploding wave front. Again, it will often be sufficient to consider a space-time diagram with two spatial dimensions suppressed (Fig. 16.6). In a **curved** space-time, the curvature manifests itself in space-time diagrams through the light cones being squashed or opened out and tipped or tilted in various ways, as we shall see below.

16.4 Space-time diagram in Schwarzschild coordinates

We first consider the class of **radial null geodesics** defined by requiring

$$ds^2 = \dot{\theta} = \dot{\phi} = 0. \tag{16.3}$$

Then, using our variational principle approach, we have

$$2K = (1 - 2m/r)\dot{t}^2 - (1 - 2m/r)^{-1}\dot{r}^2 = 0, \tag{16.4}$$

where a dot denotes differentiation with respect to an affine parameter u along the null geodesic. The Euler–Lagrange equation corresponding to $a = 0$ is

$$\frac{d}{du}[(1 - 2m/r)\dot{t}] = 0,$$

which integrates to give

$$(1 - 2m/r)\dot{t} = k, \tag{16.5}$$

where k is a constant. Substituting in (16.4) we find

$$\dot{r}^2 = k^2, \tag{16.6}$$

or

$$\dot{r} = \pm k, \tag{16.7}$$

from which it follows that r is an affine parameter (exercise). Rather than find the parametric equation of these curves, let us look directly for their equation in the form $t = t(r)$. Then

$$\frac{dt}{dr} = \frac{dt/du}{dr/du} = \frac{\dot{t}}{\dot{r}}, \tag{16.8}$$

which can be found from (16.5) and (16.7). Taking the positive sign in (16.7), we get

$$\frac{dt}{dr} = \frac{r}{r - 2m}, \qquad (16.9)$$

which can be integrated, to give (exercise)

$$t = r + 2m \ln|r - 2m| + \text{constant}. \qquad (16.10)$$

In the region I, by (16.9),

$$r > 2m \quad \Rightarrow \quad \frac{dr}{dt} > 0,$$

so that r increases as t increases. We therefore define the curves (16.10) to be a congruence of **outgoing** radial null geodesics. Similarly, the negative sign gives the congruence of **ingoing** radial null geodesics

$$t = -(r + 2m \ln|r - 2m| + \text{constant}). \qquad (16.11)$$

Notice that, under the transformation $t \to -t$, ingoing and outgoing geodesics get interchanged, as we would expect.

We can now use these congruences to draw a space-time diagram (Fig. 16.7) of the Schwarzschild solution in Schwarzschild coordinates with two dimensions suppressed (exercise). The space-time diagram is drawn for some fixed θ and ϕ. Since the diagram will be the same for all θ and ϕ, we should think of each point (t, r) in the diagram as representing a 2-sphere of area $4\pi r^2$. Notice that, as $r \to \infty$, the null geodesics make angles of $45°$ with the coordinate axes as in flat space in relativistic units, which we should

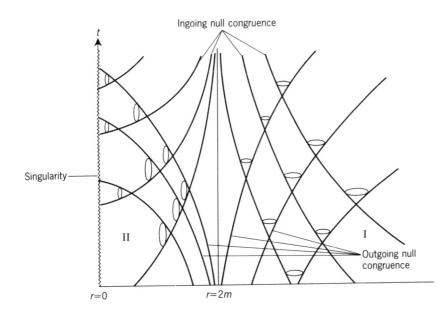

Fig. 16.7 Schwarzschild solution in Schwarzschild coordinates (two dimensions suppressed).

expect since the solution is asymptotically flat. The local light cones tip over in region II, because the coordinates t and r reverse their character. For example, the line $t = $ constant is a timelike line in region II and so must lie within the local light cone. An observer in region II cannot stay at rest, that is, at a constant value of r, but is forced to move in towards the intrinsic singularity at $r = 0$. This diagram seems to suggest that an observer in region I moving in towards the origin would take an infinite amount of time to reach the Schwarzschild radius $r = 2m$. Equally, the diagram suggests that the same is true for an ingoing light ray. However, it turns out that this space-time diagram is misleading, as we shall see.

16.5 A radially infalling particle

Let us consider the path of a radially infalling free particle. It will move on a timelike geodesic given by the equations (exercise)

$$(1 - 2m/r)\dot{t} = k, \tag{16.12}$$

$$(1 - 2m/r)\dot{t}^2 - (1 - 2m/r)^{-1}\dot{r}^2 = 1, \tag{16.13}$$

where a dot now denotes differentiation with respect to τ, the proper time along the world-line of the particle. Different choices of the constant k correspond to different initial conditions. Let us make the choice $k = 1$, which corresponds to dropping in a particle from infinity with zero initial velocity (exercise), so that, for large r, we have $\dot{t} \simeq 1$, that is, asymptotically $t \simeq \tau$. Then (16.12) and (16.13) give

$$\left(\frac{d\tau}{dr}\right)^2 = \frac{r}{2m} \tag{16.14}$$

Taking the negative square root (why?) and integrating, we find (exercise)

$$\tau - \tau_0 = \frac{2}{3(2m)^{\frac{1}{2}}}(r_0^{\frac{3}{2}} - r^{\frac{3}{2}}), \tag{16.15}$$

where the particle is at r_0 at proper time τ_0. This is, perhaps rather surprisingly, precisely the same as the classical result. No singular behaviour occurs at the Schwarzschild radius and the body falls continuously to $r = 0$ in a finite proper time.

If, instead, we describe the motion in terms of the Schwarzschild coordinate time t, then

$$\frac{dt}{dr} = \frac{\dot{t}}{\dot{r}} = -\left(\frac{r}{2m}\right)^{\frac{1}{2}}\left(1 - \frac{2m}{r}\right)^{-1}. \tag{16.16}$$

Integrating, we obtain (exercise)

$$t - t_0 = -\frac{2}{3(2m)^{\frac{1}{2}}}(r^{\frac{3}{2}} - r_0^{\frac{3}{2}} + 6mr^{\frac{1}{2}} - 6mr_0^{\frac{1}{2}})$$

$$+ 2m \ln \frac{[r^{\frac{1}{2}} + (2m)^{\frac{1}{2}}][r_0^{\frac{1}{2}} - (2m)^{\frac{1}{2}}]}{[r_0^{\frac{1}{2}} + (2m)^{\frac{1}{2}}][r^{\frac{1}{2}} - (2m)^{\frac{1}{2}}]}. \tag{16.17}$$

For situations where r_0 and r are much larger than $2m$, the results (16.15) and (16.17) are approximately the same, as we should expect. If, however, r is very

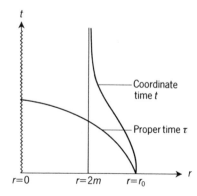

Fig. 16.8 Radially infalling particle in times τ and t.

near to $2m$, then we find (exercise)

$$r - 2m = (r_0 - 2m)e^{-(t-t_0)/2m}, \qquad (16.18)$$

from which it is clear that

$$t \to \infty \quad \Rightarrow \quad r - 2m \to 0$$

so that $r = 2m$ is approached but never passed. The two situations are illustrated in Fig. 16.8.

The coordinate t is useful and physically meaningful asymptotically since it corresponds to the proper time measured by an observer at rest far away from the origin. From the point of view of such an observer, it takes an infinite amount of time for a test body to reach $r = 2m$. However, as we have seen, from the point of view of the test body itself, it reaches both $r = 2m$ and $r = 0$ in finite proper time. Clearly, then, the Schwarzschild time coordinate t is inappropriate for describing this motion. Moreover, the coordinate system goes bad at $r = 2m$, as is evident from the behaviour of the line element there. In the next section, we shall introduce a new time coordinate which is adapted to radial infall, and in the process we shall remove the coordinate singularity at $r = 2m$.

16.6 Eddington–Finkelstein coordinates

The idea is very simple: we change to a new time coordinate in which the ingoing radial null geodesics become straight lines. It follows immediately from (16.10) that the appropriate change is given by

$$t \to \bar{t} = t + 2m \ln(r - 2m) \qquad (16.19)$$

for $r > 2m$, because in the new $(\bar{t}, r, \theta, \phi)$ coordinate system (16.11) becomes

$$\bar{t} = -r + \text{constant}, \qquad (16.20)$$

which is a straight line making an angle of $-45°$ with the r-axis. Differentiating (16.19), we get

$$d\bar{t} = dt + \frac{2m}{r - 2m}\, dr, \qquad (16.21)$$

and, substituting for dt in the Schwarzschild line element (14.47), we find the **Eddington–Finkelstein** form (exercise)

$$ds^2 = \left(1 - \frac{2m}{r}\right)d\bar{t}^2 - \frac{4m}{r}\, d\bar{t}\, dr - \left(1 + \frac{2m}{r}\right)dr^2 - r^2(d\theta^2 + \sin^2\theta\, d\phi^2).$$
$$(16.22)$$

This solution is now regular at $r = 2m$; indeed, it is regular for the whole range $0 < r < 2m$. Thus, in some sense, the transformation (16.19) extends the coordinate range from $2m < r < \infty$ to $0 < r < \infty$. The process is rather reminiscent of analytically continuing a function in complex analysis and, because of this, (16.22) is called an **analytic extension** of (14.47) (see Fig. 16.9).

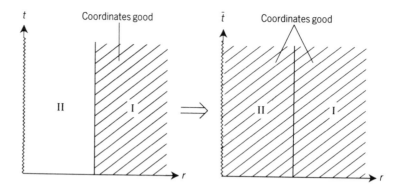

Fig. 16.9 Analytic extension of the Schwarzschild solution.

One could object that the coordinate transformation (16.19) cannot be used at $r = 2m$ because it becomes singular. However, (16.19) is just a convenient device to get us from (14.47) to (16.22). Our starting point is really the two line elements (14.47) and (16.22). Given these solutions, we then ask the question, What is the largest range of the coordinates for which each solution is regular? The answer is the patch $2m < r < \infty$ (together with, of course, $-\infty < t < \infty$, $0 \leqslant \theta \leqslant \pi$, and $-\pi < \phi \leqslant \pi$, apart from the usual problem with the coordinates on the axis $\theta = 0, \pi$) for (14.47) and the patch $0 < r < \infty$ for (16.22). In the **overlap** region $(2m < r < \infty)$, the two solutions are related by (16.19), and hence they must represent the **same** solution in this region. Note that the solution in Eddington–Finkelstein coordinates is no longer

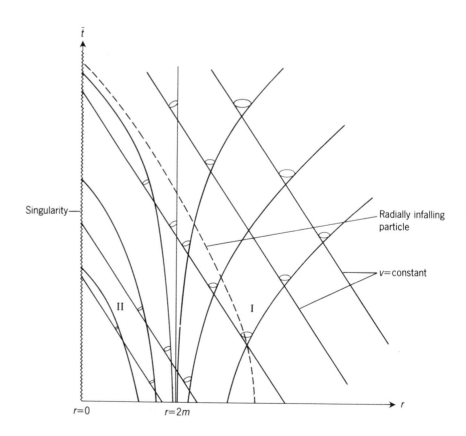

Fig. 16.10 Schwarzschild solution in advanced Eddington–Finkelstein coordinates.

time-symmetric. We can obtain a time-reversed solution by introducing a different time coordinate

$$t \to t^* = t - 2m \ln(r - 2m),$$

which straightens out **outgoing** radial null geodesics.

We can write (16.22) in a simpler form by introducing a null coordinate

$$v = \bar{t} + r, \tag{16.23}$$

which for historical reasons is called an **advanced time parameter**. The resulting line element is (exercise)

$$ds^2 = (1 - 2m/r)\,dv^2 - 2\,dv\,dr - r^2(d\theta^2 + \sin^2\theta\,d\phi^2). \tag{16.24}$$

It is then easy to show that the congruence of ingoing radial null geodesics is given by $v = $ constant, which should be evident from (16.20). The space-time diagram for the Schwarzschild solution in Eddington–Finkelstein coordinates is given in Fig. 16.10. As before, the light cones open out to 45° cones as $r \to \infty$. The left-hand edge of the light cones are all at $-45°$ to the r-axis. The right-hand edge starts at 45° to the r-axis at infinity and tips up as r decreases, becoming vertical at $r = 2m$, and tipping inwards for $r < 2m$. Notice that at $r = 2m$ radially outgoing photons 'stay where they are'. We can get a three-dimensional picture (in the equatorial plane $\theta = 0$, say) by rotating Fig. 16.10 about the $\bar{t}$-axis. Figure 16.10 now illustrates correctly what happens to a radially infalling particle.

16.7 Event horizons

Figure 16.10 suppresses the angular information in the Schwarzschild solution. This can best be depicted in the equatorial plane in a spatial diagram, as shown in Fig. 16.11. A long way from the origin, the spatial picture is similar to the special relativity picture (Fig. 16.3). As we move close to the origin, the spherical wave fronts are attracted inwards, so that the points from which they emanate are no longer at the centre. This becomes more marked until, on the surface $r = 2m$, only radial outgoing photons stay where they are, whereas all the rest are dragged inwards. In region II, all photons, even radially 'outgoing' ones, are dragged inwards towards the singularity.

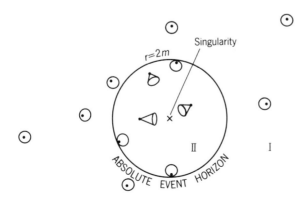

Fig. 16.11 Spatial diagram of Schwarzschild solution in advanced Eddington–Finkelstein coordinates.

It is clear from this picture that the surface $r = 2m$ acts as a **one-way membrane**, letting future-directed timelike and null curves cross only from the outside (region I) to the inside (region II). Moreover, no future-directed null or timelike curve can escape from region II to region I. The surface $r = 2m$ is called an **event horizon** because it represents the boundary of all events which can be observed in principle by an external inertial observer. The situation is reminiscent of the event horizons of hyperbolic motions in §3.8. However, they were observer-dependent. The Schwarzschild event horizon is **absolute**, since it seals off all internal events from **every** external observer.

If, instead, we use the null coordinate

$$w = t^* - r, \qquad (16.25)$$

called a **retarded time parameter**, then the line element becomes

$$ds^2 = (1 - 2m/r)\, dw^2 + 2\, dw\, dr - r^2(d\theta^2 + \sin^2\theta\, d\phi^2). \qquad (16.26)$$

This solution is again regular for $0 < r < \infty$ and corresponds to the time reversal of the advanced Eddington–Finkelstein solution (16.22) (Fig. 16.12). The surface $r = 2m$ is again a null surface which acts as a one-way membrane. However, this time it acts in the other direction of time, letting only past-directed timelike or null curves cross from the outside to the inside.

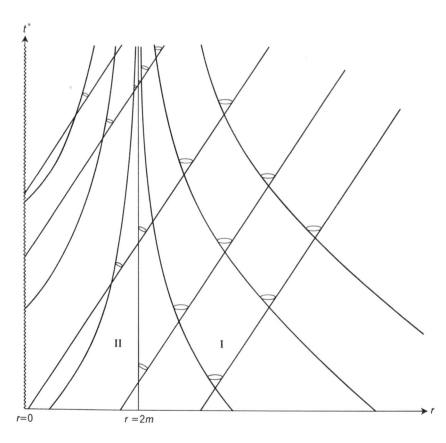

Fig. 16.12 Schwarzschild solution in retarded Eddington–Finkelstein coordinates.

16.8 Black holes

The theory of stellar evolution tells us that stars whose masses are of the order of the sun's mass can reach a final equilibrium state as a white dwarf or a neutron star. But, for much larger masses, no such equilibrium is possible, and in such a case the star will contract to such an extent that the gravitational effects will overcome the internal pressure and stresses which will not be able to halt further contraction. General relativity predicts that a spherically symmetric star will necessarily contract until all matter contained in the star arrives at a singularity at the centre of symmetry.

We imagine a situation in which the collapse of a spherically symmetric non-rotating star takes place and continues until the surface of the star approaches its Schwarzschild radius. To get an idea of the magnitude of the Schwarzschild radius, we note that the Schwarzschild radius for the Earth is about 1.0 cm and that of the Sun is 3.0 km. As long as the star remains spherically symmetric, its external field remains that given by the Schwarzschild vacuum solution. Figure 16.13 is a two-dimensional space-time diagram of the gravitational collapse, where the Schwarzschild vacuum solution is taken to be in Eddington–Finkelstein coordinates. As is clear from the diagram, an observer can follow a collapsing star through its Schwarzschild radius. If signals are sent out from an observer on the surface of the star at regular intervals according to that observer's clock, then as the surface of the star reaches the Schwarzschild radius, a distant observer will receive these signals with an ever-increasing time gap between them. The signal at $r = 2m$ will never escape from $r = 2m$, and all successive signals will ultimately be dragged back to the singularity at the centre. In fact, no matter how long the distant observer waits, it will only be possible to see the surface of the star as it was just before it plunged through the Schwarzschild radius. In practice, however, the distant observer would soon see nothing of the star's surface, since the observed intensity would die off very fast owing to the infinite red shift at the Schwarzschild radius. The star would quickly fade from view leaving behind a 'black hole' in space, waiting to gobble up anything which ventured too close.

If this were not bizarre enough, the theory apparently allows for the possibility of the time reversal of a black hole, which is called a **white hole**. A

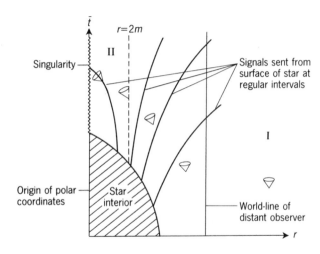

Fig. 16.13 Gravitational collapse (two spatial dimensions suppressed).

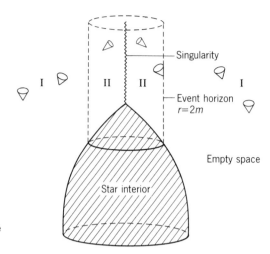

Fig. 16.14 Gravitational collapse (one spatial dimension suppressed).

white hole consists of a visible singularity (?) which, for no apparent reason, suddenly errupts into a star whose radius increases inexorably through its Schwarzschild radius. For completeness, we conclude this section with a three-dimensional space-time diagram of gravitational collapse (Fig. 16.14), which is obtained essentially by rotating Fig. 16.13 about the $\bar{t}$-axis.

16.9 A classical argument

The idea of a black hole, in the restricted sense of the gravitational field of a star being so strong that light cannot escape to distant regions, is in fact a consequence of Newtonian theory, if we adopt a particle theory of light. Consider a particle of mass m moving away radially from a spherically symmetric distribution of matter of radius R, uniform density ρ, and total mass M (Fig. 16.15). If the particle possesses a velocity v at a distance r from the centre, then conservation of energy E gives

$$E = \text{kinetic energy} + \text{potential energy}$$

$$= \tfrac{1}{2}mv^2 - GMm/r. \qquad (16.27)$$

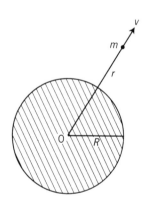

Fig. 16.15 Escape velocity in Newtonian gravitation.

We define the **escape velocity** v_0 to be the velocity at the surface of the distribution of matter which enables the particle to escape to infinity with zero velocity. This requires $v \to 0$ as $r \to \infty$, which by (16.27) results in $E = 0$. Solving for v, we find $v^2 = 2GM/r$, and hence the escape velocity is

$$v_0^2 = 2GM/R. \qquad (16.28)$$

Then, if a particle has a radial velocity less than v_0 at the surface, it will eventually be pulled back by the gravitational attraction of the distribution. If light has velocity c, then it will just escape to infinity if it is related to the mass and radius of the distribution by

$$c^2 = 2GM/R. \qquad (16.29)$$

Thus, if the mass M were increased (keeping the radius constant) or, equivalently, the radius R decreased (keeping the mass constant), then it follows that light could no longer escape. This was recognized by Laplace in 1798 who pointed out that a body of about the same density as the Sun but 250 times its radius would prevent light from escaping. Note that the limiting condition (16.29) in terms of the radius R is

$$R = 2GM/c^2, \qquad (16.30)$$

or $R = 2m$ in relativistic units, which is the Schwarzschild radius.

16.10 Tidal forces in a black hole

Consider a distribution of non-interacting particles falling freely towards the Earth in Newtonian theory, where initially the distribution is spherical (see Exercise 9.6). Each particle moves on a straight line through the centre of the Earth, but those nearer the Earth fall faster because the gravitational attraction is stronger. The sphere no longer remains a sphere but is distorted into an ellipsoid with the same volume (Fig. 16.16). Thus, the gravitation produces a **tidal force** in the sphere of particles. The tidal effect results in an elongation of the distribution in the direction of motion and a compression of the distribution in transverse directions. The same effect occurs in a body falling towards a spherical object in general relativity, but if the object is a black hole the effect becomes infinite as the singularity is reached. We can gain some idea of this by considering the equation of geodesic deviation (see (10.38) and (10.39)) in the form

$$\frac{D^2\eta^\alpha}{D\tau^2} - R^a{}_{bcd}\, e^\alpha{}_a\, v^b\, v^c\, e_\beta{}^d\, \eta^\beta = 0 \qquad (16.31)$$

for the spacelike components of the orthogonal connecting vector η^a connecting two neighbouring particles in freefall. Let the frame $e_i{}^a$ be defined in Schwarzschild coordinates as

$$e_0{}^a \overset{*}{=} (1 - 2m/r)^{-\frac{1}{2}}\,(1, 0, 0, 0), \qquad (16.32)$$

$$e_1{}^a \overset{*}{=} (1 - 2m/r)^{\frac{1}{2}}\,(0, 1, 0, 0), \qquad (16.33)$$

$$e_2{}^a \overset{*}{=} r^{-1}\,(0, 0, 1, 0), \qquad (16.34)$$

$$e_3{}^a \overset{*}{=} (r\sin\theta)^{-1}\,(0, 0, 0, 1), \qquad (16.35)$$

and let us denote the components of η^a by

$$\eta^\alpha = (\eta^1, \eta^2, \eta^3) = (\eta^r, \eta^\theta, \eta^\phi). \qquad (16.36)$$

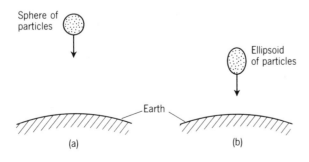

Fig. 16.16 Newtonian tidal force: (a) before; (b) after.

Then (16.31) reduces in Schwarzschild space-time in the above frame to the equations (exercise)

$$\frac{D^2\eta^r}{D\tau^2} = +\frac{2m}{r^3}\eta^r, \tag{16.37}$$

$$\frac{D^2\eta^\theta}{D\tau^2} = -\frac{m}{r^3}\eta^\theta, \tag{16.38}$$

$$\frac{D^2\eta^\phi}{D\tau^2} = -\frac{m}{r^3}\eta^\phi. \tag{16.39}$$

The positive sign in (16.37) indicates a tension or stretching in the radial direction and the negative signs in (16.38) and (16.39) indicate a pressure or compression in the transverse directions (see Misner, Thorne, and Wheeler (1973) for further details). Moreover, the equations reveal that the effect becomes infinite at the singularity $r = 0$.

Consider an intrepid astronaut falling feet first into a black hole (Fig. 16.17). The astronaut's feet are attracted to the centre by an infinitely mounting gravitational force, while the astronaut's head is accelerated downward by a smaller though ever-rising force. The difference between the two forces becomes greater and greater as the astronaut reaches the centre, where the difference becomes infinite. At the same time as the head–foot stretching, the astronaut is pulled by the gravitational field into regions with ever-decreasing circumference and so the astronaut is squashed on all sides. Again the squashing becomes infinite at the centre. Indeed, not only do the tidal effects tear the astronaut to pieces, but the very atoms of which the astronaut is composed must ultimately share the same fate!

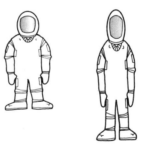

Fig. 16.17 Successive times in the astronaut's fall.

16.11 Observational evidence for black holes

Observing a black hole directly is impossible, unless one were lucky enough to see a star disappear. However, searches have been made for double stars with one invisible component. The belief is that, if a black hole is sucking off matter from its visible partner, then this will form an accretion disc around the black hole and the hot inner regions will produce intense bursts of X-rays formed by synchrotron radiation shortly before the spiralling matter disappears down the hole (Fig. 16.18). It was the discovery in 1971 of the rapid variations of the X-ray source Cygnus XI by telescopes aboard the Uhuru satellite that provided the first evidence of the likely existence of black holes. The visible component is a supergiant star, and detailed study of the X-rays

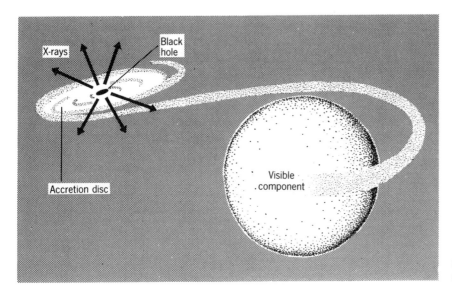

Fig. 16.18 A binary star with one visible and one black hole component.

led to the conclusion that the unseen body is a compact object with a mass in excess of 9 solar masses. Since the maximum masses of white dwarfs and neutron stars are believed to be approximately 1.4 and 4 solar masses, respectively, then the simplest conclusion is that the object is a black hole. Since 1971, a number of other black hole candidates have been found in X-ray binaries. In addition, studies of the central regions of some galaxies (including our own) and globular clusters have indicated the possible existence of supermassive black holes.

There might also be quite a number of very much smaller black holes scattered around the universe, formed not by the collapse of stars but by the collapse of highly compressed regions in the hot dense medium that is believed to have existed shortly after the big bang in which the universe originated. Such 'primordial' or 'mini' black holes are of greatest interest for their possible quantum characteristics. A black hole weighing a billion tons (about the size of a mountain) would have a radius of about 10^{-13} cm, which is the size of a neutron or proton. It could orbit, for example, around the sun or the centre of the galaxy.

16.12 Theoretical status of black holes

A crucial part of the interpretation of an X-ray binary rests on the assumption that the maximum mass of a neutron star is less than 4 solar masses. One uncertainty is the equation of state, and some arguments suggest that a high-density equation of state could lead to an increased maximum mass of 5 solar masses. Another uncertainty relates to rotation, which could increase the maximum again to 6 solar masses. A third uncertainty relates to the theory of gravitation employed. Most alternative gravity theories give similar results for weak gravitational fields, but they differ markedly in strong field regions such as occur in neutron stars. Indeed, some theories predict no maximum mass for neutron stars.

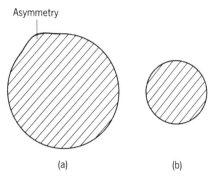

Asymmetry

(a)   (b)

Fig. 16.19 Asymmetry radiated away: (a) before; (b) eventually.

At the theoretical level, there is the objection that the solution is too special in being spherically symmetric. For example, no account has been taken of charge or rotation. In Chapters 17 and 18, we shall consider the Reissner–Nordstrøm and Kerr solutions, which deal with charged and rotating black holes, respectively. We shall see that, although the story changes in detail, the chief characteristics of a black hole, namely, the existence of absolute event horizons and singularities, persist. The next objection is that asymmetries have been excluded. It is not surprising, it can be argued, that if all the matter is moving in radially towards the centre then it will ultimately result in a singularity there. However, perturbations of the Schwarschild solution have been considered and appear to suggest that all asymmetries are eventually radiated away and that, asymptotically in time, the system settles down to a Schwarzschild black hole (Fig. 16.19).

Another objection relates to the particular set of field equations used, namely, those of general relativity. However, Penrose and Hawking have managed to prove some remarkable theorems, the so-called **singularity theorems**, which suggest that many of the qualitative features of this collapse picture remain in a more general situation. Their results do not depend on the particular field equations of general relativity, but on much weaker assumptions such as the geometrical interpretation of gravity and the consequent curvature of space-time, relativistic causality, and the dominant energy conditions (§19.12). The theorems prove that, with these very reasonable assumptions, there exist geodesics which come to an end, that is to say, that cannot be extended any further. This is usually taken to mean that they are ending on a singularity. Quite where the singularities are located and what their structure is like are issues which the theorems do not directly address. Of course, even these very weak assumptions may not apply in extremely strong gravitational fields. It could be possible, for example, that such fields result in the emission of tachyons which would violate causality. The general belief, however, is that the theorems provide strong evidence that singularities are, in fact, generic features of relativistic theories of gravitation.

There is another problem which has not yet been resolved. In order to discuss in detail the stability of a collapse situation, we need to understand what is going on inside the star. That is, we need realistic **interior** solutions which can be matched on to the known exterior solutions. However, all attempts at finding a realistic interior Kerr solution, and there have been many of them, appear to have failed. This is somewhat disturbing, because the attempts seem to suggest that the matching cannot be done. Were we to have an interior solution, it is conceivable that the motion might be unstable leading finally to fragmentation rather than collapse. Finally, we point out that gravitational collapse deals with situations of high densities and that these are really the province of quantum theory. It seems likely that a classical theory like general relativity might be modified profoundly by quantum effects. Indeed, some theories of quantum gravity suggest that the collapse is halted before a singularity is reached and a bounce takes place. Penrose has pointed out that we do not need high densities to create event horizons. It is possible to take an average-size galaxy of about 10^{11} stars, and then, if one could arrange to steer them all in the right direction, they could all fit together inside their collective gravitational radius without ever coming into contact with each other, and yet the resulting overall density would only be similar to that of the density of water.

Exercises

16.1 (§**16.1**) Interpret the solution

$$ds^2 = \left(1 - \frac{2m}{\phi}\right)d\theta^2 - \left(1 - \frac{2m}{\phi}\right)^{-1}d\phi^2$$
$$- \phi^2\,dt^2 - \phi^2\sin^2 t\,dr^2.$$

16.2 (§**16.1**) Apply the transformations

$$t = \bar{t}^2\bar{r}, \qquad\qquad r = \bar{r}\cos\bar{\theta} + 2m$$
$$\theta = \sin^{-1}(\bar{r}\bar{\theta}), \qquad \phi = \cos(\bar{\phi}\bar{t})$$

to the Schwarzschild line element (14.47) and find the coefficient of $d\bar{t}^2$.

16.3 (§**16.1**) What is the character of the coordinates of
(i) (t, ρ, z, ϕ) in

$$ds^2 = \rho^{-2m}dt^2 - \rho^{-2m}[\rho^{2m^2}(d\rho^2 + dz^2) + \rho^2\,d\phi^2];$$

(ii) (u, r, x, y) in

$$ds^2 = x^2\,du^2 - 2\,du\,dr + 4rx^{-1}\,du\,dx$$
$$- r^2\,dx^2 - x^2\,dy^2.$$

16.4 (§**16.1**) Dingle's metric is the most general diagonal metric

$$ds^2 = A\,dt^2 - B\,dx^2 - C\,dy^2 - D\,dz^2,$$

where A, B, C, and D are functions of all four coordinates. What does this solution become if $\partial/\partial x$, $\partial/\partial y$, $\partial/\partial z$ are commuting vector fields and the solution is adapted to these Killing vector fields?

16.5 (§**16.2**) Write the Schwarzschild line element (14.47) in coordinates (t, x, y, z) where x, y, z are defined by

$$x = r\sin\theta\cos\phi, \qquad y = r\sin\theta\sin\phi, \qquad z = r\cos\theta.$$

16.6 (§**16.3**) Draw a two-dimensional space-time diagram of null geodesics in special relativity. Draw the world-line of an observer moving into the origin and out again.

16.7 (§**16.4**) Integrate (16.6). Deduce that r is an affine parameter. Integrate (16.9) to obtain (16.10).

16.8 (§**16.4**) Confirm Fig. 16.7 by first drawing the graphs of
(i) $y = \ln x \ (x > 0)$
(ii) $y = \ln|x|$

(iii) $y = 2m\ln|x|$
(iv) $y = x + 2m\ln|x|$,
in turn, translating the y-axis to $x = 2m$, and then drawing the graphs of
(v) $y = x - 2m + 2m\ln|x - 2m| \ (x > 0)$
(vi) $y = x + 2m\ln|x - 2m| + c \ (x > 0)$
for different values of the constant c. What is the slope of the radial null geodesics at $r = 0$?

16.9 (§**16.5**) Establish (16.12) and (16.13) for the equations of a radially infalling particle. Show that the choice $k = 1$ corresponds to the particle having zero velocity at spatial infinity $(r = \infty)$.

16.10 (§**16.5**) Integrate (16.14) to obtain (16.15). Show that this is the same result as that for a particle falling radially from r_0 to r in Newtonian theory under the influence of a point particle situated at the origin of mass M, where the particle has zero velocity at infinity.

16.11 (§**16.5**) Integrate (16.16) to obtain (16.17).

16.12 (§**16.5**) If r is near $2m$, set $\varepsilon = 1 - r/2m$ and show that the dominant term in (16.16) is $1/\varepsilon$. Hence deduce (16.18).

16.13 (§**16.6**) Show that (16.19) transforms the Schwarzschild line element (14.47) into the form (16.22). Use (16.23) to express the resulting line element in the form (16.24).

16.14 (§**16.6**) Draw the Schwarzschild solution in advanced Eddington–Finkelstein coordinates with one spatial dimension suppressed in the equatorial plane $\theta = \frac{1}{2}\pi$. [Hint: rotate Fig. 16.10 about the $\bar{t}$-axis.]

16.15 (§**16.7**) Show that (16.25) leads to the form (16.26). Find the equations for radial null geodesics and establish Fig. 16.12.

16.16 (§**16.8**) Draw the white hole analogue of Fig. 16.13 and describe its appearance to an external observer.

16.17 (§**16.10**) Show that (16.32)–(16.35) defines an orthonormal frame in Schwarzschild space-time. Show that the spatial part of the equation of geodesic deviation leads to (16.37)–(16.39). Give a qualitative argument which reveals that η^r increases without bound as $r \to 0$.

17 Maximal extension and conformal compactification

17.1 Maximal analytic extensions

We saw in the last chapter that the Schwarzschild solution for $2m < r < \infty$ can be extended either into the advanced Eddington–Finkelstein solution (16.24) or the retarded Eddington–Finkelstein solution (16.26), where $0 < r < \infty$. That this is possible is indicated by the fact that a radial timelike geodesic can be extended through $r = 2m$ down to $r = 0$. The question naturally arises, Is it possible to extend these solutions further?

We need to make this question more precise, which we do by introducing a couple of definitions. A manifold endowed with an affine or metric geometry is said to be **maximal** if every geodesic emanating from an arbitrary point of the manifold either can be extended to infinite values of the affine parameter along the geodesic in both directions or terminates on an intrinsic singularity. If, in particular, all geodesics emanating from any point can be extended to infinite values of the affine parameters in both directions, the manifold is said to be **geodesically complete**. Clearly, a geodesically complete manifold is maximal, but the converse is not true in general. Minkowski space-time provides a trivial example of a geodesically complete manifold. Neither the Schwarzschild nor the Eddington–Finkelstein advanced or retarded extensions is in fact maximal. However, Kruskal has found the maximal analytic extension of the Schwarzschild solution and, moreover, this extension is unique. The Kruskal solution, although maximal, is again not complete because of the existence of intrinsic singularities. The Kruskal solution can be obtained by simultaneously straightening out both incoming and outgoing radial null geodesics. We shall sketch the original procedure of Kruskal in the next section.

17.2 The Kruskal solution

We start by introducing both an advanced null coordinate v and a retarded null coordinate w, in which case, in the coordinates (v, w, θ, ϕ), the Schwarzschild line element becomes (exercise)

$$ds^2 = (1 - 2m/r)\, dv\, dw - r^2(d\theta^2 + \sin^2\theta\, d\phi^2), \tag{17.1}$$

where r is a function of v and w determined implicitly by

$$\tfrac{1}{2}(v - w) = r + 2m \ln(r - 2m). \tag{17.2}$$

The 2-space $\theta = \text{constant}$, $\phi = \text{constant}$ has metric

$$ds^2 = (1 - 2m/r)\, dv\, dw, \tag{17.3}$$

and hence by the second theorem in §6.13 must be conformally flat. To make this evident, we define

$$t = \tfrac{1}{2}(v + w), \qquad x = \tfrac{1}{2}(v - w),$$

and then (17.3) becomes

$$ds^2 = (1 - 2m/r)(dt^2 - dx^2).$$

The most general coordinate transformation which leaves the 2-space (17.3) expressed in such conformally flat double null coordinates is

$$v \rightarrow v' = v'(v), \qquad w \rightarrow w' = w'(w),$$

where v' and w' are arbitrary, which leads to

$$ds^2 = (1 - 2m/r)\frac{dv}{dv'}\frac{dw}{dw'}\,dv'\,dw'.$$

Introducing

$$t' = \tfrac{1}{2}(v' + w'), \qquad x' = \tfrac{1}{2}(v' - w'),$$

we can write (17.3) in the general form

$$ds^2 = F^2(t', x')(dt'^2 - dx'^2).$$

A particular choice of v' and w' will then determine the precise form of the line element.

The choice which Kruskal made was

$$v' = \exp(v/4m), \tag{17.4}$$

$$w' = -\exp(-w/4m). \tag{17.5}$$

The radial coordinate r is to be considered a function of t' and x' determined implicitly by the equation

$$t'^2 - x'^2 = -(r - 2m)\exp(r/2m) \tag{17.6}$$

and F is given by

$$F^2 = \frac{16m^2}{r}\exp(-r/2m).$$

Then the line element becomes

$$ds^2 = \frac{16m^2}{r}\exp\left(-\frac{r}{2m}\right)dt'^2 - \frac{16m^2}{r}\exp\left(-\frac{r}{2m}\right)dx'^2$$
$$- r^2(d\theta^2 + \sin^2\theta\,d\phi^2) \tag{17.7}$$

subject to (17.6).

A two-dimensional space-time diagram of the Kruskal solution is shown in Fig. 17.1. As we indicated, all the light cones are now $45°$ cones and the incoming and outgoing radial null geodesics are straight lines. Figure 17.1 shows a radial timelike geodesic which starts from ($r = 4m$, $t' = 0$) and falls into the event horizon $r = 2m$, ending up on the future singularity at $r = 0$. The figure includes some of the signals sent out from this geodesic and illustrates the trapped nature of the signals sent inside the event horizon.

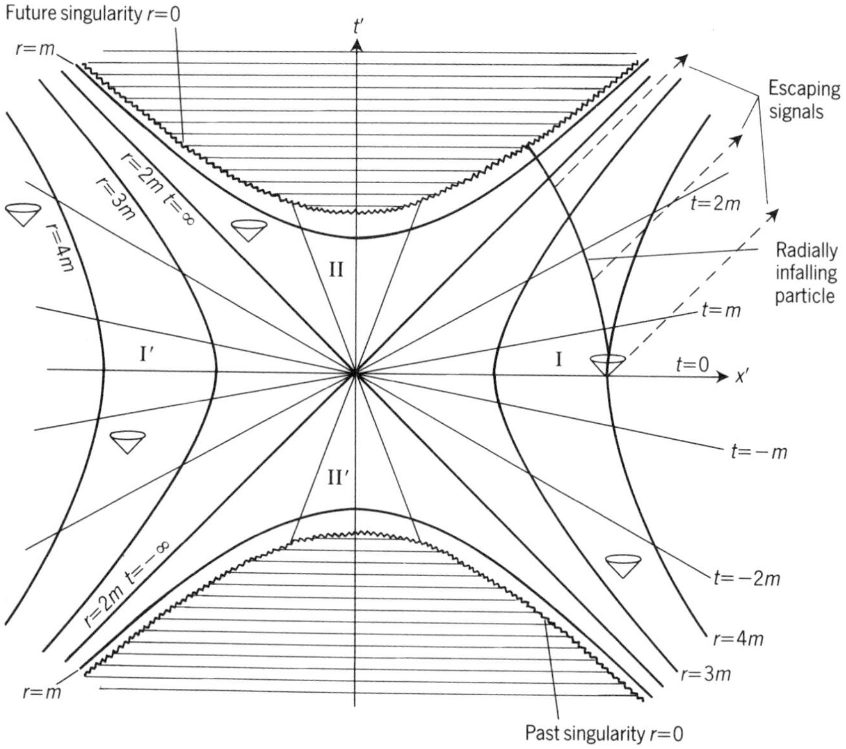

Fig. 17.1 Space-time diagram of the Kruskal solution.

Notice from (17.6), which is quadratic in t' and x', that one value of r determines two hypersurfaces. In two dimensions, the space-time is bounded by two hyperbolae representing the intrinsic singularity at $r = 0$. They are termed the **past** singularity and **future** singularity, respectively. The future singularity is **spacelike** and hence unavoidable in region II. The asymptotes of the hyperbolae represent the event horizons corresponding to $r = 2m$. These asymptotes divide the space-time into four regions labelled I, II, I′, and II′. The regions I and II correspond to the advanced Eddington–Finkelstein solution (see Fig. 16.10) with region I corresponding to the Schwarzschild solution for $r > 2m$ and region II corresponding to the black hole solution. The regions I and II′ correspond to the retarded Eddington–Finkelstein solution (see Fig. 16.12) with region II′ corresponding to the white hole solution. What is surprising is that there is a new region called I′ which is geometrically identical to the asymptotically flat exterior Schwarzschild solution region I. The topology connecting I and I′ is rather complicated and we consider it next.

17.3 The Einstein–Rosen bridge

Remember that each point in the diagram represents a 2-sphere. We can gain some intuitive idea of the overall four-dimensional structure if we consider first the submanifold $t' = 0$. Then from (17.7) the line element induced on this hypersurface is given by

$$ds^2 = -F^2 dx'^2 - r^2(d\theta^2 + \sin^2\theta \, d\phi^2). \tag{17.8}$$

As we move along the x'-axis from $+ \infty$ to $- \infty$, the value of r decreases to a minimum $2m$ at $x' = 0$ and then increases again as x' goes to $-\infty$. We can draw a cross-section of this manifold corresponding to the equatorial plane $\theta = \frac{1}{2}\pi$, in which case (17.8) reduces further to

$$ds^2 = -(F^2\,dx'^2 + r^2\,d\phi^2). \tag{17.9}$$

To interpret this we consider a two-dimensional surface possessing this line element embedded in a flat three-dimensional space. The surface appears as in Fig. 17.2. Thus, at $t' = 0$, the Kruskal manifold can be thought of as being formed by two distinct but identical asymptotically flat Schwarzschild manifolds joined at the 'throat' $r = 2m$. As t' increases, the same qualitative picture holds but the throat narrows down, the universes joining at a value of $r < 2m$. At $t' = 1$, the throat pinches off completely and the two universes touch at the singularity $r = 0$. For larger values of t', the two universes, each containing a singularity at $r = 0$, are completely separate. The Kruskal solution is time-symmetric with respect to t', and so the same thing happens if we run time backwards from $t' = 0$. The full time evolution is shown schematically in Fig. 17.3, where each diagram should be rotated about the central vertical axis to get the two-dimensional picture analogous to that shown in Fig. 17.2.

The intriguing question of whether or not the mathematical procedure for extending the solution which results in the 'new universe' I′ has any physical significance is still an open one. Although Einstein's equations fix the local geometry of space-time, they do not fix its global geometry or its topology. In Fig. 17.4, we see an embedding of the slice $t' = $ constant which is geometrically identical but topologically different. This embedding leads to a Schwarzschild 'wormhole' which connects two distant regions of a single asymptotically flat universe. We shall not pursue the idea further.

Although Fig. 17.1 is very informative, it does not indicate what happens to points at 'infinity'. We shall see that the process of conformal compactification allows us to investigate the structure of these points and leads to another picture called a **Penrose diagram**.

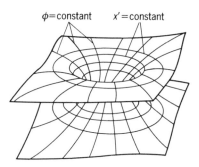

Fig. 17.2 The Einstein–Rosen bridge.

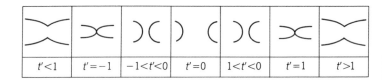

$t' < 1$	$t' = -1$	$-1 < t' < 0$	$t' = 0$	$1 < t' < 0$	$t' = 1$	$t' > 1$

Fig. 17.3 Time evolution of the Einstein–Rosen bridge.

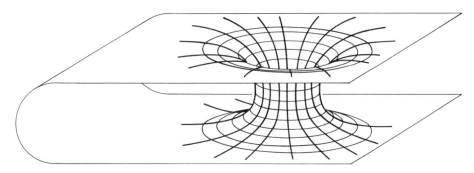

Fig. 17.4 A Schwarzschild wormhole.

17.4 Penrose diagram for Minkowski space-time

We shall introduce the idea of a Penrose diagram by first of all considering the procedure for Minkowski space-time. This will provide a prototype for other solutions. The essential idea is to start off with a metric g_{ab}, which we call the **physical** metric, and introduce another metric $\bar{g}_{ab}$, called the **unphysical** metric, which is conformally related to g_{ab}, that is,

$$\bar{g}_{ab} = \Omega^2 g_{ab}, \tag{17.10}$$

where Ω is the conformal factor. Then, by a suitable choice of Ω^2, it may be possible to 'bring in' the points at infinity to a finite position and hence study the causal structure of infinity. As we found in Exercise 6.28, the null geodesics of conformally related metrics are the same. The null geodesics determine the light cones, which in turn define the causal structure. The essential idea for bringing in the points at infinity is to use coordinate transformations involving functions like $\tan^{-1} x$, which, for example, maps the infinite interval $(-\infty, \infty)$ onto the finite interval $(-\frac{1}{2}\pi, \frac{1}{2}\pi)$ (Fig. 17.5). We introduce double null coordinates defined by

$$v = t + r, \tag{17.11}$$

$$w = t - r, \tag{17.12}$$

in which case the line element of Minkowski space-time becomes (exercise)

$$ds^2 = dv\,dw - \tfrac{1}{4}(v - w)^2(d\theta^2 + \sin^2\theta\,d\phi^2). \tag{17.13}$$

From (17.11) and (17.12), it follows that $r = \frac{1}{2}(v - w)$, and so the coordinate range $(-\infty < t < \infty, 0 \leqslant r < \infty)$ becomes $(-\infty < v < \infty, -\infty < w < \infty)$, with the requirement

$$r \geqslant 0 \quad \Rightarrow \quad v - w \geqslant 0 \quad \Rightarrow \quad v \geqslant w. \tag{17.14}$$

The space-time diagram for Minkowski space-time is shown in Fig. 17.6. We next define new coordinates p and q by

$$p = \tan^{-1} v, \tag{17.15}$$

$$q = \tan^{-1} w, \tag{17.16}$$

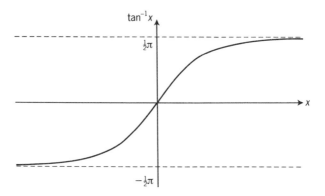

Fig. 17.5 The function $\tan^{-1} x$ maps $(-\infty, \infty)$ on to $(-\frac{1}{2}\pi, \frac{1}{2}\pi)$.

with the coordinate ranges $-\frac{1}{2}\pi < p < \frac{1}{2}\pi$ and $-\frac{1}{2}\pi < q < \frac{1}{2}\pi$, and where by (17.14)

$$p \geqslant q. \tag{17.17}$$

Then (17.13) becomes (exercise)

$$ds^2 = g_{ab}\,dx^a\,dx^b = \tfrac{1}{4}\sec^2 p\sec^2 q[4\,dp\,dq - \sin^2(p-q)(d\theta^2 + \sin^2\theta\,d\phi^2)], \tag{17.18}$$

and the line element of the unphysical metric is

$$d\bar{s}^2 = \bar{g}_{ab}\,d\bar{x}^a\,d\bar{x}^b = 4\,dp\,dq - \sin^2(p-q)(d\theta^2 + \sin^2\theta\,d\phi^2), \tag{17.19}$$

with the conformal factor

$$\Omega = \tfrac{1}{4}\sec^2 p\sec^2 q.$$

Finally, we introduce the coordinates

$$t' = p + q, \tag{17.20}$$

$$r' = p - q, \tag{17.21}$$

where the coordinate range is

$$-\pi < t' + r' < \pi, \tag{17.22}$$

$$-\pi < t' - r' < \pi, \tag{17.23}$$

$$r' \geqslant 0, \tag{17.24}$$

the last condition resulting from (17.17). The unphysical line element is now

$$d\bar{s}^2 = dt'^2 - dr'^2 - \sin^2 r'(d\theta^2 + \sin^2\theta\,d\phi^2) \tag{17.25}$$

subject to the coordinate range (17.22)–(17.24).

The line element (17.25) is that of the **Einstein static universe** which we introduced in §13.3 and which we shall meet in more detail in Part F. The topology of this solution is **cylindrical** with the time coordinate running along the generators of the cylinder. A cross-section of the cylinder, $t' = $ constant, has the topology of a 3-sphere S^3. Then the coordinate range of the manifold is

$$-\infty < t' < \infty, \quad 0 \leqslant r' \leqslant \pi, \quad 0 \leqslant \theta \leqslant \pi, \quad -\pi < \phi \leqslant \pi, \tag{17.26}$$

where $r' = 0, \pi$ and $\theta = 0, \pi$ are coordinate singularities. We shall discuss this further in Part F, but for the moment it is sufficient to think of an S^3 as a three-dimensional generalization of a 2-sphere S^2. In fact, the Einstein static universe can be embedded as the cylinder

$$x^2 + y^2 + z^2 + w^2 = 1$$

in a five-dimensional flat space of signature -3 with line element

$$ds^2 = dt^2 - dx^2 - dy^2 - dz^2 - dw^2.$$

(Suppressing two dimensions, this is the more familiar equation of a cylinder,

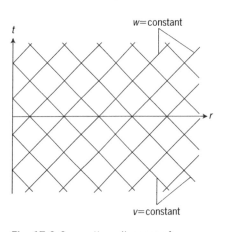

Fig. 17.6 Space-time diagram of Minkowski space-time.

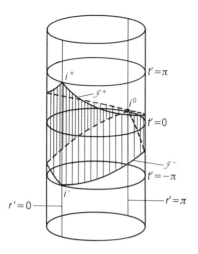

Fig. 17.7 Compactified Minkowski space-time (two dimensions suppressed).

namely, $x^2 + y^2 = 1$, in a three-dimensional space, but with a Minkowski-type geometry $ds^2 = -dt^2 + dx^2 + dy^2$.) The Einstein static universe then has line element (17.25) and coordinate range (17.26). We have shown that Minkowski space-time is conformal to that part of the Einstein static universe defined by the coordinate range (17.22)–(17.24). This is depicted in Fig. 17.7. The coordinate range (17.22)–(17.24) defines the diamond-shape region of the cylinder indicated. Thus, the whole of Minkowski space-time has been shrunk or compacted into this finite region. The process is called **conformal compactification** and the region is called **compactified Minkowski space-time**. The boundary of this region represents the conformal structure of infinity for Minkowski space-time. In terms of the coordinates p and q, it consists of the following:

a null surface $p = \frac{1}{2}\pi$ called $\mathscr{I}^+$,

a null surface $q = -\frac{1}{2}\pi$ called $\mathscr{I}^-$,

a point $(p = \frac{1}{2}\pi, q = \frac{1}{2}\pi)$ called i^+,

a point $(p = \frac{1}{2}\pi, q = -\frac{1}{2}\pi)$ called i^0,

a point $(p = -\frac{1}{2}\pi, q = -\frac{1}{2}\pi)$ called i^-,

where $\mathscr{I}$ is pronounced 'scri' — short for script i. Then it can be shown that all timelike geodesics originate at i^- and terminate at i^+. Similarly, null geodesics originate at points of $\mathscr{I}^-$ and end at points of $\mathscr{I}^+$, while spacelike geodesics both originate and end at i^0 (but these rules are not satisfied by non-geodesic curves). Thus, one may regard i^+ and i^- as representing **future** and **past timelike infinity**, $\mathscr{I}^+$ and $\mathscr{I}^-$ as representing **future** and **past null infinity**, and i^0 as representing **spacelike infinity**. This is illustrated in Fig. 17.8.

A **Penrose diagram** is a space-time diagram of a conformally compactified space-time. The Penrose diagram for Minkowski space-time is shown in Fig. 17.9. The diagram shows the curves $r = \text{constant}$ which correspond to the histories of 2-spheres $r = \text{constant}$, and the curves $t = \text{constant}$ which correspond to timelike slices. Ingoing and outgoing radial null geodesics are

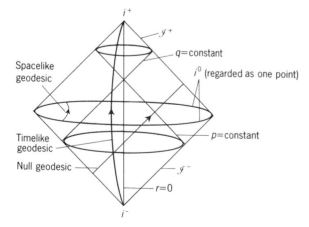

Fig. 17.8 Origin and termination of geodesics in compactified Minkowski space-time (one dimension suppressed).

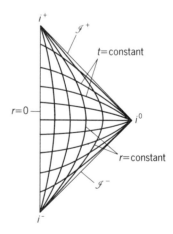

Fig. 17.9 Penrose diagram of Minkowski space-time (two dimensions suppressed).

represented by the straight lines $p = $ constant and $q = $ constant making angles $-45°$ and $45°$, respectively. A large class of asymptotically flat space-times, which Penrose calls **simple** space-times, can be analysed in a similar manner.

17.5 Penrose diagram for the Kruskal solution

The conformal compactification of the Kruskal solution may be obtained by defining new advanced and retarded null coordinates in terms of the null coordinates v' and w' of §17.2

$$v'' = \tan^{-1}[v'/(2m)^{\frac{1}{2}}], \qquad w'' = \tan^{-1}[w'/(2m)^{\frac{1}{2}}]$$

for the coordinate range

$$-\tfrac{1}{2}\pi < v'' < \tfrac{1}{2}\pi,$$

$$-\tfrac{1}{2}\pi < w'' < \tfrac{1}{2}\pi,$$

$$-\pi < v'' + w'' < \pi.$$

We omit the calculational details and simply present the Penrose diagram in Fig. 17.10. Again null geodesics and light cones have angles $\pm 45°$ in the figure. Both regions I and I′ have their own future, past, and null infinities. For any point outside $r = 2m$, an outward radial null geodesic ends up at $\mathscr{I}^+$ but an inward radial null geodesic ends up at the future singularity. For any point lying inside $r = 2m$, both outward and inward radial null geodesics end up on the future singularity.

We now take into account the fact that each point in the diagram represents a 2-sphere. Consider a 2-sphere S_0 situated in region I which is illuminated at some time. Then the photons at each point of S_0 move out in a 2-sphere and the envelope of these 2-spheres is again two 2-spheres S_1 and S_2 as shown in Fig. 17.11. The area of S_2 will be greater than S_0, which in turn will have a greater area than S_1. However, if S_0 lies in region II, both wave fronts are imploding and the areas of S_1 and S_2 will both be less than S_0. Such

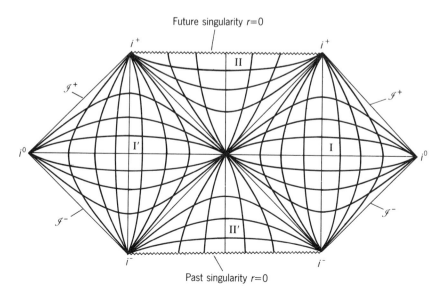

Fig. 17.10 Penrose diagram of the Kruskal solution.

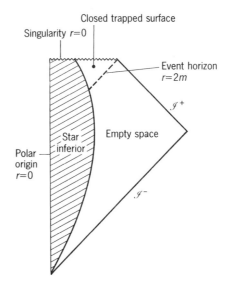

Fig. 17.12 Penrose diagram of spherically symmetric gravitational collapse.

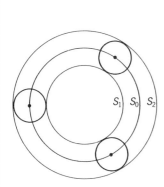

Fig. 17.11 Spherical wave fronts of an illuminated 2-sphere S_0.

a 2-surface is called a **closed trapped surface.** Similarly, each point in region II′ represents a time-reversed closed trapped surface. It turns out that it is precisely the existence of closed trapped surfaces which lead in the singularity theorems to the existence of singularities.

In Fig. 17.12, we show the Penrose diagram for a collapsing spherical star (compare with Figs. 16.13 and 16.14).

Exercises

17.1 (§17.2) Show that Schwarzschild space-time can be written in the form (17.1) subject to (17.2) in double null coordinates. [Hint: use (16.23) and (16.25)].

17.2 (§17.2) Show that (17.4) and (17.5) lead to the form (17.7) subject to (17.6).

17.3 (§17.2) Show that radial null geodesics make angles of $\pm 45°$ with the x'-axis in the Kruskal space-time diagram.

17.4 (§17.2) Where can observers from universes I and I′ meet in the Kruskal solution? What is their ultimate fate?

17.5 (§17.4) Show that Minkowski space-time takes the form (17.13) in double null coordinates.

17.6 (§17.4) Show that Minkowski space-time takes the form (17.18) under the coordinate transformations (17.15) and (17.16).

17.7 (§17.4) Draw a diagram of the region in the (t', r')-plane described by the inequalities (17.22) and (17.23). What subregion satisfies (17.24) as well?

17.8 (§17.4) Write down the transformation from the usual Minkowski coordinates (t, r) to (t', r') given in (17.20) and (17.21). Find the equations for the curves $t =$ constant and $r =$ constant in terms of t' and r' and draw them in the Penrose diagram of Minkowski space-time.

17.9 (§17.5) Draw the analogue of Fig. 17.11 for a closed trapped surface. Draw the corresponding figure for a 2-sphere in region II′.

17.10 (§17.5) Consider the transition from Fig. 17.10 to Fig. 17.12. What has happened to regions I′ and II′?

Charged black holes

18

18.1 The field of a charged mass point

In this chapter, we shall obtain and investigate the Reissner–Nordstrøm solution for a **charged** mass point. The importance of this solution is that its structure is in many ways similar to that of the more complicated Kerr solution describing rotating black holes which we shall meet in the next chapter. The approach we adopt is to look for a static, asymptotically flat, spherically symmetric solution of the Einstein–Maxwell field equations. The Einstein–Maxwell equations are

$$G_{ab} = 8\pi T_{ab}, \tag{18.1}$$

where T_{ab} is the Maxwell energy–momentum tensor, which in source-free regions is given by (12.49). In Exercise 13.3, we saw that this tensor is trace-free, which, by (18.1), implies that the Ricci scalar vanishes (exercise). We can therefore also work with the equivalent equations to (18.1), namely,

$$R_{ab} = 8\pi T_{ab}. \tag{18.2}$$

In addition, the Maxwell tensor F_{ab} must satisfy Maxwell's equations in source-free regions

$$\nabla_b F^{ab} = 0, \tag{18.3}$$

$$\partial_{[a} F_{bc]} = 0. \tag{18.4}$$

The assumption of spherical symmetry means that we can introduce coordinates (t, r, θ, ϕ) in which the line element reduces to the canonical form (14.33), namely,

$$ds^2 = e^{\nu} dt^2 - e^{\lambda} dr^2 - r^2(d\theta^2 + \sin^2\theta \, d\phi^2), \tag{18.5}$$

where ν and λ are functions of t and r. If we next impose the condition that the solution is static, then this requires that ν and λ are functions of r only, namely,

$$\nu = \nu(r), \qquad \lambda = \lambda(r). \tag{18.6}$$

The assumption that the field is due to a charged particle, which we take to be situated at the origin of coordinates, means that the line element and the Maxwell tensor will become singular there. Moreover, the charged particle will give rise to an electrostatic field which is purely radial (Fig. 18.1). This

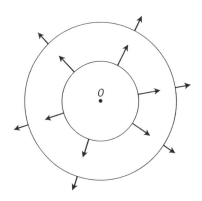

Fig. 18.1 Radial electrostatic field of charged point particle.

means that the Maxwell tensor must take on the form (exercise)

$$F_{ab} = E(r) \begin{bmatrix} 0 & -1 & 0 & 0 \\ 1 & 0 & 0 & 0 \\ 0 & 0 & 0 & 0 \\ 0 & 0 & 0 & 0 \end{bmatrix}. \tag{18.7}$$

Plugging the assumptions (18.5)–(18.7) in (18.3) and (18.4), we find (exercise) that (18.4) is satisfied automatically and (18.3) reduces to one equation, namely,

$$(e^{-\frac{1}{2}(v+\lambda)} r^2 E)' = 0, \tag{18.8}$$

where the prime indicates differentiation with respect to r. This integrates to give

$$E = e^{\frac{1}{2}(v+\lambda)} \varepsilon / r^2, \tag{18.9}$$

where ε is a constant of integration. Our assumption that the solution is asymptotically flat requires

$$v, \lambda \to 0 \quad \text{as} \quad r \to \infty, \tag{18.10}$$

and so $E \sim \varepsilon / r^2$ asymptotically. This latter result is exactly the same as the classical result for the electric field of a point particle of charge ε situated at the origin. We therefore interpret ε as the **charge** of the particle.

We now use (18.5) to (18.9) together with (12.49) to compute the Maxwell energy momentum tensor T_{ab}. Plugging this into the field equations (18.2), we find that the 00 and 11 equations lead to

$$\lambda' + v' = 0, \tag{18.11}$$

which by (18.10) results in $\lambda = -v$. The 22 equation is the one remaining independent equation and it leads to

$$(re^v)' = 1 - \varepsilon^2 / r^2, \tag{18.12}$$

which integrates immediately to give

$$e^v = 1 - 2m/r + \varepsilon^2 / r^2, \tag{18.13}$$

where m is a constant of integration. We have finally obtained the **Reissner–Nordstrøm solution**

$$ds^2 = \left(1 - \frac{2m}{r} + \frac{\varepsilon^2}{r^2}\right) dt^2 - \left(1 - \frac{2m}{r} + \frac{\varepsilon^2}{r^2}\right)^{-1} dr^2 - r^2(d\theta^2 + \sin^2\theta \, d\phi^2). \tag{18.14}$$

When $\varepsilon = 0$, this reduces to the Schwarzschild line element (14.47), and so we again identify m as the **geometric mass**. In deriving this solution, we have, in addition to assuming spherical symmetry, also assumed the solution is static and asymptotically flat. In fact, as in the case of the Schwarzschild solution, it is **not** necessary to adopt these last two assumptions: they are forced on you. The full calculation is similar to the Schwarzschild case but rather longer, which is why we have omitted it. There is therefore an analogue to Birkhoff's theorem.

Theorem: A spherically symmetric solution of the Einstein–Maxwell field equations is necessarily static.

18.2 Intrinsic and coordinate singularities

Consider the coefficients

$$g_{00} = -(g_{11})^{-1} = 1 - 2m/r + \varepsilon^2/r^2 = Q/r^2,$$

where

$$Q = r^2 - 2mr + \varepsilon^2. \tag{18.15}$$

The discriminant of the quadratic Q is

$$\Delta = m^2 - \varepsilon^2,$$

and, if this is negative, i.e. $\varepsilon^2 > m^2$, the quadratic has no real roots and is positive for all values of r. Hence, it follows that the line element (18.14) is non-singular for all values of r except at the origin $r = 0$. The solution possesses an intrinsic singularity at $r = 0$ — as can be shown by calculating the Riemann invariant $R^{abcd}R_{abcd}$ — which is not surprising since this is where the point charge producing the field is located. The more interesting case occurs when $\varepsilon^2 \leqslant m^2$, for then the metric has singularities when Q vanishes, namely, at $r = r_+$ and $r = r_-$, where

$$r_\pm = m \pm (m^2 - \varepsilon^2)^{\frac{1}{2}} \tag{18.16}$$

In Fig. 18.2, we plot g_{00} in the case $\varepsilon^2 < m^2$ and compare it with the Schwarzschild coefficient $_sg_{00} = 1 - 2m/r$.

The line element (18.14) is regular in the regions:

    I. $r_+ < r < \infty$,

   II. $r_- < r < r_+$,

  III. $0 < r < r_-$.

If $\varepsilon^2 = m^2$, then only the regions I and III exist. The regions are separated by the null hypersurfaces $r = r_+$ and $r = r_-$. The situation at $r = r_+$ is rather

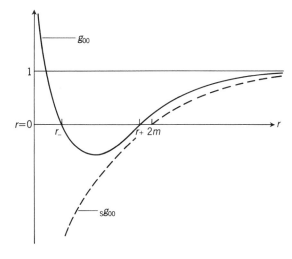

Fig. 18.2 Graphs of g_{00} for Reissner–Nordstrøm and Schwarzschild solutions.

similar to the Schwarzschild case at $r = 2m$. The coordinates t and r are timelike and spacelike, respectively, in the regions I and III, but interchange their character in region II. Thus, regions I and III are static, but region II is not. As in the case of the Schwarzschild solution, these coordinates suggest that the regions I, II, and III appear totally disconnected because the light cones have totally different orientations on either side of the null hyper-surfaces $r = r_\pm$. We will not pursue the structure of the solution in these coordinates further, but rather proceed as we did with the Schwarzschild solution and look for the analogue of the Eddington–Finkelstein coordinates.

18.3 Space-time diagram of Reissner–Nordstrøm solution

In the next two sections, we restrict our attention to the important case $\varepsilon^2 < m^2$. We first find the equation for the congruence of ingoing radial null geodesics (exercise). Then, defining for $r > r_+$ the new time coordinate

$$\bar{t} = t + \frac{r_+^2}{r_+ - r_-} \ln(r - r_+) - \frac{r_-^2}{r_+ - r_-} \ln(r - r_-), \qquad (18.17)$$

the line element takes on the form (exercise)

$$ds^2 = (1 - f)d\bar{t}^2 - 2f\,d\bar{t}\,dr - (1 + f)dr^2 - r^2(d\theta^2 + \sin^2\theta\,d\phi^2), \qquad (18.18)$$

where, for convenience, we define

$$f = 1 - g_{00} = 2m/r - \varepsilon^2/r^2. \qquad (18.19)$$

This form is regular for all positive values of r and again has an intrinsic singularity at $r = 0$. The conditions for radial null geodesics are

$$\dot{\theta} = \dot{\phi} = ds^2 = 0. \qquad (18.20)$$

These lead to (exercise) the ingoing family of null geodesics

$$\bar{t} + r = \text{constant} \qquad (18.21)$$

and the outgoing family whose differential equation is

$$\frac{d\bar{t}}{dr} = \frac{1 + f}{1 - f}. \qquad (18.22)$$

We do not, in fact, need to solve this equation exactly since our aim is to draw a space-time diagram, in which case it is sufficient to use the equation to obtain qualitative information about the slope for different values of r. The graphs of $1 + f$ and $1 - f$ are shown in Fig. 18.3. At infinity, f vanishes and so the slope is $45°$, as we would expect for an asymptotically flat solution. As we come in from infinity, $1 + f$ increases and $1 - f$ decreases and so the slope increases until, at $r = r_+$, $1 - f$ vanishes and the slope becomes infinite. In region II, the slope increases from $-\infty$ at $r = r_+$ to some maximum negative value at $r = \varepsilon^2/m$, and then decreases again to $-\infty$ as r approaches r_-. In region III, the slope decreases from $+\infty$ to 1, where the graphs cross, and continues decreasing to zero, where the graph $1 + f$ crosses

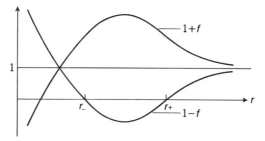

Fig. 18.3 Graphs of the functions $1 + f$ and $1 - f$.

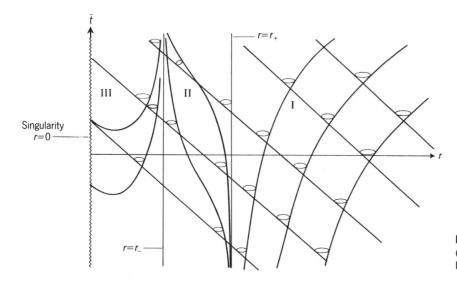

Fig. 18.4 Reissner–Nordstrøm solution ($\varepsilon^2 < m^2$) in advanced Eddington–Finkelstein type coordinates.

the r-axis. The slope then decreases through negative values until it reaches -1 at the origin. With this information, we can draw the space-time diagram in Fig. 18.4. It is clear from the light cones at $r = r_+$ that no light signal can escape from region II to region I. Thus, the surface $r = r_+$ is an event horizon. In region II, the light cones are inclined towards the singularity $r = 0$, and hence any particle entering region II will move necessarily towards the centre until it either crosses $r = r_-$ or reaches it asymptotically. In the region III, the light cones are no longer inclined towards the centre and consequently particles need not fall into the singularity. In fact, the opposite occurs in that neutral particles **cannot** reach the singularity, as we shall next show.

18.4 Neutral particles in Reissner–Nordstrøm space-time

To consider the motion of a neutral test particle, we shall investigate a radial timelike geodesic, the conditions for which are

$$\dot{\theta} = \dot{\phi} = \dot{s}^2 - 1 = 0, \tag{18.23}$$

where dot denotes differentiation with respect to the proper time τ. Defining the covariant 4-velocity $u_a = g_{ab}\,\mathrm{d}x^b/\mathrm{d}\tau$, we find that the geodesic equations lead to a first integral of the motion (exercise)

$$u_0 = \text{constant} \tag{18.24}$$

and a remaining equation which can be written in the form

$$\dot{r}^2 = A, \tag{18.25}$$

where

$$A = u_0^2 - g_{00}. \tag{18.26}$$

We can investigate what qualitative forms of motion are possible by plotting the curve of g_{00} against r and drawing in lines parallel to the r-axis a distance u_0^2 from it, for various values of u_0^2 (Fig. 18.5). Consider first the case when $u_0^2 < 1$. Then the line intersects the graph of g_{00} at two points P and Q in regions I and III, respectively. At these points, A vanishes by (18.26), and (18.25) then shows that $\dot{r} = 0$. Moreover, from (18.25), the left-hand side of the equation is positive, from which it follows that A must be positive. Therefore, by (18.26), motion is only possible when $u_0^2 \geqslant g_{00}$. It follows that the motion is bounded between the two values $r = r_P$ and $r = r_Q$. Similar arguments show that if $u_0^2 > 1$ then unbounded motion is possible, but there is a minimum distance of approach $r = r_R$ in region III. Thus, the point charge at the origin produces a potential barrier, which means that a neutral free particle can only approach within a certain distance before being repelled.

According to Fig. 18.4, once a particle is in region III, it cannot cross $r = r_-$ but can only reach it asymptotically. However, it can be shown that if a particle reaches $r = r_-$ then it does so in **finite** proper time. The diagram is misleading in exactly the same way as the Schwarzschild diagram in Schwarzschild coordinates (Fig. 16.7) is misleading in describing what happens to a radially infalling free particle in region I. Thus, the manifold described by the line element (18.18) is not maximal and needs extending in analogy with the Kruskal case.

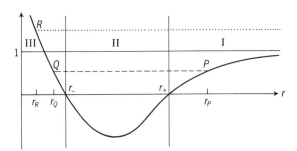

Fig. 18.5 Bounded and unbounded motion of a neutral particle.

18.5 Penrose diagrams of the maximal analytic extensions

If we introduce double null coordinates

$$v = \bar{t} + r, \qquad w = 2t - v, \tag{18.27}$$

then the line element (18.18) takes on the form (exercise)

$$ds^2 = \left(1 - \frac{2m}{r} + \frac{\varepsilon^2}{r^2}\right) dv\, dw - r^2(d\theta^2 + \sin^2\theta\, d\phi^2). \tag{18.28}$$

In the case $\varepsilon^2 < m^2$, we define new coordinates

$$v'' = \tan^{-1}\left(\exp\frac{r_+ - r_-}{4r_+^2} v\right), \qquad w'' = \tan^{-1}\left(-\exp\frac{r_- - r_+}{4r_+^2} w\right),$$
(18.29)

which transforms the line element into the form

$$ds^2 = -64\left(1 - \frac{2m}{r} + \frac{\varepsilon^2}{r^2}\right)\frac{r_+^4}{(r_+ - r_-)^2}\,\mathrm{cosec}\,2v''\,\mathrm{cosec}\,2w''\,dv''\,dw''$$
$$- r^2(d\theta^2 + \sin^2\theta\,d\phi^2),$$
(18.30)

where r is defined implicitly by

$$\tan v''\tan w'' = -\exp\left(\frac{r_+ - r_-}{2r_+^2}r\right)(r - r_+)^{\frac{1}{2}}(r - r_-)^{-r_-^2/2r_+^2}$$
(18.31)

This line element is the analogue of the Kruskal solution and represents the maximal analytic extension of the Reissner–Nordstrøm solution for $\varepsilon^2 < m^2$. The Penrose diagram for this maximal extension is shown in Fig. 18.6.

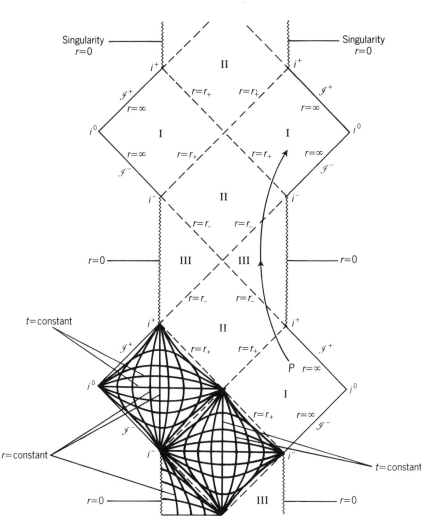

Fig. 18.6 Penrose diagram for maximal analytic extension ($\varepsilon^2 < m^2$).

This time, the maximal extension gives rise to an infinity of 'new universes'. There are an infinite number of asymptotically flat regions I where $r > r_+$. These are connected by intermediate regions II and III where $r_- < r < r_+$ and $0 < r < r_-$, respectively. Region III possesses an intrinsic singularity at $r = 0$ but, unlike the Kruskal solution, it is **timelike** and so can be avoided by a future-directed timelike curve from a region I which crosses $r = r_+$. A timelike curve is drawn in Fig. 18.6 which starts in a particular region I, passes through regions II, III, and II and re-emerges into another asymptotically flat region I. This gives rise to the highly speculative possibility that it may be possible to travel to other universes by passing through the 'wormholes' produced by charges. Unfortunately, it would seem as though it would not be possible to return. However, there is the possibility of identifying regions I (giving rise to a more complicated topology), so that a particle could then re-emerge from the black hole through the horizon $r = r_+$. Whether or not the particle emerges into the same part or a different part of the universe will depend on how the identification is made. A particle crossing the event horizon $r = r_+$ would appear to suffer an infinite red shift to an observer who remains in region I. In region II, each point represents a closed trapped surface. The extended solution possesses a very bizarre property in that any observer crossing the surface $r = r_-$ would see the whole of the remaining history of the asymptotically flat region I in a finite time! The line element (18.30) has a coordinate singularity at $r = r_-$. It is therefore necessary to introduce new null coordinates (in fact, an infinity of such coordinates) in order to 'patch' the manifold together. We shall not pursue this further.

The case $\varepsilon^2 = m^2$ can be extended similarly and the Penrose diagram is shown in Fig. 18.7. The remaining case $\varepsilon^2 > m^2$ is already inextendible in the original coordinates and the Penrose diagram is shown in Fig. 18.8.

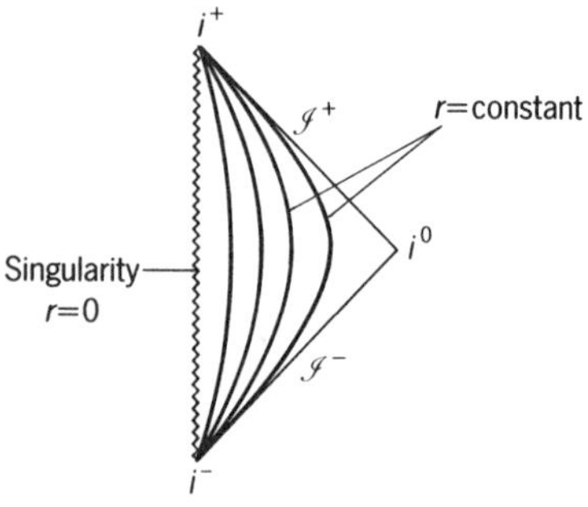

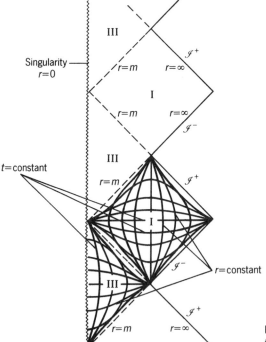

Fig. 18.8 Penrose diagram for the case $\varepsilon^2 > m^2$.

Fig. 18.7 Penrose diagram for the case $\varepsilon^2 = m^2$.

Exercises

18.1 (§**18.1**) Show that Einstein–Maxwell equations can be written in the equivalent form (18.2) in source-free regions.

18.2 (§**18.1**) Given the definition (12.20) of the Maxwell tensor in Minkowski coordinates (t, x, y, z), find its components in spherical polar coordinates (t, r, θ, ϕ). Hence confirm the ansatz (18.7).

18.3 (§**18.1**) Show that the assumptions (18.5)–(18.7) lead to the result that (18.4) is satisfied automatically and (18.3) reduces to (18.8).

18.4 (§**18.1**) Use (18.5)–(18.8) and (12.49) to compute the energy–momentum tensor T_{ab}. Show that the two independent Einstein–Maxwell field equations are (18.11) and (18.12). Hence obtain the Reissner–Nordstrøm solution.

18.5 (§**18.2**) Establish Fig. 18.2.

18.6 (§**18.2**) Establish the character of the coordinates t and r in (18.14) for $\varepsilon^2 < m^2$ in the regions I, II, and III. Find the surfaces of infinite red shift.

18.7 (§**18.2**) Draw a retarded Eddington–Finkelstein space-time diagram for the Reissner–Nordstrøm solution.

18.8 (§**18.3**) Find the equation for the congruence of in-going radial null geodesics for the line element (18.14) in the case $\varepsilon^2 < m^2$.

18.9 (§**18.3**) Draw a space-time diagram for Reissner–Nordstrøm solution in the coordinates of (18.14) for $\varepsilon^2 < m^2$. What happens to the diagram when $\varepsilon^2 = m^2$?

18.10 (§**18.3**) Show that (18.17) transforms (18.14) into the form (18.18). Show that the two families of radial null geodesics are given by (18.21) and (18.22).

18.11 (§**18.3**) Show that the transformations

$$\bar{t} = t + \frac{r_+^2}{r_+ - r_-} \ln(r_+ - r) - \frac{r_-^2}{r_+ - r_-} \ln(r - r_-)$$

$$\text{for } r_- < r < r_+$$

$$\bar{t} = t + \frac{r_+^2}{r_+ - r_-} \ln(r_+ - r) - \frac{r_-^2}{r_+ - r_-} \ln(r_- - r)$$

$$\text{for } 0 < r < r_-$$

transform (18.14) for $\varepsilon^2 < m^2$ into the form (18.18).

18.12 (§**18.3**) Show that the transformations

$$\bar{t} = t + m \ln(r - m)^2 - \frac{m^2}{r - m} \quad \text{if } \varepsilon^2 = m^2$$

$$\bar{t} = t + m \ln(r^2 - 2mr + \varepsilon^2) + \frac{2m^2 - \varepsilon^2}{(\varepsilon^2 - m^2)^{\frac{1}{2}}} \tan^{-1} \frac{r - m}{(\varepsilon^2 - m^2)^{\frac{1}{2}}}$$

$$\text{if } \varepsilon^2 > m^2$$

transform (18.14) into the form (18.18).

18.13 (§**18.3**) Find the advanced Eddington–Finkelstein form of the Reissner–Nordstrøm solution.

18.14 (§**18.3**) Consider the graphs of $1 + f$ and $1 - f$ in Fig. 18.3. Where is the slope $d\bar{t}/dr$ a maximum in region II? Where is the slope zero in the region III? What is the slope at the origin?

18.15 (§**18.4**) Show that the equation for a radial timelike geodesic for the solution (18.18) in the case $\varepsilon^2 < m^2$ leads to (18.24) and (18.25).

18.16 (§**18.5**) Show that the transformation (18.27) transforms the line element (18.18) into the form (18.28). Show that the transformation (18.29) transforms (18.28) into the form (18.30) subject to (18.31).

18.17 (§**18.5**) Consider the world-line of the observer in Fig. 18.6 emanating from the point P and the histories of all timelike geodesics in the region I containing P. Hence show that the observer will see all the remaining history of these geodesics as the horizon $r = r_-$ is crossed.

19 Rotating black holes

19.1 Null tetrads

In this chapter, we shall investigate the Kerr solution which describes rotating black holes. It turns out to be a rather long process to solve Einstein's vacuum equations directly for the Kerr solution. We shall, instead, describe a 'trick' of Newman and Janis for obtaining the Kerr solution from the Schwarzschild solution. This same trick can then be applied to the Reissner–Nordstrøm solution to obtain the Kerr–Newman solution, the most general solution for a charged rotating black hole. In order to discuss this approach, we start by introducing the very important idea of a null tetrad.

In §10.4, we met the idea of a tetrad $e_i{}^a$ of one timelike and three spacelike vectors. In fact, these tetrads, or frames, possess a formalism of their own called the frame formalism, which has proved extremely useful in many applications in general relativity. The most important case is when the tetrad vectors are taken to be **null** vectors. The systematic use of null tetrads is the basis of the 'Newman–Penrose' formalism, which has been used extensively in the study of gravitational radiation, among other topics. In this section, we shall restrict our attention to the definition of a null tetrad.

We start with four linearly independent vector fields $e_i{}^a$, where $\mathbf{i}$ serves to label the vectors. Then, working at a point, we define a matrix of scalars g_{ij}, called the **frame metric**, by

$$g_{ij} = g_{ab} e_i{}^a e_j{}^b. \tag{19.1}$$

Since $e_i{}^a$ are linearly independent and g_{ab} is non-singular, it follows that the matrix g_{ij} is non-singular and hence invertible. We therefore define its inverse g^{ij}, the contravariant frame metric, by the relation

$$g_{ij} g^{jk} = \delta_i^k. \tag{19.2}$$

We then use the frame metric to raise and lower frame indices in the same way that we use the metric tensor to raise and lower tensor indices. It is then easy to verify that the inverse relationship to (19.1) is

$$g_{ab} = g_{ij} e^i{}_a e^j{}_b. \tag{19.3}$$

In §10.4, we took the tetrad to consist of one timelike vector v^a and three spacelike vectors i^a, j^a, and k^a, say, in which case the orthonormality relations lead to

$$g_{ij} = \eta_{ij} = \mathrm{diag}(1, \ -1, \ -1, \ -1),$$

where the frame metric is the Minkowski metric η_{ij}. We now take

$$e_0{}^a = l^a = \frac{1}{\sqrt{2}}(v^a + i^a), \tag{19.4}$$

$$e_1{}^a = n^a = \frac{1}{\sqrt{2}}(v^a - i^a), \tag{19.5}$$

in which case l^a and n^a are **null** vectors (Fig. 19.1), that is,

$$l^a l_a = n^a n_a = 0 \tag{19.6}$$

and satisfy the normalization condition

$$l^a n_a = 1. \tag{19.7}$$

Then, if we take $e_2{}^a = j^a$ and $e_3{}^a = k^a$, the orthonormality relations (19.1) lead to the frame metric

$$g_{ij} = \begin{bmatrix} 0 & 1 & 0 & 0 \\ 1 & 0 & 0 & 0 \\ 0 & 0 & -1 & 0 \\ 0 & 0 & 0 & -1 \end{bmatrix}. \tag{19.8}$$

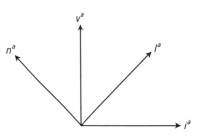

Fig. 19.1 The null vectors n^a and l^a.

Finally, it is advantageous to introduce a **complex** null vector defined by

$$m^a = \frac{1}{\sqrt{2}}(j^a + ik^a), \tag{19.9}$$

together with its complex conjugate

$$\bar{m}^a = \frac{1}{\sqrt{2}}(j^a - ik^a). \tag{19.10}$$

It is then easy to verify (exercise) that the vectors are null,

$$m^a m_a = \bar{m}^a \bar{m}_a = 0, \tag{19.11}$$

and satisfy the normalizing condition

$$m^a \bar{m}_a = -1. \tag{19.12}$$

If we choose

$$(e_0{}^a, e_1{}^a, e_2{}^a, e_3{}^a) = (l^a, n^a, m^a, \bar{m}^a), \tag{19.13}$$

then this defines a **null tetrad** with frame metric

$$g_{ij} = \begin{bmatrix} 0 & 1 & 0 & 0 \\ 1 & 0 & 0 & 0 \\ 0 & 0 & 0 & -1 \\ 0 & 0 & -1 & 0 \end{bmatrix}. \tag{19.14}$$

Thus, writing out (19.3) explicitly, we have decomposed g_{ab} into products of the null tetrad vectors according to

$$g_{ab} = l_a n_b + l_b n_a - m_a \bar{m}_b - m_b \bar{m}_a. \tag{19.15}$$

The contravariant form of this equation is

$$g^{ab} = l^a n^b + l^b n^a - m^a \bar{m}^b - m^b \bar{m}^a. \qquad (19.16)$$

19.2 The Kerr solution from a complex transformation

The Schwarzschild solution in advanced Eddington–Finkelstein coordinates is given by (16.24). The non-zero components of the contravariant metric g^{ab} are found to be (exercise)

$$g^{01} = -1, \quad g^{11} = -\left(1 - \frac{2m}{r}\right), \quad g^{22} = -\frac{1}{r^2}, \quad g^{33} = -\frac{1}{r^2 \sin^2\theta}. \qquad (19.17)$$

It is straightforward to check, using (19.16), that the contravariant metric may be written in terms of the following null tetrad:

$$\left. \begin{aligned} l^a &= (0,\ 1,\ 0,\ 0) = \delta_1^a, \\ n^a &= \left(-1,\ -\tfrac{1}{2}(1 - 2m/r),\ 0,\ 0\right) = -\delta_0^a - \tfrac{1}{2}(1 - 2m/r)\,\delta_1^a, \\ m^a &= \frac{1}{\sqrt{2}\,r}\left(0,\ 0,\ 1,\ \frac{i}{\sin\theta}\right) = \frac{1}{\sqrt{2}\,r}\left(\delta_2^a + \frac{i}{\sin\theta}\,\delta_3^a\right). \end{aligned} \right\} \qquad (19.18)$$

The 'trick' starts by allowing the coordinate r to take on complex values and the tetrad is rewritten in the form

$$\left. \begin{aligned} l^a &= \delta_1^a, \\ n^a &= -\delta_0^a - \tfrac{1}{2}[1 - m(r^{-1} + \bar{r}^{-1})]\,\delta_1^a, \\ m^a &= \frac{1}{\sqrt{2}\,\bar{r}}\left(\delta_2^a + \frac{i}{\sin\theta}\,\delta_3^a\right), \end{aligned} \right\} \qquad (19.19)$$

where throughout this procedure we keep l^a and n^a real and m^a and $\bar{m}^a$ complex conjugate to each other. We next formally perform the complex coordinate transformations

$$v \to v' = v + ia\cos\theta, \quad r \to r' = r + ia\cos\theta, \quad \theta \to \theta', \quad \phi \to \phi' \qquad (19.20)$$

on the null tetrad. Then, if we require that v' and r' are real, we obtain the following tetrad (exercise):

$$\left. \begin{aligned} l'^a &= \delta_1^a, \\ n'^a &= -\delta_0^a - \frac{1}{2}\left(1 - \frac{2mr'}{r'^2 + a^2\cos^2\theta}\right)\delta_1^a. \\ m'^a &= \frac{1}{\sqrt{2}(r' + ia\cos\theta)}\left(-ia\sin\theta(\delta_0^a + \delta_1^a) + \delta_2^a + \frac{i}{\sin\theta}\,\delta_3^a\right). \end{aligned} \right\} \qquad (19.21)$$

This is the promised Kerr solution and the contravariant components can be read off using (19.16).

19.3 The three main forms of the Kerr solution

The procedure of the last section gives rise to the following line element:

$$ds^2 = \left(1 - \frac{2mr}{\rho^2}\right)dv^2 - 2\,dv\,dr + \frac{2mr}{\rho^2}(2a\sin^2\theta)\,dv\,d\bar\phi + 2a\sin^2\theta\,dr\,d\bar\phi$$
$$- \rho^2\,d\theta^2 - \left((r^2+a^2)\sin^2\theta + \frac{2mr}{\rho^2}(a^2\sin^4\theta)\right)d\bar\phi^2, \qquad (19.22)$$

where

$$\rho^2 = r^2 + a^2\cos^2\theta \qquad (19.23)$$

and, for later convenience, we have replaced ϕ by $\bar\phi$. This is obtained by complexifying the advanced Eddington–Finkelstein form of the Schwarzschild solution, and so we shall term (19.22) the **advanced Eddington–Finkelstein** form of Kerr's solution. To obtain the analogue of the Schwarzschild form, we carry out the coordinate transformation (to be explained later) from the old coordinates $(v, r, \theta, \bar\phi)$ to new coordinates (t, r, θ, ϕ). It turns out to be easier to work with the coordinate differentials rather than the coordinates themselves, in which case the transformation is given by

$$dv = d\bar t + dr = dt + \frac{2mr + \Delta}{\Delta}\,dr \qquad (19.24)$$

$$d\bar\phi = d\phi + \frac{a}{\Delta}\,dr \qquad (19.25)$$

where

$$\Delta = r^2 - 2mr + a^2 \qquad (19.26)$$

and r and θ remain unchanged. This leads to the form of Kerr's solution called the **Boyer–Lindquist** form, namely,

$$ds^2 = \frac{\Delta}{\rho^2}(dt - a\sin^2\theta\,d\phi)^2 - \frac{\sin^2\theta}{\rho^2}[(r^2+a^2)d\phi - a\,dt]^2$$
$$- \frac{\rho^2}{\Delta}\,dr^2 - \rho^2\,d\theta^2. \qquad (19.27)$$

In fact neither (19.22) nor (19.27) was the form in which Kerr originally discovered the solution. He used Cartesian-type coordinates $(\bar t, x, y, z)$ to obtain the **Kerr** form

$$ds^2 = d\bar t^2 - dx^2 - dy^2 - dz^2$$
$$- \frac{2mr^3}{r^4 + a^2z^2}\left(d\bar t + \frac{r}{a^2 + r^2}(x\,dx + y\,dy)\right.$$
$$\left. + \frac{a}{a^2 + r^2}(y\,dx - x\,dy) + \frac{z}{r}\,dz\right)^2, \qquad (19.28)$$

where

$$\left.\begin{aligned}
\bar{t} &= v - r, \\
x &= r\sin\theta\cos\phi + a\sin\theta\sin\phi, \\
y &= r\sin\theta\sin\phi - a\sin\theta\cos\phi, \\
z &= r\cos\theta.
\end{aligned}\right\} \tag{19.29}$$

This line element has the general form

$$ds^2 = \eta_{ab}dx^a dx^b - \lambda l_a l_b\, dx^a\, dx^b, \tag{19.30}$$

where the vector l^a is null with respect to the Minkowski metric η_{ab}, that is,

$$\eta_{ab}l^a l^b = 0. \tag{19.31}$$

In the particular case of the Kerr form (19.28), we have

$$\lambda = \frac{2mr^3}{r^4 + a^2 z^2} \tag{19.32}$$

and

$$l_a = \left(1, \frac{rx + ay}{a^2 + y^2}, \frac{ry - ax}{a^2 + y^2}, \frac{z}{r}\right). \tag{19.33}$$

In the special case of the Schwarzschild solution this reduces to

$$\lambda = 2m/r \tag{19.34}$$

and

$$l_a = (1, x/r, y/r, z/r). \tag{19.35}$$

Indeed, it was precisely by considering metrics of the form (19.30) subject to (19.31) that Kerr originally found the solution; see Adler et al. (1975) for the details. We shall now attempt to gain some physical insight into the Kerr solution, and in so doing we shall make use of all three forms, namely, the Eddington–Finkelstein, Boyer–Lindquist, and Kerr versions of the solution, which is why we have collected them together in this section.

19.4 Basic properties of the Kerr solution

The Boyer–Lindquist form is the most useful one for investigating the elementary properties of the Kerr solution. First of all, it is clear that the solution depends on the two parameters m and a. If we set $a = 0$, we regain the Schwarzschild solution in Schwarzschild coordinates and so m is identified as the **geometric mass**. The metric coefficients in (19.27) are independentof both t and ϕ, and hence the solution is both **stationary** and **axially symmetric**. In other words, both $\partial/\partial t$ and $\partial/\partial\phi$ are Killing vector fields. To say that a solution is axially symmetric means that there exists an invariantly defined axis (which in coordinate terms we take to be the z-axis or $\theta = 0$) such that the solution is invariant under rotation about this axis. Or, equivalently, the orbits of the Killing vector field $\partial/\partial\phi$, namely, the curves $t = $ constant, $r = $ constant, $\theta = $ constant, are circles. These are the only continuous symmetries. As for discrete symmetries, the solution is not symmetric separately under time reflection or ϕ reflection (reflection in the (x, z)-plane), but it is invariant under the simultaneous inversion of t and ϕ, that is, under the transformation

$$t \to -t, \qquad \phi \to -\phi. \tag{19.36}$$

This suggests that the Kerr field may arise from a spinning source, since running time backwards with a negative spin direction is equivalent to running time forwards with a positive spin direction. Again, the line element is invariant under

$$t \to -t, \qquad a \to -a,$$

which suggests that a specifies a spin direction.

A third property which lends support to the spinning source interpretation is the presence in these canonical (t, ϕ)-coordinates of a cross term involving $d\phi \, dt$ (the only cross term present). Let us consider in Newtonian theory two frames $Oxyz$ and $Ox'y'z'$ whose origins and z-axes coincide, in which the primed frame is rotating relative to the unprimed frame with constant angular velocity $a\mathbf{k}$ (Fig. 19.2). Then a point P has cylindrical coordinates (r, ϕ, z) and (r', ϕ', z') relative to the two frames, where

$$r' = r, \qquad \phi' = \phi - at, \qquad z' = z. \tag{19.37}$$

If we take $Oxyz$ to be inertial, then this represents a transformation to a rotating frame. Now write flat space in cylindrical polar coordinates (t, r, ϕ, z), namely,

$$ds^2 = dt^2 - (dr^2 + r^2 \, d\phi^2 + dz^2), \tag{19.38}$$

and carry out the coordinate transformation (19.37) to a 'rotating frame' (leaving t unchanged). The line element (19.38) becomes (exercise)

$$ds^2 = (1 - a^2 r^2) \, dt^2 - 2ar^2 \, d\phi' \, dt - (dr^2 + r^2 \, d\phi'^2 + dz^2), \tag{19.39}$$

which, as we see, also possesses a cross term in $d\phi' \, dt$. This is somewhat imprecise since we have not discussed rigid rotation in special relativity (nor shall we). The argument presented is merely suggestive of rotation. Nor have we said anything precise yet about the coordinates r and θ in (19.27). Indeed, r is **not** the usual spherical polar radial coordinate except asymptotically (although we shall retain r to agree with standard notation). For, if we take (x, y, z) in (19.28) to be the usual Cartesian coordinates, then the standard spherical polar coordinate R is defined by

$$R^2 = x^2 + y^2 + z^2, \tag{19.40}$$

and hence, from (19.29),

$$R^2 = r^2 + a^2 \sin^2 \theta. \tag{19.41}$$

However, for $r \gg a$ (exercise),

$$R = r + \frac{a^2 \sin^2 \theta}{2r} + \cdots, \tag{19.42}$$

which shows that R and r coincide asymptotically. They also coincide in the Schwarzschild limit $a \to 0$. Further, it follows from the Kerr form (19.28) that

$$g_{ab} \to \eta_{ab} \quad \text{as} \quad R \to \infty,$$

so that the Kerr solution is **asymptotically flat**.

If we return to the idea that the Kerr solution represents a vacuum field exterior to a spinning source, then there are a number of independent arguments to suggest that a is related to the **angular velocity** and ma to the **angular momentum** (as measured at infinity). One argument involves comparing the Kerr solution with a solution due to Lens and Thirring for the

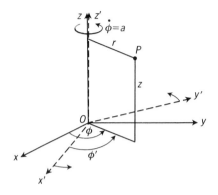

Fig. 19.2 Primed frame rotating about z-axis of unprimed frame.

gravitational field exterior to a spinning sphere of constant density in the weak-field limit. Another argument is based on the definition of the multipole moments of an isolated source. There are difficulties associated with this latter work because a number of different definitions have been proposed (indeed an infinitude of them); however, they all lead to the angular momentum being proportional to ma for the Kerr metric. We have already seen that in the weak-field limit the $1/R$ term in g_{00} determines the total mass of the field. It is also possible to show that, in certain circumstances, the $1/R$ terms in $g_{0\alpha}$ ($\alpha = 1, 2, 3$) determine the components of the angular momentum. Expanding the Kerr solution (19.28) in powers of $1/R$, we find

$$ds^2 = \left(1 - \frac{2m}{R} + \cdots\right)dt^2 - \frac{4ma}{R^3}(x\,dy - y\,dz)\,dt + \cdots \quad (19.43)$$

which again suggests that the total angular momentum is proportional to ma.

19.5 Singularities and horizons

Calculation of the Riemann invariant $R^{abcd}R_{abcd}$ reveals that the Kerr metric has only one intrinsic singularity and that is when $\rho = 0$. Since

$$\rho^2 = r^2 + a^2\cos^2\theta = 0,$$

it follows that $r = \cos\theta = 0$ and, from (19.29), (19.40), and (19.41), this occurs when

$$x^2 + y^2 = a^2, \qquad z = 0. \qquad (19.44)$$

This singularity is, rather surprisingly, a **ring** of radius a lying in the equatorial plane $z = 0$. We have only considered how to calculate the gravitational red shift in a static space-time, but it can be shown that the surfaces of infinite red shift in the Kerr solution are again given by the vanishing of the coefficient g_{00}. From (19.27), we find

$$g_{00} = (r^2 - 2mr + a^2\cos^2\theta)/\rho^2, \qquad (19.45)$$

and so the surfaces of infinite red shift are

$$r = r_{s_\pm} = m \pm (m^2 - a^2\cos^2\theta)^{\frac{1}{2}}. \qquad (19.46)$$

In the Schwarzschild limit, $a \to 0$ and the surface S_+ reduces to $r = 2m$ and S_- to $r = 0$. The surfaces are axially symmetric, with S_+ possessing a radius $2m$ at the equator and (assuming $a^2 < m^2$) a radius $m + (m^2 - a^2)^{\frac{1}{2}}$ at the poles, and the surface S_- being completely contained inside S_+. We shall primarily be concerned with the physically more interesting case $a^2 < m^2$, when the spin is small compared with the mass.

The existence of these infinite red-shift surfaces imply the existence of a null event horizon as follows. The Killing vector field

$$X^a = (1, 0, 0, 0)$$

has magnitude

$$X^2 = X_a X^a = g_{ab} X^a X^b = g_{00}.$$

It follows from (19.45) and (19.46) that X^a is timelike outside S_+ and inside S_-, null on S_+ and S_-, and spacelike between S_+ and S_-. In analogy with the Schwarzschild solution, we search for the event horizon by looking for the hypersurfaces where $r = $ constant becomes null, that is, where g^{11} vanishes. From the Boyer–Lindquist form (19.27), we find (exercise)

$$g^{11} = -\frac{\Delta}{\rho^2} = -\frac{r^2 - 2mr + a^2}{r^2 + a^2 \cos^2 \theta},$$

and hence g^{11} vanishes when

$$\Delta = r^2 - 2mr + a^2 = 0,$$

which results in **two null event horizons** (assuming $a^2 < m^2$)

$$r = r_\pm = m \pm (m^2 - a^2)^{\frac{1}{2}}. \tag{19.47}$$

Then, in a similar way in which the Reissner–Nordstrøm solution is regular in three regions, the Kerr solution is regular in the three regions:

   I. $r_+ < r < \infty$,

  II. $r_- < r < r_+$,

 III. $0 < r < r_-$.

In the Schwarzschild limit, $a \to 0$, and the two event horizons reduce to $r = 2m$ and $r = 0$, from which it follows that in the Schwarzschild solution the surfaces of infinite red shift and the event horizons coincide. The event horizon $r = r_+$ lies entirely within S_+, giving rise to a region between the two called the **ergosphere**, the properties of which we shall discuss in §19.11. The various surfaces and the ring singularity are illustrated in Fig. 19.3.

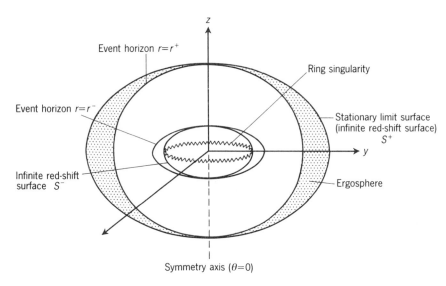

Fig. 19.3 The event horizons, stationary limit surface, and ring singularity of the Kerr solution.

We end this section by summarizing the properties we have met so far. The Kerr solution:

(1) is stationary;
(2) is axisymmetric;
(3) is invariant under the discrete transformations

$$t \rightarrow -t, \quad \phi \rightarrow -\phi \quad \text{and} \quad t \rightarrow -t, \quad a \rightarrow -a;$$

(4) has geometric mass m;
(5) represents the field exterior to a spinning source where the spin of the field is related to a and the angular momentum to ma;
(6) is asymptotically flat;
(7) has a ring singularity at

$$x^2 + y^2 = a, \quad z = 0;$$

(8) has two surfaces of infinite red shift

$$r = m \pm (m^2 - a^2 \cos^2 \theta)^{\frac{1}{2}};$$

(9) in the case $a^2 < m^2$, has two event horizons

$$r = m \pm (m^2 - a^2)^{\frac{1}{2}}.$$

19.6 The principal null congruences

The Kerr solution is no longer spherically symmetric and so we no longer expect that there are any curves corresponding to radial null geodesics. This is because, in a loose sense, we expect a rotating source to 'drag' space around with it and consequently drag the geodesics with it. The situation is very different from what happens in Newtonian theory, where, if one was investigating a source rotating about the z-axis, say, one could transfer to a frame rotating with the source and so reduce it to rest. However, one cannot do this in general relativity because it is **not** possible to find a coordinate system which reduces the Kerr solution to the Schwarzschild solution. Put another way, the nonlinear field equations couple the source to the exterior field.

Since the metric is axially symmetric, we might expect to obtain null geodesics which lie in the hypersurface $\theta = $ constant. We therefore search for null geodesics for which

$$\dot{\theta} = \mathrm{d}s^2 = 0, \tag{19.48}$$

where the dot denotes differentiation with respect to an affine parameter and where, throughout, θ is kept constant. We use the Boyer–Lindquist form (19.27), and then the fact that the metric coefficients are independent of t and ϕ means that the Euler–Lagrange equations immediately lead to first integrals of the motion. These are

$$\frac{\Delta}{\rho^2}(\dot{t} - a\sin^2\theta \, \dot{\phi}) + \frac{a\sin^2\theta}{\rho^2}[(r^2 + a^2)\dot{\phi} - a\dot{t}] = l, \tag{19.49}$$

$$\frac{a\Delta\sin^2\theta}{\rho^2}(\dot{t} - a\sin^2\theta \, \dot{\phi}) + \frac{(r^2 + a^2)\sin^2\theta}{\rho^2}[(r^2 + a^2)\dot{\phi} - a\dot{t}] = n, \tag{19.50}$$

where l and n are constants of integration. We have another first integral from the condition $ds^2 = 0$, namely,

$$\frac{\Delta}{\rho^2}(\dot{t} - a\sin^2\theta\,\dot{\phi})^2 - \frac{\sin^2\theta}{\rho^2}[(r^2 + a^2)\dot{\phi} - a\dot{t}]^2 - \frac{\rho^2\dot{r}^2}{\Delta} = 0. \quad (19.51)$$

Finally, we have the Euler–Lagrange equation corresponding to $x^2 = \theta$ and, using the fact that $\ddot{\theta} = 0$ from (19.48), this becomes

$$\frac{a^2\Delta}{\rho^4}(\dot{t} - a\sin^2\theta\,\dot{\phi})^2 - \frac{2a\Delta\dot{\phi}}{\rho^2}(\dot{t} - a\sin^2\theta\,\dot{\phi})$$

$$- \frac{r^2 + a^2}{\rho^4}[(r^2 + a^2)\dot{\phi} - a\dot{t}]^2 + \frac{a^2\dot{r}^2}{\Delta} = 0. \quad (19.52)$$

Since (19.49), (19.50), (19.51), and (19.52) represent four equations in the three unknowns $\dot{t}$, $\dot{r}$, and $\dot{\phi}$, it follows that there must exist some constraint between l and n. Some algebra reveals that the constraint is (exercise)

$$(n + al\sin^2\theta)(n - al\sin^2\theta) = 0, \quad (19.53)$$

where θ is constant. Restricting attention to the condition

$$n - al\sin^2\theta = 0, \quad (19.54)$$

the system of equations can be solved for $\dot{t}$, $\dot{r}$, and $\dot{\phi}$, to give

$$\dot{t} = (r^2 + a^2)l/\Delta, \quad (19.55)$$

$$\dot{r} = \pm l, \quad (19.56)$$

$$\dot{\phi} = al/\Delta. \quad (19.57)$$

We have therefore found two null congruences corresponding to the two signs in (19.56). Moreover, (19.56) shows that r is an affine parameter along each congruence. Choosing $\dot{r} = +l$, we get

$$\frac{dt}{dr} = \frac{\dot{t}}{\dot{r}} = \frac{r^2 + a^2}{\Delta} \quad (19.58)$$

and

$$\frac{d\phi}{dr} = \frac{\dot{\phi}}{\dot{r}} = \frac{a}{\Delta}. \quad (19.59)$$

If we restrict our attention to the case $a^2 < m^2$, these equations can be immediately integrated to give (exercise)

$$t = r + \left(m + \frac{m^2}{(m^2 - a^2)^{\frac{1}{2}}}\right)\ln|r - r_+|$$

$$+ \left(m - \frac{m^2}{(m^2 - a^2)^{\frac{1}{2}}}\right)\ln|r - r_-| + \text{constant} \quad (19.60)$$

and

$$\phi = \frac{a}{2(m^2 - a^2)^{\frac{1}{2}}}\ln\left|\frac{r - r_+}{r - r_-}\right| + \text{constant}. \quad (19.61)$$

Using the fact that $\Delta > 0$ in regions I and III and $\Delta < 0$ in region II, then it follows from (19.58) that $dr/dt > 0$ in region I, and so this congruence is called the **principal congruence of outgoing null geodesics**. The solution corresponding to $\dot{r} = -l$ is again given by (19.60) and (19.61) if we simply replace t by $-t$ and ϕ by $-\phi$, and so is called the **principal congruence of ingoing null geodesics**. The solutions reduce to the Schwarzschild congruences (16.10) and (16.11), respectively, in the limit $a \to 0$, as we should expect.

These two congruences play the same role as the null radial congruences do in the Schwarzschild solution. They give information about the radial variation of the light cone structure in that the most outgoing and most ingoing null lines — those for which $|dr/dt|$ is a maximum at any point — are members of the principal null congruences. We can draw a space-time diagram of the light cones using these equations and we find in region I a diagram analogous to Fig. 16.7 with the light cones narrowing down as $r \to r_+$. On $r = r_+$, both t and ϕ become infinite, suggesting, as in the Schwarzschild solution, that $r = r_+$ is a coordinate singularity. We therefore proceed as we did in the Schwarzschild solution and look for the analogue of the Eddington–Finkelstein coordinate system.

19.7 Eddington–Finkelstein coordinates

We use the principal null congruences to obtain a coordinate transformation which extends the solution through $r = r_+$. We could work explicitly with the equations of the congruence (19.60) and (19.61), but it turns out to be simpler to work with them in the differential form (19.58) and (19.59), that is,

$$dt = -\frac{r^2 + a^2}{\Delta}\, dr, \tag{19.62}$$

$$d\phi = -\frac{a}{\Delta}\, dr, \tag{19.63}$$

for the ingoing congruence. In the Schwarzschild case, we looked for a transformation to new coordinates $(\bar{t}, r, \theta, \phi)$ in which the equations for the ingoing radial null congruence take on the simpler differential form

$$d\bar{t} = -dr, \qquad d\theta = d\phi = 0. \tag{19.64}$$

Proceeding similarly in the Kerr case, we search for a transformation to new coordinates $(\bar{t}, r, \theta, \bar{\phi})$ in which the principal ingoing congruence reduces to

$$d\bar{t} = -dr, \qquad d\theta = d\bar{\phi} = 0. \tag{19.65}$$

Using (19.58) and (19.59), the requisite transformations are (exercise)

$$t \to \bar{t} \quad \text{where} \quad d\bar{t} = dt + \frac{2mr}{\Delta}\, dr, \tag{19.66}$$

$$\phi \to \bar{\phi} \quad \text{where} \quad d\bar{\phi} = d\phi + \frac{a}{\Delta}\, dr. \tag{19.67}$$

If we define an advanced time coordinate

$$v = \bar{t} + r, \tag{19.68}$$

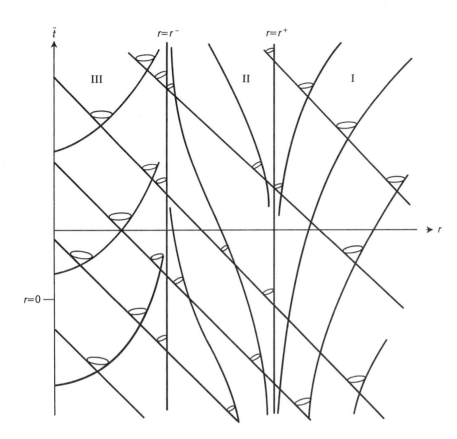

Fig. 19.4 Kerr solution ($a^2 < m^2$) in advanced Eddington–Finkelstein type coordinates.

then the Boyer–Lindquist line element is transformed into (19.22), the advanced Eddington–Finkelstein form of the Kerr solution. The two-dimensional space-time diagram for this solution is given in Fig. 19.4 (compare this with the Reissner–Nordstrøm space-time diagram, Fig. 18.4).

19.8 The stationary limit

Consider the set of null curves in the region I given by

$$\mathrm{d}r = \mathrm{d}\theta = \mathrm{d}s^2 = 0. \tag{19.69}$$

Then the Boyer–Lindquist line element reduces to

$$\frac{\Delta}{\rho^2}(\mathrm{d}t - a\sin^2\theta\,\mathrm{d}\phi)^2 - \frac{\sin^2\theta}{\rho^2}[(r^2 + a^2)\mathrm{d}\phi - a\,\mathrm{d}t]^2 = 0,$$

and solving for $\mathrm{d}\phi/\mathrm{d}t$ produces

$$\frac{\mathrm{d}\phi}{\mathrm{d}t} = \frac{a\sin\theta \pm \Delta^{\frac{1}{2}}}{(r^2 + a^2)\sin\theta \pm a\Delta^{\frac{1}{2}}\sin^2\theta}. \tag{19.70}$$

These curves are not geodesics, but are tangent to world-lines of photons initially constrained to orbit the source with fixed r and θ. The positive sign in (19.70) leads to $\mathrm{d}\phi/\mathrm{d}t > 0$, that is, the photon orbits the source in the same direction as the rotation of the source. We now investigate when it is possible for $\mathrm{d}\phi/\mathrm{d}t \leqslant 0$, in which case we must restrict attention to the negative sign in

(19.70). In region I,

$$r > r_+ \quad \Leftrightarrow \quad (r^2 + a^2)\sin^2\theta - a\Delta^{\frac{1}{2}}\sin^2\theta > 0,$$

so that the denominator of (19.70) is positive. Hence (exercise),

$$\frac{\mathrm{d}\phi}{\mathrm{d}t} \leqslant 0 \quad \Leftrightarrow \quad a\sin\theta - \Delta^{\frac{1}{2}} \leqslant 0 \quad \Leftrightarrow \quad r \geqslant r_{\mathrm{s}_+}. \tag{19.71}$$

Thus, on S_+, the derivative $\mathrm{d}\phi/\mathrm{d}t$ is zero, and hence any particle on this hypersurface attempting to orbit the source against its direction of rotation must travel with the local speed of light just to remain stationary (that is, to be precise, stationary relative to a stationary observer at infinity). In the ergosphere, the light cones tip over in the direction of ϕ increasing to such an extent that photons and particles are forced to orbit the source in the direction of its rotation. It is because of this that the infinite red-shift surface S_+ is also termed the **stationary limit surface**. The stationary limit surface is a timelike surface except at the two points on its axis, where it is null and where it coincides with the event horizon $r = r_+$. Where the surface is timelike, the light cone structure reveals that it can be crossed by particles in either the ingoing or outgoing direction. These properties are most clearly revealed in a spatial diagram of the Kerr solution ($a^2 < m^2$) in the equatorial plane (Fig. 19.5).

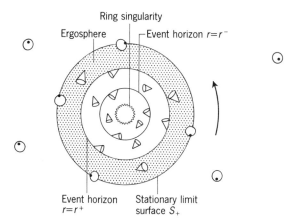

Fig. 19.5 Spatial diagram of Kerr solution ($a^2 < m^2$) in the equatorial plane.

19.9 Maximal extension for the case $a^2 < m^2$

The Kerr metric can be extended by using advanced and retarded Eddington–Finkelstein coordinates

$$\mathrm{d}u_\pm = \mathrm{d}t \pm \frac{r^2 + a^2}{\Delta}\,\mathrm{d}r, \qquad \mathrm{d}\phi_\pm = \mathrm{d}\phi \pm \frac{a}{\Delta}\,\mathrm{d}r$$

in a manner analogous to the Reissner–Nordstrøm case, where the maximal extension is built up by a combination of these extensions. The global structure is very similar to that of the Reissner–Nordstrøm solution except that now one can continue through the ring singularity to negative values of r. Figure 19.6 shows the conformal structure of the solution along the symmetry axis for the case $a^2 < m^2$. The regions I ($r_+ < r < \infty$) are stationary asymptotically flat regions exterior to the outer event horizon. The

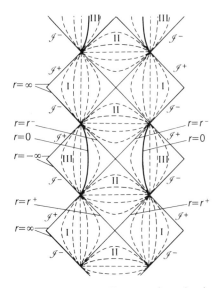

Fig. 19.6 Penrose diagram of maximal extension of Kerr solution ($a^2 < m^2$) along symmetry axis.

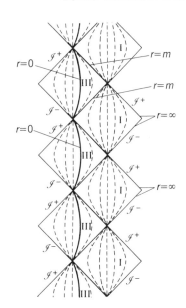

Fig. 19.7 Penrose diagram of maximal extension of Kerr solution ($a^2 = m^2$) along symmetry axis.

regions II $(r_- < r < r_+)$ are non-stationary and each point in one is a closed trapped surface. The regions III $(-\infty < r < r_-)$ contain the ring singularity which is **timelike** and hence avoidable. This region also contains **closed timelike curves**. Such curves violate causality and would seem highly unphysical since, if they represent world-lines of observers, then these observers would travel back and meet themselves in the past! There is no causality violation in the regions I and II. In the limiting case $a^2 = m^2$, the event horizons r_+ and r_- coincide and there are no regions II. The maximal extension is similar to that of the Reissner–Nordstrøm solution when $\varepsilon^2 = m^2$ and its conformal structure along the symmetry axis is shown in Fig. 19.7.

19.10 Maximal extension for the case $a^2 > m^2$

In the case $a^2 > m^2$, we find that $\varDelta > 0$ and the Boyer–Lindquist form of the Kerr solution (19.27) is regular everywhere except at $r = 0$, where there is a ring singularity. The coordinate r, by (19.29), can be determined in terms of x, y, z from

$$r^4 - (x^2 + y^2 + z^2 - a^2)r^2 - a^2z^2 = 0.$$

For $r \neq 0$, the surfaces $r = $ constant are confocal ellipsoids in a slice $t = $ constant which degenerate to the disc $x^2 + y^2 \leqslant a^2$, $z = 0$ when $r = 0$. The ring singularity is the boundary of this disc. The function r can be analytically continued from positive to negative values through the interior of the disc to obtain a maximal analytic extension of the solution. To do this, one attaches another surface with coordinates (x', y', z'), where a point on the top side of the disc is identified with a point with the same x- and y-coordinates on the bottom side of the corresponding disc in the (x', y', z')-surface, and similarly for points on the bottom of the disc (see Fig. 19.8). The

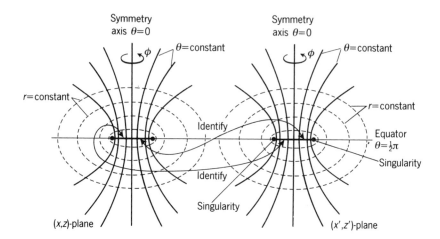

Fig. 19.8 Maximal extension of Kerr solution ($a^2 > m^2$).

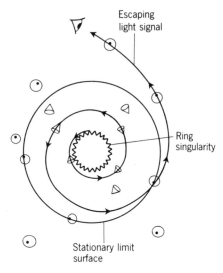

Fig. 19.9 The Kerr solution ($a^2 > m^2$) as a naked singularity.

line element (19.27) then extends to this larger manifold and has the same form on the (x', y', z')-region, but r is now negative. Then on circling twice round the ring singularity, for example, one passes from the (x, y, z)-region, where r is positive, to the (x', y', z')-region, where r is negative, and back to the (x, y, z)-region. At large negative values of r, the space is again asymptotically flat, but this time it has **negative** mass.

For a small value of r near the singularity, the vector $\partial/\partial\phi$ is timelike so the circles $t = $ constant, $r = $ constant, $\theta = $ constant are timelike curves. These closed timelike curves can be deformed to pass through any point of the extended space, so that the solution badly violates causality. The solution is geodesically incomplete at the ring singularity, but the only timelike and null geodesics which reach this singularity are those in the equatorial plane on the positive-r side. This leads to another bizarre property of the solution. The event horizons have now disappeared, but an intrinsic space-time singularity still exists at the ring and now it is possible for information to escape from the singularity to the outside world, provided it spirals around sufficiently (Fig. 19.9). In short, the singularity is visible, in all its nakedness, to the outside world. Such a singularity is called a **naked singularity**. If naked singularities exist, then they open up a whole new realm for wild speculation, so much so that Penrose has suggested the existence of the **cosmic censorship hypothesis**, which would forbid the existence of naked singularities but would only allow singularities to be hidden behind event horizons. Attempts to establish under what conditions, if any, the cosmic censorship hypothesis holds have been an area of active research in recent years and the source of considerable controversy.

19.11 Rotating black holes

We consider the ideal case of a rotating star whose exterior field is given by the Kerr solution for $0 < a^2 < m^2$. Intuitively we may think of the source as a rotating sphere or ellipsoid of matter, but as we have indicated before there is as yet, despite considerable efforts, no known physically realistic interior Kerr solution. (Perhaps the existence of a ring singularity suggests that we might be able to fill in the Kerr solution with a toroidal rather than a spherical source.) Nonetheless, we envisage this source collapsing through the event horizon $r = r_+$ to give rise to a black hole. As before, any observer following the collapse through $r = r_+$ will be unable to return to their original region I. The collapse will necessarily continue through $r = r_-$ and any observer in region II must follow the collapse through to region III. A difference arises in the rotating case, as compared with the non-rotating case, in that the collapse may now halt. The maximal extension then suggests that an observer in region III is able to escape into a new asymptotically flat region I. We shall return to the question of a more physically realistic collapse situation later.

Penrose has suggested that it might be possible to extract energy from a rotating black hole as follows. A particle is fired into the ergosphere, where it decays into two products, one falling into the black hole and the other escaping outside the stationary limit. Calculations reveal that the escaping component can contain **more** mass-energy than the original particle. This is possible because the angular momentum of the black hole is reduced in the process. This leads to a fanciful suggestion that an advanced civilization could live near a rotating black hole and develop some mechanism for extracting their energy requirements from the black hole's rotation (Fig. 19.10).

In order to obtain the most general black hole solution, we apply the Newman–Janis trick of §19.2 to the Reissner–Nordstrøm solution in advanced Eddington–Finkelstein coordinates (see Exercise 18.13), namely,

$$ds^2 = \left(1 - \frac{2m}{r} + \frac{\varepsilon^2}{r^2}\right) dv^2 - 2 dv\, dr - r^2(d\theta^2 + \sin^2\theta\, d\phi^2).$$

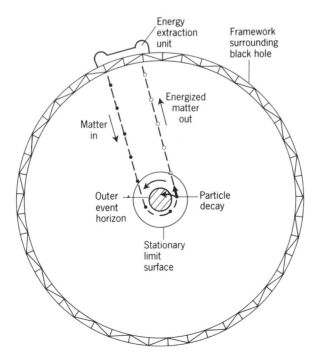

Fig. 19.10 Living off a rotating black hole.

We find the result (exercise)

$$ds^2 = \left(1 - \frac{2mr}{\rho^2} + \frac{\varepsilon^2}{\rho^2}\right) dv^2 - 2\,dv\,dr + \frac{2a}{\rho^2}(2mr - \varepsilon^2)\sin^2\theta\,dv\,d\bar{\phi}$$
$$+ 2a\sin^2\theta\,dr\,d\bar{\phi} - \rho^2\,d\theta^2$$
$$- [(r^2 + a^2)^2 - (r^2 - 2mr + a^2 + \varepsilon^2)a^2\sin^2\theta]\frac{\sin^2\theta}{\rho^2}\,d\bar{\phi}^2, \qquad (19.72)$$

which is the **Kerr–Newman** solution in advanced Eddington–Finkelstein coordinates. The solution clearly depends on the three parameters m, a, ε, defining the mass, spin, and charge, respectively. It is stationary and axisymmetric, and possesses a stationary limit surface

$$r = m + (m^2 - \varepsilon^2 - a^2\cos^2\theta)^{\frac{1}{2}} \qquad (19.73)$$

and, provided that $a^2 + \varepsilon^2 \leqslant m^2$, an outer event horizon

$$r = m + (m^2 - \varepsilon^2 - a^2)^{\frac{1}{2}}. \qquad (19.74)$$

It has properties analogous to the Kerr solution, but we shall not pursue the details further.

If we consider a realistic collapse of a charged rotating black hole, then the Kerr–Newman solution will not represent the true geometry exterior to the star at early times. This is because, if the star has not gone far down the road to collapse, it will not possess the symmetries of stationarity and axisymmetry. Gravitational moments will arise from mountains and other asymmetries. However, if an event horizon develops, then these asymmetries will be radiated away. In fact, a remarkable theorem has been proved which states that, if an event horizon develops in an asymptotically flat space-time, then the solution exterior to this horizon necessarily approaches a Kerr–Newman solution asymptotically in space-time. Thus, we have remarkably complete information as to the asymptotic state of affairs resulting from a gravitational collapse.

Detailed considerations of gravitational collapse suggest the following picture. A body, or collection of bodies, collapses down to a size comparable to its Schwarzschild radius, after which a trapped surface can be found in the region surrounding the matter. Some way outside the trapped surface there is another surface which will ultimately form the event horizon. But at present this surface is still expanding somewhat. Its exact location is a complicated affair and it depends on how much more matter or radiation falls in. We assume only a finite amount falls in. Then the expansion of the absolute event horizon gradually slows down to stationarity. Thus, when a black hole is created by gravitational collapse, it rapidly settles down to a stationary state that is characterized by the three parameters m, ε, and a. Apart from these three properties, the black hole preserves no other details of the object that collapsed. Wheeler has termed this the theorem that 'a black hole has no hair'. If you've seen one, you've seen them all! Wheeler depicts this rather humorously by a picture in which a vase of flowers and a television set fall into a black hole (Fig. 19.11). Once the system has settled down the only quantities which may have altered are m, ε, and a. All details of the objects swallowed up are obliterated. Considering the time reversal of this situation, we see that if you happen to be an astronaut travelling in space and you

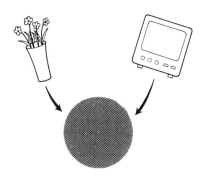

Fig. 19.11 A black hole has no hair.

suddenly see a vase of flowers and a television set pop out of nowhere then you know you are in the vicinity of a **white** hole — a rather 'hairy' prospect!

19.12 The singularity theorems

In this section, we consider briefly the question, Are singularities a necessary consequence of gravitational collapse in general relativity? As we have stated before, the singularity theorems of Hawking and Penrose state that the presence of a trapped surface always implies the presence of some form of space-time singularity. There are also versions of the theorems which apply in cosmological situations, that is, when the cosmological constant Λ is included (see (13.5)). As we shall see in Part F, some of the cosmological models involve singularities at the big bang and the big crunch. We shall refer to these cosmological singularities as Friedmann singularities. Again, gravitational waves seem to have singularities associated with them, as we shall see in Part E. The main significance of the theorems is that they show that the presence of space-time singularities in exact models is not just a feature of their high symmetry, but can be expected in generically perturbed models. This is not to say that all solutions are singular; in fact, many exact solutions are known which are complete, that is, maximal and singularity-free. But those which closely resemble the Kerr–Newman collapse models, or the Friedmann cosmological models containing a big bang or big crunch, or colliding plane gravitational waves, must be expected to be singular. The theorems do not, however, say that the singularities need look like those of Kerr–Newman, Friedmann, or colliding plane gravitational waves; in fact, there is some evidence that generic singularities may have a much more complicated structure, but little is known about this.

The main assumption that the theorems depend upon is the **dominant energy condition** (12.56), which can be written in the more general form

$$t^a t_a = 1 \quad \Leftrightarrow \quad T^{ab} t_a t_b \geqslant 0 \quad (T^{ab} t_b \text{ non-spacelike}), \tag{19.75}$$

so that the vector $w^a = T^{ab} t_b$ must be timelike or null. The significance of the energy condition lies in the effect discovered by Raychaudhuri (see (21.44)) which states that, whenever a system of timelike geodesics normal to a spacelike hypersurface starts converging, then this convergence inevitably increases along the geodesics until finally the geodesics focus (assuming the geodesics are complete). There is a corresponding focusing effect in the case of **null** geodesics. This depends on the **weak energy condition** which can be expressed in the form

$$v^a \text{ non-spacelike} \quad \Rightarrow \quad T^{ab} v_a v_b \geqslant 0. \tag{19.76}$$

If the energy–momentum tensor has energy density μ and principal stresses p_α ($\alpha = 1, 2, 3$), then the weak energy condition can be expressed equivalently as

$$\mu \geqslant 0, \quad \mu + p_\alpha \geqslant 0. \tag{19.77}$$

There is a strong physical basis for this requirement, but the basis is not so strong as it is for (12.56), although all known matter satisfies it. Most theorems require the dominant energy condition.

Most of the theorems require as an additional assumption the non-existence of closed timelike curves, so that causality is not violated. In addition, some of the theorems require a **genericity** condition, namely,

$$v_{[a}R_{b]cd[e}v_{f]}v^c v^d \neq 0,$$

somewhere along every timelike or null geodesic, where v^a is the tangent vector. It is only in very special cases that we might expect this condition to be violated.

None of the theorems leads directly to the existence of singularities. Instead, one obtains the result that space-time is not geodesically complete in timelike or null directions and, furthermore, cannot be extended to a geodesically complete space-time. The most reasonable explanation would seem to be that space-time is confronted with infinite curvature at its boundary. But the theorems do not quite say this and other types of space-time singularities may be possible.

19.13 The Hawking effect

This book is concerned with classical relativity theory and quantum considerations are beyond its brief. However, we shall make an exception and finish our treatment of black holes by describing in simple terms a quantum effect which suggests that black holes are not the permanent structures that the classical theory suggests. The surface area of the event horizon of a black hole has the remarkable property that it always increases when additional matter or radiation falls into the hole. Moreover, if two black holes collide and merge to form a single hole, the area of the new horizon is greater than the sum of the areas of the colliding holes. These properties suggest there is a resemblance between the area of the event horizon of a black hole and the concept of entropy in thermodynamics. (Entropy can be regarded simply as a measure of the disorder of a system or, equivalently, as a lack of knowledge of its precise state. The second law of thermodynamics states that entropy always increases with time.) Indeed, Hawking and collaborators discovered that the laws of thermodynamics have exact analogues in the properties of black holes. The first law relates the change in mass of a black hole to a change in area of the event horizon. The factor of proportionality involved is a quantity called the surface gravity, which is a measure of the strength of the gravitational field at the event horizon. This suggests that surface gravity is analogous to temperature, and indeed it is a constant at all points on the event horizon, just as the temperature is the same everywhere in a body at thermal equilibrium.

How, more precisely, can the area of a black hole be related to the concept of entropy? Well, the no-hair theorem implies that a large amount of information is lost in a gravitational collapse. A black hole of given mass, angular momentum, and charge could have been formed by the collapse of any one of large numbers of different configurations of matter. If one now takes into account quantum effects, the uncertainty principle requires that the number of configurations, although very large, must be finite. The logarithm of this number is the measure of the entropy of the hole and thus measures the information that was irretrievably lost during the collapse through the event horizon when the black hole was created. It follows that if this number is finite then the black hole must have a finite temperature (proportional to its surface gravity), and so it could be in thermal equilibrium with thermal radiation at some temperature other than zero. Yet, according to classical concepts, no such equilibrium is possible, since the black hole would absorb any thermal radiation that fell on it, but by definition would not be able to emit anything in return. This paradox was eventually resolved by Hawking who discovered that black holes seem to emit particles at a steady rate: this is the 'Hawking effect'.

Quantum mechanics implies that the whole of space is filled with pairs of 'virtual' particles and antiparticles that are constantly materializing in pairs, separating, and then coming together again and annihilating each other. These particles are called virtual because they cannot be observed directly with a particle detector (although they can be measured indirectly by the 'Lamb shift' in the spectrum of hydrogen). Now, in the presence of a black hole, the gravitational attraction will cause one member of a pair to fall into the hole, leaving the other member without a partner with which to undergo annihilation. This particle may also fall into the hole, but it may also escape to infinity, where it appears to be radiation emitted by the black hole. Equivalently, one may regard the member which falls into the hole (the antiparticle, say) as being really a particle travelling backwards in time. Then the motion of the antiparticle can be interpreted as a particle coming out of the hole (travelling backwards in time), and when it reaches the point at which the particle–antiparticle pair originally materialized it is scattered by the gravitational field, so that it travels forward in time. Thus quantum

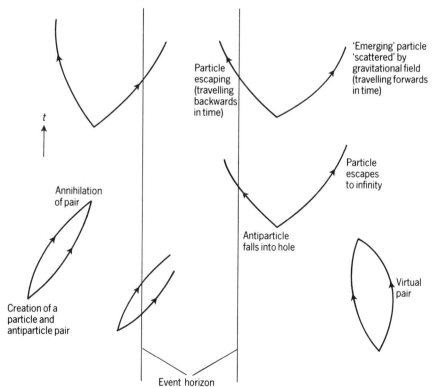

Fig. 19.12 Hawking radiation.

mechanics does allow, in this interpretation, an escape of particles from the hole — a form of quantum-mechanical 'tunnelling' (see Fig. 19.12).

As a black hole emits particles, its mass and size steadily decrease. This makes it easier for particles to tunnel out, and so the emission will continue at an ever-increasing rate until eventually the black hole radiates itself out of existence. In the long run, every black hole in the universe will evaporate in this way. For large black holes, it will take a very long time indeed (about 10^{66} years for a black hole the mass of the Sun). On the other hand, a primordial black hole, as discussed in §16.11, should have almost completely evaporated in the 10 billion years that have elapsed since the big bang. Thus mini-black holes may be exploding now and may be the source of highly energetic gamma rays. Attempts have been made to quantify this rate of production and to compare the predictions with terrestrial observations of incident gamma radiation, but the results are inconclusive.

Exercises

19.1 (§**19.1**) Show that the definitions (19.4) and (19.5) lead to (19.6), (19.7), and (19.8). Show also that the definitions (19.9) and (19.10) lead to (19.11), (19.12), (19.14), and (19.15) (see Exercise 8.3).

19.2 (§**19.2**) Find the covariant metric g_{ab} and contravariant metric g^{ab} for the Schwarzschild line element (16.24) in advanced Eddington–Finkelstein coordinates. Hence confirm (19.17) and (19.18).

19.3 (§**19.2**) Show that the transformations (19.20) applied to (19.19) lead to (19.21) (keeping v' and r' real). Deduce the line element (19.22) subject to (19.23).

19.4 (§**19.3**) Apply the transformations (19.24) and (19.25), subject to (19.26), to the line element (19.22), to obtain the form (19.27).

19.5 (§**19.3**) Apply the transformations (19.29) to (19.28), and then the transformation (19.24) to the result, to obtain the form (19.27).

19.6 (§**19.3**) Show that (19.28) can be written in the form (19.30), subject to (19.31), where λ and l_a are defined by (19.32) and (19.33). Show that, in the Schwarzschild limit, λ and l_a become (19.34) and (19.35). [Hint: $l_a l_b \mathrm{d}x^a \mathrm{d}x^b = (l_a \mathrm{d}x^a)^2$.]

19.7 (§**19.4**) Show that the transformations (19.37) together with $t' = t$ convert (19.38) into (19.39).

19.8 (§**19.4**) Show that the definition (19.40) leads to (19.41) and (19.42). Deduce that the Kerr solution is asymptotically flat.

19.9 (§**19.5**) Confirm Fig. 19.3.

19.10 (§**19.5**) Show that the stationary limit surface is timelike everywhere except at its poles.

19.11 (§**19.5**) Find g^{11} for the Boyer–Lindquist form of the Kerr solution (19.27).

19.12 (§**19.6**) Confirm equations (19.49)–(19.52). Show that they lead to (19.53). Why is it sufficient to consider the condition (19.54)? Check the deductions (19.55), (19.56), and (19.57), and show that r is an affine parameter. Obtain the geodesic equations (19.60) and (19.61).

19.13 (§**19.7**) Confirm the differential form (16.64) for the ingoing null congruence of the Schwarzschild solution in Eddington–Finkelstein coordinates. Check that the transformations (19.66) and (19.67) map the congruences (19.62) and (19.63) onto (19.65).

19.14 (§**19.7**) Confirm Fig. 19.4 and draw the retarded time version of it.

19.15 (§**19.8**) Show that (19.69) leads to (19.70), and hence deduce (19.71). [Hint: take second term in line element over to the right-hand side and take square roots.]

19.16 (§**19.11**) Use the Newman–Janis trick to obtain the Kerr–Newman solution (19.72) from the Reissner–Nordstrøm solution. Investigate the surfaces of infinite red shift and the event horizons (where present).

E. Gravitational Waves

Plane gravitational waves

20

20.1 The linearized field equations

Our consideration of gravitational radiation or gravitational waves (gravity waves for short) starts from the pioneering work of Einstein and is based on the **linearized** form of the field equations. In this approximation, we shall see that plane wave solutions lead to the result that gravitational waves are **transverse** and possess **two** polarization states. Put another way, the gravitational field has two radiation degrees of freedom. In the linearized approximation of the field equations, general relativity is recast as a Lorentz-covariant theory. Considerable caution has to be exercised in doing this because there are associated with it a number of serious difficulties and limitations (the details of which are beyond the brief of this book), but nonetheless it does throw some important light on the general theory.

We begin by assuming that the metric differs only slightly from the Minkowski metric in Minkowski coordinates, that is,

$$g_{ab} = \eta_{ab} + \varepsilon h_{ab}, \tag{20.1}$$

where ε is a small dimensionless parameter and, throughout, **we shall neglect terms of second order or higher in ε.** In addition, we adopt the boundary conditions that space-time is asymptotically flat, that is, if r denotes a radial parameter, then

$$\lim_{r \to \infty} h_{ab} = 0. \tag{20.2}$$

Defining

$$h^{ab} \equiv \eta^{ac} \eta^{bd} h_{cd}. \tag{20.3}$$

then

$$(\eta_{ab} + \varepsilon h_{ab})(\eta^{bc} - \varepsilon h^{bc}) = \delta_a^c, \tag{20.4}$$

from which we get

$$g^{ab} = \eta^{ab} - \varepsilon h^{ab}. \tag{20.5}$$

Since η_{ab} is constant, we also have (exercise)

$$\begin{aligned}
\Gamma_{bc}^a &= \tfrac{1}{2} g^{ad}(g_{dc,b} + g_{db,c} - g_{bc,d}) \\
&= \tfrac{1}{2} \varepsilon \eta^{ad}(h_{dc,b} + h_{db,c} - h_{bc,d}) \\
&= \tfrac{1}{2} \varepsilon (h^a{}_{c,b} + h^a{}_{b,c} - h_{bc,}{}^a),
\end{aligned} \tag{20.6}$$

where we make use of the result that, since this term is of order ε, we can, using (20.1) and (20.5), raise and lower indices with the Minkowski metric. The Riemann tensor then becomes

$$R_{abcd} = \tfrac{1}{2}\varepsilon(h_{ad,bc} + h_{bc,ad} - h_{ac,bd} - h_{bd,ac}). \tag{20.7}$$

The Bianchi identities

$$R_{ab[cd;e]} \equiv 0 \tag{20.8}$$

are

$$R_{ab[cd,e]} \equiv 0 \tag{20.9}$$

and are identically satisfied by (20.7).

The Ricci tensor is (exercise)

$$R_{ab} = \eta^{cd} R_{cadb} = \tfrac{1}{2}\varepsilon(h^c{}_{a,bc} + h^c{}_{b,ac} - \Box h_{ab} - h_{,ab}), \tag{20.10}$$

where

$$h \equiv \eta^{cd} h_{cd} = h^c{}_c \tag{20.11}$$

and $\Box$ is the d'Alembertian operator

$$\begin{aligned}
\Box &= \eta^{ab}\partial_a \partial_b \\
&= \partial^a \partial_a \\
&= \frac{\partial^2}{\partial t^2} - \nabla^2 \\
&= \frac{\partial^2}{\partial t^2} - \left(\frac{\partial^2}{\partial x^2} + \frac{\partial^2}{\partial y^2} + \frac{\partial^2}{\partial z^2}\right),
\end{aligned}$$

defined previously in (12.32). The Ricci scalar is

$$R = \varepsilon(h^{cd}{}_{,cd} - \Box h) \tag{20.12}$$

and finally the Einstein tensor is

$$G_{ab} = \tfrac{1}{2}\varepsilon(h^c{}_{a,bc} + h^c{}_{b,ac} - \Box h_{ab} - h_{,ab} - \eta_{ab}h^{cd}{}_{,cd} + \eta_{ab}\Box h). \tag{20.13}$$

In fact, the Einstein tensor can be found directly from the quadratic Lagrangian

$$\mathcal{L}(h^{ab}{}_{,c}) = \tfrac{1}{2}\varepsilon(h^{ab}{}_{,b}h^c{}_{c,a} - h^{ab,c}h_{cb,a} + \tfrac{1}{2}h^{cd,a}h_{cd,a} - \tfrac{1}{2}h^c{}_{c,a}h^d{}_{d}{}^{,a}), \tag{20.14}$$

using (exercise)

$$\begin{aligned}
G_{ab} &= \frac{\delta\mathcal{L}}{\delta h^{ab}} \\
&= \frac{\partial\mathcal{L}}{\partial h^{ab}} - \left(\frac{\partial\mathcal{L}}{\partial h^{ab}{}_{,c}}\right)_{,c} \\
&= -\left(\frac{\partial\mathcal{L}}{\partial h^{ab}{}_{,c}}\right)_{,c}. \tag{20.15}
\end{aligned}$$

20.2 Gauge transformations

Let us consider what happens to the linearized equations under a coordinate transformation of the form

$$x^a \rightarrow x'^a = x^a + \varepsilon \xi^a. \tag{20.16}$$

Then

$$\frac{\partial x'^a}{\partial x^b} = \delta_b^a + \varepsilon \xi^a_{,b} \tag{20.17}$$

and, applying this to the transformation formula for g_{ab}, (7.4), we find the consequent transformation of h_{ab} (see exercise 11.1), namely,

$$h_{ab} \rightarrow h'_{ab} = h_{ab} - 2\xi_{(a,b)}. \tag{20.18}$$

By analogy with electromagnetic theory (see (12.30)), this is called a **gauge transformation** of h_{ab}. It is easy to establish (exercise) that both the linearized curvature tensor (20.7) and its contractions are **gauge-invariant quantities**, that is, unchanged to first order in ε by transformations of the form (20.18).

Just as in electrodynamics, we may impose further conditions to fix the gauge. Going back to the field equations, we observe that if **new** variables ψ_{ab} are defined by

$$\psi_{ab} \equiv h_{ab} - \tfrac{1}{2}\eta_{ab}h, \tag{20.19}$$

then (20.10) becomes

$$R_{ab} = \tfrac{1}{2}\varepsilon(\psi^c_{a,bc} + \psi^c_{b,ac} - \Box h_{ab}), \tag{20.20}$$

and consequently

$$R = \tfrac{1}{2}\varepsilon(2\psi^{cd}_{,cd} - \Box h) \tag{20.21}$$

and

$$G_{ab} = \tfrac{1}{2}\varepsilon(\psi^c_{a,bc} + \psi^c_{b,ac} - \Box\psi_{ab} - \eta_{ab}\psi^{cd}_{,cd}). \tag{20.22}$$

This suggests that our field equations will reduce to **wave equations** if we impose the condition

$$\psi^a_{b,a} = 0, \tag{20.23}$$

or, in terms of h_{ab},

$$h^a_{b,a} - \tfrac{1}{2}h_{,b} = 0, \tag{20.24}$$

which is called variously the **Einstein**, **de Donder**, **Hilbert**, or **Fock** gauge. A straightforward calculation (exercise) reveals that, under the gauge transformation (20.16),

$$\psi_{ab} \rightarrow \psi'_{ab} = \psi_{ab} - \xi_{a,b} - \xi_{b,a} + \eta_{ab}\xi^c_{,c}, \tag{20.25}$$

from which we find

$$\psi'^a_{b,a} = \psi^a_{b,a} - \Box\xi_b. \tag{20.26}$$

It follows from (20.26) that the gauge transformation (20.16) will transform the equations into the Einstein gauge, that is,

$$\psi'^a_{b,\,a} = 0$$

if we choose ξ_a to satisfy

$$\Box \xi_a = \psi^b_{a,\,b}. \tag{20.27}$$

In other words, if we treat the ξ_a as unknowns then the problem involves solving wave equations with a source term. Then, by (20.22), Einstein's full field equations reduce to (dropping primes)

$$\tfrac{1}{2}\varepsilon\,\Box\,\psi_{ab} = -\kappa T_{ab}. \tag{20.28}$$

The gauge is not completely fixed by (20.27) because we can always carry out additional transformations with

$$\Box \xi_a = 0, \tag{20.29}$$

which leaves $\psi^a_{b,\,a}$ unaltered.

The vacuum field equations in the Einstein gauge reduce to

$$\Box\,\psi_{ab} = 0 \tag{20.30}$$

and, taking the trace,

$$\eta^{ab}\,\Box\,\psi_{ab} = \Box\,(\eta^{ab}\,\psi_{ab}) = \Box\,(h - 2h) = -\,\Box h = 0, \tag{20.31}$$

by (20.19). Combining this result with (20.30) and (20.19), we find that h_{ab} must also satisfy

$$\Box h_{ab} = 0 \tag{20.32}$$

in the Einstein gauge (20.24), which in terms of h_{ab} is

$$h^a_{b,\,a} - \tfrac{1}{2}h_{,\,b} = 0. \tag{20.33}$$

20.3 Linearized plane gravitational waves

Before we attempt to solve the linearized field equations, let us consider what theoretical motivation there might be which suggests that gravitational waves exist. We have seen that the linearized vacuum field equations reduce to the wave equations

$$\Box h_{ab} = 0 \tag{20.34}$$

in the Einstein gauge, from which we might be tempted to conclude that gravitational effects propagate as waves with the velocity of light. However, this is open to the objection that the perturbation h_{ab} is linked to an arbitrary coordinate system and therefore the existence of a non-zero h_{ab} is not an invariant indication of the existence of a gravitational field. A better argument is based on the fact that if (20.34) holds then, by (20.7),

$$\Box R_{abcd} = 0. \tag{20.35}$$

Thus, the Riemann tensor, which gives an absolute criterion for the existence of a gravitational field, itself obeys the wave equation. It follows that, in the linearized theory, gravitational effects propagate with the velocity of light. This does not of itself, however, prove whether or not gravitational **radiation** exists, since radiation involves energy transfer. We return to this question in the next chapter.

We look for a simple solution of the linearized vacuum field equations which represents an infinite plane wave propagating in the x-direction. We start by introducing the coordinates

$$(x^0, x^1, x^2, x^3) = (t, x, y, z)$$

and adopt the ansatz

$$h_{ab} = h_{ab}(t, x), \tag{20.36}$$

which requires

$$h_{ab,2} = h_{ab,3} = 0. \tag{20.37}$$

This assumption means that the Riemann tensor is highly degenerate and, from (20.7), we find that the 20 independent components fall into the following three groups of terms (exercise):

$$R_{0123} = R_{0223} = R_{0323} = R_{1223} = R_{1323} = R_{2323} = 0; \tag{20.38}$$

$$\left. \begin{array}{l} R_{0101} = \tfrac{1}{2}\varepsilon(2h_{01,01} - h_{00,11} - h_{11,00}), \\[4pt] R_{0102} = \tfrac{1}{2}\varepsilon(h_{02,01} - h_{12,00}), \\[4pt] R_{0103} = \tfrac{1}{2}\varepsilon(h_{03,01} - h_{13,00}), \\[4pt] R_{0112} = \tfrac{1}{2}\varepsilon(h_{02,11} - h_{12,01}), \\[4pt] R_{0113} = \tfrac{1}{2}\varepsilon(h_{03,11} - h_{13,01}); \end{array} \right\} \tag{20.39}$$

$$\left. \begin{array}{l} R_{0202} = -\tfrac{1}{2}\varepsilon h_{22,00}, \\[4pt] R_{0203} = -\tfrac{1}{2}\varepsilon h_{23,00}, \\[4pt] R_{0212} = -\tfrac{1}{2}\varepsilon h_{22,01}, \\[4pt] R_{0213} = -\tfrac{1}{2}\varepsilon h_{23,01}, \\[4pt] R_{0303} = -\tfrac{1}{2}\varepsilon h_{33,00}, \\[4pt] R_{0313} = -\tfrac{1}{2}\varepsilon h_{33,01}, \\[4pt] R_{1212} = -\tfrac{1}{2}\varepsilon h_{22,11}, \\[4pt] R_{1213} = -\tfrac{1}{2}\varepsilon h_{23,11}, \\[4pt] R_{1313} = -\tfrac{1}{2}\varepsilon h_{33,11}. \end{array} \right\} \tag{20.40}$$

We now impose the linearized vacuum field equations in the form $R_{ab} = 0$.

Then, for example,

$$R_{13} = R^a{}_{1a3} = R_{0103} = 0, \tag{20.41}$$

so that one of the independent components of (20.39) vanishes. In fact, the vacuum field equations result in all the group (20.39) vanishing (exercise). Thus, only the components in the group (20.40) are non-zero and these only involve the components h_{22}, h_{23}, and h_{33}. This means that we can decompose h_{ab} into two parts:

$$h_{ab} = h_{ab}^{(1)} + h_{ab}^{(2)}, \tag{20.42}$$

where

$$h_{ab}^{(1)} = \begin{bmatrix} 0 & 0 & 0 & 0 \\ 0 & 0 & 0 & 0 \\ 0 & 0 & h_{22} & h_{23} \\ 0 & 0 & h_{23} & h_{33} \end{bmatrix} \tag{20.43}$$

and

$$h_{ab}^{(2)} = \begin{bmatrix} h_{00} & h_{01} & h_{02} & h_{03} \\ h_{01} & h_{11} & h_{12} & h_{13} \\ h_{02} & h_{12} & 0 & 0 \\ h_{03} & h_{13} & 0 & 0 \end{bmatrix}. \tag{20.44}$$

The vacuum field equations then lead to the result that the curvature tensor of $h_{ab}^{(2)}$ is identically zero. This suggests that there may exist a coordinate system in which h_{ab} has only h_{22}, h_{23}, and h_{33} components; that is, h_{ab} is a pure $h_{ab}^{(1)}$-type solution. We shall show that we can exploit the gauge freedom to achieve this in the case of a plane wave.

We sharpen our ansatz (20.36) by requiring

$$h_{ab} = h_{ab}(t - x), \tag{20.45}$$

so that it clearly represents a solution propagating in the x-direction with the speed of light (see Fig. 20.1). The Einstein gauge conditions (20.33) then

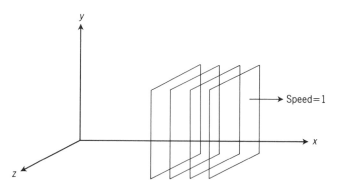

Fig. 20.1 The ansatz (20.45).

become

$$\left.\begin{array}{rl} h_{00,0} - h_{01,1} - \tfrac{1}{2}h_{,0} &= 0, \\ h_{01,0} - h_{11,1} - \tfrac{1}{2}h_{,1} &= 0, \\ h_{02,0} - h_{12,1} &= 0, \\ h_{03,0} - h_{13,1} &= 0, \end{array}\right\} \tag{20.46}$$

or, letting a prime denote differentiation with respect to the argument $t - x$, these can be written

$$\begin{array}{rl} h'_{00} + h'_{01} - \tfrac{1}{2}h' &= 0, \\ h'_{01} + h'_{11} + \tfrac{1}{2}h' &= 0, \\ h'_{02} + h'_{12} &= 0, \\ h'_{03} + h'_{13} &= 0. \end{array}$$

These integrate to give

$$\left.\begin{array}{rl} h_{00} + h_{01} - \tfrac{1}{2}h &= f_1, \\ h_{01} + h_{11} + \tfrac{1}{2}h &= f_2, \\ h_{02} + h_{12} &= f_3, \\ h_{03} + h_{13} &= f_4, \end{array}\right\} \tag{20.47}$$

where the f's are all functions of y and z only. However, since the h_{ab} all vanish at spatial infinity by (20.2), it follows that

$$f_1 = f_2 = f_3 = f_4 = 0.$$

Then (20.47) gives

$$h_{12} = -h_{02}, \qquad h_{13} = -h_{03}, \qquad h_{01} = -\tfrac{1}{2}(h_{00} + h_{11}), \qquad h_{33} = -h_{22},$$

that is,

$$h_{ab} = \begin{bmatrix} h_{00} & -\tfrac{1}{2}(h_{00} + h_{11}) & h_{02} & h_{03} \\ -\tfrac{1}{2}(h_{00} + h_{11}) & h_{11} & -h_{02} & -h_{03} \\ h_{02} & -h_{02} & h_{22} & h_{23} \\ h_{03} & -h_{03} & h_{23} & -h_{22} \end{bmatrix}. \tag{20.48}$$

We still have the remaining gauge freedom (20.18), where ξ_a satisfies (20.29). Let us try and choose this so that

$$h'_{00} = h'_{02} = h'_{03} = h'_{11} = 0. \tag{20.49}$$

Then, by (20.18), this requires

$$\left.\begin{array}{rl} h_{00} - 2\xi_{0,0} &= 0, \\ h_{02} - \xi_{0,2} - \xi_{2,0} &= 0, \\ h_{03} - \xi_{0,3} - \xi_{3,0} &= 0, \\ h_{11} - 2\xi_{1,1} &= 0. \end{array}\right\} \tag{20.50}$$

If we assume that

$$\xi_a = \xi_a(t - x), \tag{20.51}$$

then (20.29) is automatically satisfied. We choose

$$(\xi_0, \xi_1, \xi_2, \xi_3) = \big(F_0(t-x), F_1(t-x), F_2(t-x), F_3(t-x)\big), \qquad (20.52)$$

where, setting $u = t - x$, we see that the functions F_0, F_1, F_2, and F_3 are all functions of u only and are determined by the ordinary differential equations

$$\frac{\mathrm{d}F_0}{\mathrm{d}u} = \tfrac{1}{2}h_{00}(u), \quad \frac{\mathrm{d}F_1}{\mathrm{d}u} = -\tfrac{1}{2}h_{11}(u), \quad \frac{\mathrm{d}F_2}{\mathrm{d}u} = h_{02}(u), \quad \frac{\mathrm{d}F_3}{\mathrm{d}u} = h_{03}(u). \quad (20.53)$$

This choice satisfies (20.51) and (20.50), and, moreover, it leaves h_{22}, h_{23}, and h_{33} unchanged. Hence, dropping primes, we have shown that h_{ab} may be transformed into the **canonical form**

$$h_{ab} = \begin{bmatrix} 0 & 0 & 0 & 0 \\ 0 & 0 & 0 & 0 \\ 0 & 0 & h_{22} & h_{23} \\ 0 & 0 & h_{23} & -h_{22} \end{bmatrix}. \qquad (20.54)$$

Clearly, h_{ab} only depends on **two** functions, namely,

$$h_{22}(t-x) \quad \text{and} \quad h_{23}(t-x).$$

We consider the physical significance of these two independent functions in the next section.

20.4 Polarization states

In the case $h_{23} = 0$, the line element becomes

$$\mathrm{d}s^2 = \mathrm{d}t^2 - \mathrm{d}x^2 - [1 - \varepsilon h_{22}(t-x)]\,\mathrm{d}y^2 - [1 + \varepsilon h_{22}(t-x)]\,\mathrm{d}z^2. \qquad (20.55)$$

We shall call this an 'h_{22}-wave'. Let us suppose that h_{22} is some oscillatory function of u so that there are values when $h_{22} > 0$ and values when $h_{22} < 0$. Let us investigate what happens when an h_{22}-wave is incident on a distribution of test particles. First of all, consider two neighbouring particles in the (y, z)-plane which initially have coordinates (y_0, z_0) and $(y_0 + \mathrm{d}y, z_0)$ in the plane. Then, using (20.55), the proper distance between them is given by

$$\mathrm{d}s^2 = -(1 - \varepsilon h_{22})\,\mathrm{d}y^2. \qquad (20.56)$$

The proper distance is a coordinate-independent quantity, and hence if initially h_{22} changes from zero to $h_{22} > 0$ the particles move closer together and, conversely, if h_{22} changes from zero to $h_{22} < 0$ the particles move further apart. The opposite happens if we consider free particles with coordinates (y_0, z_0) and $(y_0, z_0 + \mathrm{d}z)$ in the plane, since now

$$\mathrm{d}s^2 = -(1 + \varepsilon h_{22})\,\mathrm{d}z^2. \qquad (20.57)$$

Thus, if an oscillatory plane gravitational wave propagating in the x-direction is incident on a ring of dust particles situated in the yz-plane, then

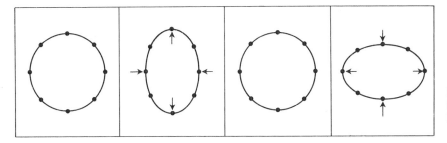

Fig. 20.2 Time sequence showing the transverse effect of an oscillatory linear plane gravitational wave with + polarization.

the ring is distorted into a pulsating ellipse whose major axis is in turn parallel to the *y*- and *z*-axes (see Fig. 20.2). The **transverse** character of an h_{22}-wave is clear from this. We refer to this state as a wave with **+ polarization**.

Let us turn attention to an 'h_{23}-wave', that is, the case when $h_{22} = 0$, and the line element becomes

$$ds^2 = dt^2 - dx^2 - dy^2 + 2\varepsilon h_{23}(t - x)\,dy\,dz - dz^2. \qquad (20.58)$$

Let us perform a rotation through 45° in the (*y*, *z*) plane given by (see (3.9) and (3.10))

$$y \to \bar{y} = \frac{1}{\sqrt{2}}(y + z), \qquad z \to \bar{z} = \frac{1}{\sqrt{2}}(-y + z), \qquad (20.59)$$

so that the line element becomes (exercise)

$$ds^2 = dt^2 - dx^2 - [1 - \varepsilon h_{23}(t - x)]\,d\bar{y}^2 - [1 + \varepsilon h_{23}(t - x)]\,d\bar{z}^2. \qquad (20.60)$$

Comparing this with (20.55) we see that an h_{23}-wave produces exactly the same effect as an h_{22}-wave but with the axes rotated through 45° (see Fig. 20.3). The transverse character of an h_{23}-wave is again clear and we refer to the state as a wave with **× polarization**.

Clearly, a general wave is a superposition of these two polarization states. The fact that the two polarization states are at 45° to each other contrasts with the two polarization states of an electromagnetic wave, which are at 90° to each other. (This can be shown to stem from the fact that gravity is represented by the second-rank symmetric tensor $h_{\mu\nu}$, whereas electromagnetism is represented by the vector potential A^μ.)

An alternative method for investigating these results is to consider the equation of geodesic deviation (10.21). If we introduce a local coordinate

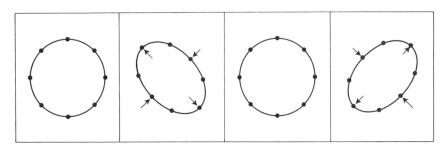

Fig. 20.3 Time sequence showing the transverse effect of an oscillatory linear plane gravitational wave with × polarization.

system adapted to the tetrad so that

$$e_i{}^a \overset{*}{=} \delta_i{}^a,$$

then, by (10.37), the equation becomes

$$\frac{D^2 \eta^\alpha}{D\tau^2} + R^\alpha{}_{0\beta 0} \eta^\beta = 0.$$

Setting

$$\eta^\beta = (x, y, z)$$

and using (20.38), (20.39), and (20.40), we get

$$\left.\begin{aligned}
\frac{D^2 x}{D\tau^2} &= 0, \\[1mm]
\frac{D^2 y}{D\tau^2} + \tfrac{1}{2}\varepsilon(h_{22,00}\,y + h_{23,00}\,z) &= 0, \\[1mm]
\frac{D^2 z}{D\tau^2} - \tfrac{1}{2}\varepsilon(h_{22,00}\,z - h_{23,00}\,y) &= 0.
\end{aligned}\right\} \tag{20.61}$$

Then, for example, an h_{22}-wave leads to

$$\frac{D^2 y}{D\tau^2} = -\tfrac{1}{2}\varepsilon h''_{22}\, y, \qquad \frac{D^2 z}{D\tau^2} = \tfrac{1}{2}\varepsilon h''_{22}\, z,$$

and, as in §16.10, the different signs in the relative accelerations lead to the behaviour we have described above.

20.5 Exact plane gravitational waves

If we introduce double null coordinates defined by

$$u = t - x, \qquad v = t + x$$

in (20.55), then an h_{22}-wave has a line element of the form

$$ds^2 = du\,dv - f^2(u)\,dy^2 - g^2(u)\,dz^2, \tag{20.62}$$

where

$$f^2(u) = 1 - \varepsilon h_{22}(u), \qquad g^2(u) = 1 + \varepsilon h_{22}(u). \tag{20.63}$$

The functions are squared to ensure the correct signature (which is justified in the linearized approximation by assuming that ε is small in (20.63)).

Let us now choose (20.62) as an ansatz and plug this line element into the full vacuum field equations to see if we can solve them. We find that the non-vanishing components of the connection are (exercise)

$$\Gamma^1_{22} = 2ff', \qquad \Gamma^1_{33} = 2gg', \qquad \Gamma^2_{02} = f'/f, \qquad \Gamma^3_{03} = g'/g, \tag{20.64}$$

where a prime denotes differentiation with respect to u. The Riemann tensor has two independent components

$$R_{0202} = ff'', \qquad R_{0303} = gg''$$

and there is only one vacuum field equation, namely,

$$f''/f + g''/g = 0. \qquad (20.65)$$

Let us denote the first term by the function $h(u)$, i.e.

$$f''/f = h. \qquad (20.66)$$

Then the field equation will be satisfied if g is chosen so that

$$g''/g = -h. \qquad (20.67)$$

These last two equations determine f and g in terms of $h(u)$ up to constants of integration. Hence any choice of the arbitrary function $h(u)$ gives rise to a vacuum solution. Such exact solutions are called **linearly polarized plane gravitational waves**. They represent plane-fronted gravitational waves, abstracted away from any sources, propagating in the x-direction.

The form of the line element (20.62) is essentially that due originally to **Rosen**. If we carry out the coordinate transformation

$$U = u, \qquad V = v + y^2 ff' + z^2 gg', \qquad Y = fy, \qquad Z = gz, \qquad (20.68)$$

then the line element is transformed into the **Brinkmann** form (exercise)

$$ds^2 = h(U)(Z^2 - Y^2)dU^2 + dU\,dV - dY^2 - dZ^2, \qquad (20.69)$$

which shows the explicit dependence on the freely specifiable function h. This function can be shown to represent the amplitude of the polarized wave.

Although such solutions are highly unphysical, being infinite in extent, it may be hoped that they represent some of the properties of real waves from bounded sources in some far zone limit. In particular, they allow us to investigate the question of the scattering of gravitational waves. For, unlike electromagnetic theory, where the linearity of the theory means that electromagnetic waves pass through each other unaltered, there is, in general, no superposition principle in general relativity. Indeed, we may expect the non-linearity of the theory to reveal itself in the interaction of two gravitational waves. However, (20.69) does reveal a limited superposition principle in that two plane waves moving in the **same** direction can be superposed simply by adding their corresponding h functions. Thus, when moving in the same direction, two such gravitational waves do not scatter one another. To exhibit scattering, we need two waves moving in different directions. If we consider two linearly polarized waves colliding at an angle, we can always find a class of observers who consider the collision to be head on (see, for example, Exercise 4.10). Hence, it is sufficient to work in a coordinate system in which the waves appear to collide head on. We shall consider this question in the limited case of impulsive gravitational waves, which we discuss next.

20.6 Impulsive plane gravitational waves

We start with a mathematical digression. The Heaviside step function $\theta(u)$ is defined by

$$\theta(u) = \begin{cases} 0 & \text{if } u \leqslant 0, \\ 1 & \text{if } u > 0. \end{cases} \qquad (20.70)$$

It is closely related to the Dirac delta function $\delta(u)$. Strictly speaking, δ is not a function but rather a distribution and lives under an integral sign. It will be sufficient for our purposes to define δ by the requirements

$$\delta(u) = 0 \qquad \text{if } u \neq 0, \tag{20.71}$$

$$\int_{-\infty}^{\infty} f(u)\delta(u)\,\mathrm{d}u = f(0), \tag{20.72}$$

for any suitably defined function $f(u)$. Then, with these definitions, we can establish the results (exercise)

$$\theta'(u) = \delta(u), \tag{20.73}$$

$$u\delta(u) = 0, \tag{20.74}$$

$$u\theta'(u) = 0. \tag{20.75}$$

We now consider a line element in the Rosen form defined by

$$f(u) = 1 + u\theta(u), \qquad g(u) = 1 - u\theta(u). \tag{20.76}$$

Then we find, using the above results, that (exercise)

$$f' = -g' = \theta(u), \quad f'' = -g'' = \delta(u), \quad f''/f = -g''/g = \delta(u), \tag{20.77}$$

which means, from (20.65), that (20.76) gives rise to a plane wave. Hence, the Ricci and Einstein tensors vanish, but the Riemann tensor (or, since the solution is vacuum, equivalently the Weyl tensor) does not vanish, having non-vanishing components

$$R_{0202} = -R_{0303} = \delta(u). \tag{20.78}$$

The solution has delta functions in the curvature and hence it is non-flat only when $u = 0$. This can be seen more clearly in the Brinkmann form of the solution, which, from (20.66), is obtained by setting

$$h(U) = \delta(U). \tag{20.79}$$

Hence, for $u = U \neq 0$, the line element reduces to

$$\mathrm{d}s^2 = \mathrm{d}U\,\mathrm{d}V - \mathrm{d}Y^2 - \mathrm{d}Z^2, \tag{20.80}$$

which is Minkowski space-time in double null coordinates. The hypersurface $u = 0$, where the field is concentrated, thus separates two flat regions. It represents a plane wave similar to that of Fig. 20.1, except that now there is just one wave front (Fig. 20.4). Such a solution is called a **shock wave** or **impulsive plane gravitational wave**. Figure 20.5 is a space-time picture (with two dimensions suppressed) of such a solution.

We define a **sandwich wave** to be a non-flat vacuum solution bounded by plane hypersurfaces outside of which the solution is flat (Fig. 20.6). An observer moving on a geodesic will 'feel' the wave passing for a finite period when moving from region I through region II and out into region III. Neighbouring test particles will be accelerated transversely to the direction of propagation of the wave. Then an impulsive gravitational wave can be viewed as a thin sandwich wave in a suitable limit as the thickness goes to zero. Although impulsive waves are yet another idealization, they do prove easier to work with than more general waves at first.

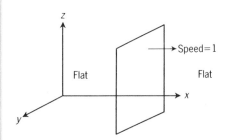

Fig. 20.4 Spatial picture of an impulsive plane gravitational wave.

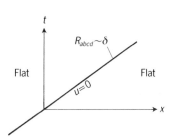

Fig. 20.5 Space-time picture (two dimensions suppressed) of an impulsive plane wave.

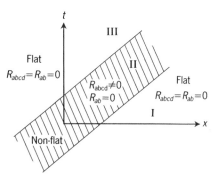

Fig. 20.6 A space-time picture of a sandwich wave.

20.7 Colliding impulsive plane gravitational waves

In the next two sections, we shall outline the pioneering work of Penrose, Khan, and Szekeres on the important problem of colliding plane gravitational waves. We start by generalizing the Rosen form (20.62) to the form

$$ds^2 = l\,du\,dv - f^2\,dy^2 - g^2\,dz^2, \tag{20.81}$$

where l, f, and g are now functions of both u and v. This form then allows us to incorporate waves moving in both directions. The explicit vacuum solution of Penrose and Khan is then given by

$$l = \frac{m^3}{rw(pq + rw)^2}, \qquad f^2 = m^2\left(\frac{r+q}{r-q}\right)\left(\frac{w+p}{w-p}\right),$$

$$g^2 = m^2\left(\frac{r-q}{r+q}\right)\left(\frac{w-p}{w+p}\right),$$

where

$$p = u\theta(u), \quad q = v\theta(v), \quad r = (1-p^2)^{\frac{1}{2}}, \quad w = (1-q^2)^{\frac{1}{2}}, \quad m = (1-p^2-q^2)^{\frac{1}{2}}.$$

The space-time diagram is shown in Fig. 20.7.

The solution is only valid in the four regions:

   I. $u < 0, v < 0,$
  II. $0 < u < 1, v < 0,$
 III. $u < 0, 0 < v < 1,$
 IV. $u > 0, v > 0, u^2 + v^2 < 1.$

Regions I, II, and III are flat, and region IV is curved. Region I is separated from region II by an incoming impulsive wave and from region III by another impulsive wave travelling in the opposite direction. They collide at the origin in the figure, and then region IV represents the interaction region between them. If we consider the world-line of the observer $x = 0$, then the two waves collide at $t = 0$, scatter each other, and leave a curved region between them,

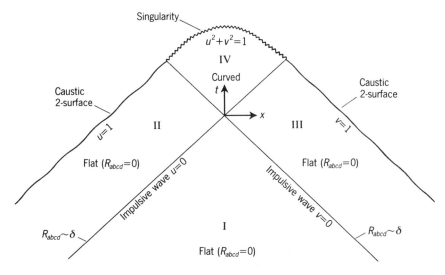

Fig. 20.7 Penrose and Khan space-time picture of two colliding impulsive plane waves.

which, in finite proper time according to the observer, develops an **intrinsic singularity**. (This is an intrinsic singularity in the usual sense that scalar invariants in the curvature tensor blow up.) There is also a coordinate singularity in region II at $(u = 1, v < 0)$ and an analogous one in region III at $(v = 1, u < 0)$. These singularities are, in fact, topological singularities, sometimes called fold singularities, and are in this case **caustic 2-surfaces** caused by each wave focusing the other, i.e. they are surfaces where the null geodesics cross. They are not intrinsic curvature singularities. The space-time diagram (Fig. 20.7) is a bit misleading at first sight since you might think it possible for an observer in region II to cross $u = 1$ and escape. However, the caustic surface is just a 'seam' in the hypersurface $v = 0$, and so the chances of hitting it are remote, and, anyway, any observer getting close will be swept up into region III and end up on the singularity. There is a finite jump in the curvature tensor at $(u = 0, 0 < v < 1)$ and at $(v = 0, 0 < u < 1)$ (sometimes called a step wave) in addition to the delta function there. Furthermore, inspection of the solution reveals that the waves no longer have planar symmetry after impact.

To summarize, two impulsive plane gravitational waves approaching each other from different directions scatter each other and cease to be plane waves. Eventually, the focusing effect of each wave on the other results in the formation of a spacelike intrinsic singularity (recall that, whereas timelike singularities are avoidable, spacelike singularities are not).

20.8 Colliding gravitational waves

The fact that two colliding impulsive waves give rise to a singularity is perhaps something of a surprise. At first (recall the situation in black holes with the Schwarzschild solution), it was thought that this may be due to the high symmetry of the solution and that a more realistic solution would remain regular. However, Peter Szekeres provided a general framework for investigating colliding gravitational waves and discovered some exact solutions which again result in singularities. The framework consists essentially of formulating the problem as a characteristic initial value problem (see §21.5), which, in double null coordinates (u, v), consists of prescribing initial data on a pair of null hypersurfaces $u = 0$, $v = 0$ intersecting in a spacelike 2-surface (Fig. 20.8). Region I is taken to be flat, and regions II and III contain two waves which are approaching from opposite directions. Region IV is then the interaction region of the two waves. The problem is well posed in that it can be shown that any given initial data gives rise to a unique solution in region IV. It is convenient to assume that two commuting spacelike Killing vectors $\partial/\partial y$ and $\partial/\partial z$ exist throughout the whole space-time. Szekeres shows that coordinates of the Rosen type exist in which the metric takes on the form

$$ds^2 = e^{-M}du\,dv - e^{-U}(e^V \cosh W\,dy^2 - 2\sinh W\,dy\,dz + e^{-V}\cosh W\,dz^2),$$

where M, U, V, and W are functions of u and v in general. However, in region II, the functions M, U, V, and W depend on u only; in region III, they depend on v only. If the waves have constant and parallel polarizations, then it can be shown that one can put $W = 0$ globally and the solution to the initial value problem reduces to a one-dimensional integral for V and two quadratures for M.

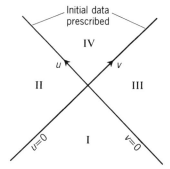

Fig. 20.8 The characteristic initial value problem for colliding waves.

Szekeres considered the more realistic case of sandwich waves in regions II and III and again found that they give rise to singularities in region IV. Since the early 1970s, when this work was first reported, there have been a large number of exact solutions found for colliding gravitational waves, including plane gravitational waves and waves coupled with electromagnetic waves, perfect fluids, and null dust (i.e. an energy–momentum tensor of the form (12.1) but where the 4-vector u^a is null). Indeed, there has been considerable controversy over what happens when two planar impulsive gravitational waves, each followed by a distribution of null dust, collide. Do the two distributions pass through each other or do they mix magically to produce a perfect fluid with a 'stiff' equation of state $p = \rho$? These ambiguities make it clear that these sorts of problems, which are a mixture of initial value and boundary value problems, need careful handling and that particular attention needs to be paid to the physical interpretation of the resulting solutions.

Although most solutions develop spacelike singularities, not all do. Some examples give rise to non-singular horizons, for which the metric can be analytically extended across the horizon to produce a maximal space-time which contains topological singularities and in some cases curvature singularities as well. Some advances in our understanding of these solutions has resulted from the work of Yurtsever. By considering a class of perturbations of the initial data producing such solutions, he has shown that these horizons are in fact unstable. Moreover, the work seems to suggest that the development of spacelike singularities by colliding gravitational waves is a **generic** phenomenon. If this is indeed the case, it leads to the puzzling question of why we do not detect singularities in our locality, since we are certainly immersed in a sea of colliding gravitational waves emanating from many sources situated both within and outside our own galaxy. However, the time taken for these singularities to form for the amplitudes of waves which are likely to exist is very large (perhaps comparable with the age of the universe). On the other hand, it could well be that the singularities are simply an artefact of the **planar** symmetry, and that if more realistic (non-planar) solutions are employed then the singularities will likely disappear.

20.9 Detection of gravitational waves

We turn to the possible detection of gravitational waves. The pioneer in this field is J. Weber whose work dates back to the 1960s. His method is based on the fact that free particles moving through a gravitational field experience relative accelerations as expressed through the equation of geodesic deviation. Weber's technique consists in measuring the deformations set up in a large aluminium cylinder by any incident gravitational radiation. Considerable controversy surrounded his claims to be detecting radiation emanating from the centre of the galaxy, since the sensitivity of the bar was considered to be too low to detect radiation at the energy which might be expected. Such signals would probably be swamped by the noise emanating from people, vehicles, aircraft, and so on, passing near the equipment. Moreover, there was also disquiet over the way the results were analysed, and the consensus is that the equipment was probably not detecting gravity waves. However, Weber has played an important part in alerting the experimentalists to the need to undertake this work.

Let us discuss briefly the possible sources of gravitational radiation. Thorne distinguishes between three sorts of radiation, namely, **bursts, periodic,** and **stochastic.** The possible sources of bursts are collapsing and bouncing cores of supernovae in our galaxy and other galaxies; the birth of black holes, especially massive ones; collisions between black holes and between black holes and neutron stars in globular clusters, galactic nuclei, and quasars; and the final spiralling in, coalescence, and destruction of compact binaries (like the binary pulsar PSR 1913 + 16). Possible sources of periodic waves include binary star systems, rotating deformed stars, rotating deformed white dwarfs, and pulsations of white dwarfs following nova outbursts. Stochastic sources include the hot big bang, inhomogeneities in the very early universe, and black holes formed from population III stars (stars born before galaxies were formed).

It is extremely difficult to obtain estimates of the energy output from the various sources, because they often depend on the details of the model employed about which little is known. Added to which, it may not be possible to carry out the algebraic computations involved either exactly or even approximately. There is a growing role here for **numerical relativity,** which is the field of using computers to solve Einstein's equations numerically from prescribed initial data. For example, numerical codes exist which suggest that a collapsing star may emit up to 1 or 2% of its mass in the form of gravitational waves.

Thorne gives some estimates of the energy output, as shown in Fig. 20.9, in which the amplitude h of the gravitational waves is plotted against their frequency v. He suggests that bursts with a frequency of once a month are likely to lie in the vertically hatched region, although there may be stronger bursts from supernovae in our own galaxy. The line labelled 'cherished beliefs' is based on the most optimistic of estimates. The horizontally hatched region gives estimates of where the strongest periodic sources may lie. On

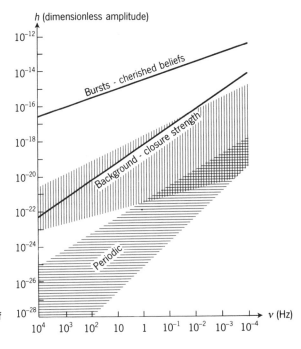

Fig. 20.9 Estimates of the strength of gravity waves bathing the Earth.

some estimates, there could be a stochastic background as large as the solid line (whose energy density would be sufficient to close the universe).

There have been three sorts of detectors proposed so far: **Weber bars, laser interferometers,** and **spacecraft tracking**. Bar detectors are currently narrow-band detectors, being tuned to a particular frequency. Laser systems, on the other hand, are broad band and can operate over a range in frequency of an order of magnitude, which means that they can be used to study the detailed time structure $h(t)$ of the wave. The bars are made of a number of materials including aluminium, silicon, and niobium and are isolated from various sources of noise and cooled to a temperature of $2\,\mathrm{K}$ or less. The oscillations are then measured by mechanical or electrical transducer devices. Early bars achieved r.m.s. noise levels of $h \simeq 10^{-16}$ for frequencies $v \simeq 10^3\,\mathrm{Hz}$ at room temperatures. There are a large number of centres which have bar detectors, and the best currently operates around $h \simeq 10^{-18}$, although 10^{-20} seems possible. Perhaps of more interest are attempts to detect coincidences in two or more detectors situated at different sites, since this would provide stronger evidence for the observation of some real external source. There appears to be little evidence of statistically significant coincidences to date.

The basic design of a laser interferometer is shown in Fig. 20.10. The first prototype system was run in 1972 with an r.m.s. sensitivity of $h \simeq 10^{-14}$ for $v \simeq 1\text{--}10\,\mathrm{kHz}$. These sort of detectors are very promising because the sensitivity goes like

$$h = \Delta(L_1 - L_2)/L_1,$$

where Δ is the change due to gravity waves, and so this sensitivity can be improved rapidly by simply scaling up the length of the arms without other major changes in instrumentation. There are many different strategies for noise reduction, which is reflected in wide differences in design. For example, some involve multiple bounces of the beam, others use an optically resonant cavity. There is a quantum-mechanical limit which suggests a possible sensitivity ultimately of $h \simeq 10^{-22}$. The sensitivity of current interferometers goes like $h \simeq 10^{-18}$ or 10^{-19}, but there are considerable technological hurdles to overcome before the quantum-mechanical limit can be reached.

At low frequencies of $v \leqslant 1\,\mathrm{Hz}$, it is difficult, if not impossible, to shield Earth-based detectors from noise. The only solution is to use detectors in space. One possibility is the Doppler tracking of spacecraft. Measurements on the Voyager spacecraft yielded a sensitivity in the range 10^{-13} to 10^{-14} for v between 10^{-4} and $10^{-2}\,\mathrm{Hz}$. Future flights may yield $h \lesssim 10^{-16}$ using

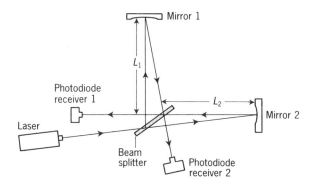

Fig. 20.10 Idealized version of a laser interferometer gravity wave detector.

very accurate onboard clocks such as hydrogen maser clocks. Using several such spacecraft to track each other, it may eventually be possible to achieve $h < 10^{-21}$ for 10^{-4} Hz $< v < 30$ Hz.

The binary pulsar PSR 1913 + 16 appears to provide indirect evidence of gravitational radiation. It is thought that this is a rotating binary system emitting gravitational radiation leading to a decrease in the energy of the system and a consequent increase in the rotation and a decrease in the period τ. The general relativistic prediction for $\dot{\tau}$ is -2.4×10^{-12}, which compares well with the observed value of $-(2.30 \pm 0.22) \times 10^{-12}$ [see Schutz (1980) for further details].

In short, there appears as yet to be no clear-cut direct observation of gravity waves by detectors. However, all the evidence is that these observations may be just around the corner and that, in addition to radio, infrared, optical, ultraviolet, X-ray and gamma-ray observations, we may soon have a window onto new phenomena in our universe through the advent of gravitational astronomy.

Exercises

20.1 (§**20.1**) Show that, if we work to order ε^2, then (20.1) implies (20.4), (20.5), (20.6), (20.7), and (20.10) (subject to (20.11)), (20.12), and (20.13).

20.2 (§**20.1**) Show that the Bianchi identities (20.8) can be written in the form (20.9) to order ε^2, and that these equations are satisfied automatically by (20.1).

20.3 (§**20.1**) Show that the quadratic Lagrangian (20.14) leads to the field equations (20.13). [Hint: the field equations must be symmetric in a and b.]

20.4 (§**20.2**) Show that h_{ab} transforms according to (20.18) to order ε^2 under the coordinate transformation (20.16). Show also that ψ_{ab} transforms according to (20.25) under this transformation.

20.5 (§**20.2**) Show that in the slow-motion approximation for a distribution of dust of proper density ρ_0 that (20.28) reduces to

$$\varepsilon \nabla^2 \psi_{00} = 16\pi\rho_0.$$

Compare this with Poisson's equation in relativistic units to deduce that

$$\varepsilon \psi_{00} = 4\phi$$

with all other components vanishing. Use (20.19) to deduce that

$$\varepsilon h_{00} = \varepsilon h_{11} = \varepsilon h_{22} = \varepsilon h_{33} = 2\phi$$

and hence that, in this approximation, the metric is

$$ds^2 = (1 + 2\phi)\,dt^2 - (1 - 2\phi)(dx^2 + dy^2 + dz^2)$$

Show that this is consistent with the Schwarzschild solution (in isotropic coordinates) in the weak-field limit.

20.6 (§**20.2**) Confirm equations (20.20), (20.21), and (20.22), and deduce (20.28), (20.30), and (20.32) in the Einstein gauge.

Show that there is an additional gauge freedom (20.18) subject to (20.29).

20.7 (§**20.3**) Show that the ansatz (20.36) leads to a Riemann tensor satisfying (20.38), (20.39), and (20.40). [Hint: use the identity (6.78) to eliminate R_{0312}.] Show that the linearized vacuum field equations lead to the vanishing of the group of equations (20.39). [Hint: Consider $R_{00} = R_{03} = R_{12} = R_{13} = R_{00} - R_{11} + R_{22} + R_{33} = 0$, and remember to raise and lower indices with η_{ab}.]

20.8 (§**20.3**) Fill in the details of the argument which shows that the ansatz (20.45) leads to the canonical form (20.54). [Hint: be careful about signs.]

20.9 (§**20.4**) Show that the transformation (20.59) transforms (20.58) to (20.60).

20.10 (§**20.4**) Show that the equation of geodesic deviation can be written in the form (20.61). Investigate the equation for an h_{22}-wave and an h_{23}-wave.

20.11 (§**20.5**) Show that the line element (20.62) leads to (20.64) and (20.65).

20.12 (§**20.5**) Show that (20.68) transforms vacuum solutions in the Rosen form into the Brinkmann form (20.69). What is the inverse form of (20.68)?

20.13 (§**20.5**) Some authors write the Rosen line element with a 2 in front of the first term, i.e.

$$ds^2 = \bar{g}_{ab}\,d\bar{x}^a\,d\bar{x}^b = 2\,d\bar{u}\,d\bar{v} - \bar{f}^2(\bar{u})\,d\bar{y}^2 - \bar{g}^2(\bar{v})\,d\bar{z}^2.$$

(i) Show that if

$$\bar{u} = (1/\sqrt{2})u, \quad \bar{v} = (1/\sqrt{2})v, \quad \bar{y} = y, \quad \bar{z} = z$$

then the line element reduces to the Rosen form (20.62).

(ii) Show that if

$$\bar{u} = u, \quad \bar{v} = v, \quad \bar{y} = \sqrt{2}y, \quad \bar{z} = \sqrt{2}z$$

then $\bar{g}_{ab} = 2g_{ab}$, where g_{ab} is the Rosen metric (20.62) and deduce that $\bar{g}_{ab}$ gives rise to the same connection, Ricci and Einstein tensors as g_{ab} does.

20.14 (§**20.6**) Show that the definitions (20.71) and (20.72)

lead to the results (20.73), (20.74), and (20.75). [Hint: use integration by parts to establish (20.73).] Deduce that $u\delta'(u) = -\delta(u)$.

20.15 (§**20.6**) Show that (20.76) leads to (20.77), (20.78), and (20.79).

Radiation from an isolated source

21

21.1 Radiating isolated sources

The extent to which the results of the linearized theory can be trusted is not clear. The non-linearity of the gravitational field is one of its most characteristic properties, and it is likely that at least some of the crucial properties of the field should show themselves through the non-linear terms. Indeed, we have met exact solutions of the Einstein vacuum field equations corresponding to plane gravitational waves and we have seen that superposition of them leads to the creation of intrinsic singularities. This result is certainly absent in the linear case, so clearly there are differences. However, even these solutions are global vacuum solutions abstracted away from sources and as such are physically unrealistic, even if they may give us important information about how waves behave in asymptotic regions. What we would really like to do is to be able to investigate gravitational waves from bounded isolated sources, since then we would be in a position to discuss **energy transfer** and it is this which determines whether or not gravitational waves behave in the same way as other forms of radiation. Such a model system consists of an isolated bounded source (preferably possessing as much symmetry as possible, so that the field equations are easier to handle) which has been quiescent for a semi-infinite period, then radiates for a finite time, and afterwards becomes quiescent again. If the resulting waves are real physical waves, in that they carry energy, then we might expect the source to lose mass (and possibly other multipole moments may change) in the process.

The simplest field due to a bounded isolated source is spherically symmetric, but Birkhoff's theorem reveals that a spherically symmetric vacuum field is necessarily static and therefore spherically symmetric solutions cannot emit waves. Spherical symmetry assumes the existence of **three** spacelike Killing vector fields. The next simplest starting assumption, therefore, is to assume that the solution possesses **two** Killing vector fields. These fields must both be spacelike, otherwise, if one is timelike, the solution is stationary and so could not accommodate any time-dependent phenomena such as mass loss. An important case which has attracted a lot of attention is that of cylindrical symmetry. A solution is **cylindrically symmetric** if it admits a symmetry axis and is invariant under both rotations about this axis and translations parallel to it. In adapted coordinates, this requires invariance under rotations, namely,

$$\phi \to \phi' = \phi + \text{constant},$$

and translations, namely,

$$z \to z' = z + \text{constant},$$

where the z-axis is the symmetry axis. Such a solution admits two commuting spacelike Killing vector fields, namely, $\partial/\partial\phi$ and $\partial/\partial z$. Investigations of simple model interior solutions joined on to radiative exterior solutions do indeed suggest that the radiation carries away mass from the source. Unfortunately, there are considerable difficulties with the interpretation of cylindrical solutions and, moreover, they are considered physically unrealistic because the source is infinite in extent (being a cylinder extending from $z = -\infty$ to $z = +\infty$). It turns out, similarly, that work on other solutions admitting two spacelike Killing vectors also leads to problems of interpretation, and so we shall not pursue such solutions further.

The next simplest assumption is to consider a system admitting just one spacelike Killing vector field together possibly with discrete reflection symmetries. This, indeed, was the starting point of Bondi in his pioneering work on gravitational radiation in the early 1960s in which he considered a source which is axially symmetric and non-rotating. The symmetry assumptions are therefore, in adapted coordinates,

$$\phi \to \phi' = \phi + \text{constant}, \tag{21.1}$$

$$\phi \to \phi' = -\phi, \tag{21.2}$$

where the reflection symmetry (21.2) prohibits the solution from rotating (why?). Although these assumptions simplify things somewhat, the mathematics is still quite difficult and ultimately recourse has to be made to asymptotic approximation methods to discuss the radiation. It is also possible to employ the additional reflection symmetry in the equatorial plane, namely,

$$\theta \to \theta' = \pi - \theta, \tag{21.3}$$

but this does not lead to any great simplification and so we shall omit it. We shall return to the definition of the other coordinates in §21.3.

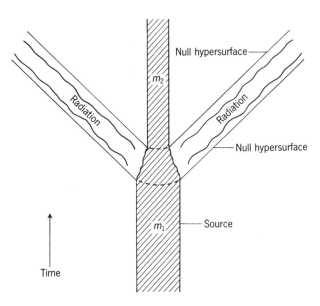

Fig. 21.1 Bondi mass loss: $m_2 < m_1$.

We therefore consider an axially symmetric non-rotating bounded isolated source which is initially static, radiates for a finite period (for example by pulsating axially symmetrically), and subsequently returns to a static configuration (Fig. 21.1). This model assumes that, once a system has radiated, it is possible subsequently for it to become quiescent again. One might expect the non-linearity to cause the waves to interfere, backscatter, and so excite the source, causing it to radiate indefinitely. This is a delicate problem, which is outside the scope of this book, and so, following Bondi, we shall assume a quiescent model is possible and restrict ourselves to outlining the proof of the mass-loss result in this case. We start by considering the surfaces which act as wave fronts in the theory.

21.2 Characteristic hypersurfaces of Einstein's equations

The field equations of general relativity form a system of hyperbolic partial differential equations. This is most easily seen in the linearized approximation, where, in an appropriate gauge, the equations are simply wave equations. As Bondi has pointed out, hyperbolic equations are very different in character to elliptic or parabolic equations since they allow for 'time-bomb' solutions, that is, solutions which are initially static but then suddenly become dynamic. Such solutions propagate their effects along privileged curves called the **bicharacteristics** of the theory. Moreover, these bicharacteristics lie on privileged surfaces called **characteristic hypersurfaces** which play the role of **wave fronts** in the propagation of these effects. Along characteristic hypersurfaces, different solutions can meet continuously and, as a consequence, they are defined as those singular hypersurfaces for which the usual Cauchy initial value problem **cannot** be solved.

To find the characteristic hypersurfaces for the vacuum field equations, recall that, in considering the Cauchy problem, we obtained the evolution equations in the form (13.14), namely,

$$g^{00}g_{\alpha\beta,\,00} = 2M_{\alpha\beta}.$$

Thus, we would be unable to solve for $g_{\alpha\beta,\,00}$ if and only if $g^{00} = 0$. As we have seen in §16.1, this is the condition for the hypersurface $x^0 = $ constant to be a **null hypersurface**. The normal vector to such a hypersurface is null and consequently it is also tangent to the hypersurface. Thus, a null hypersurface is a hypersurface that is locally tangent to the light cone (Fig. 21.2). Not only are null hypersurfaces characteristic surfaces, but they are ruled by **null geodesics** which turn out to be the **bicharacteristics** of the theory (see §21.3

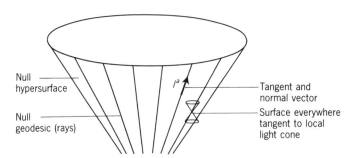

Null hypersurface

Null geodesic (rays)

l^a

Tangent and normal vector

Surface everywhere tangent to local light cone

Fig. 21.2 A null hypersurface.

below). This makes clearer the idea we met in the linearized theory, namely, that gravitational disturbances are propagated along null geodesics with the speed of light. It is clear from these considerations that null hypersurfaces play an important role in the study of gravitational radiation.

21.3 Radiation coordinates

The discussion of the last section suggests that, in order to investigate radiation, we should introduce the coordinate hypersurfaces

$$x^0 = u = \text{constant} \tag{21.4}$$

as a family of non-intersecting null hypersurfaces. The normal covariant vector field to these surfaces is therefore

$$l_a = \partial_a u = (1, 0, 0, 0) = \delta_a^0 \tag{21.5}$$

and, since it is null,

$$l_a l^a = g^{ab} \partial_a u \partial_b u = 0, \tag{21.6}$$

and the vector field is both tangent and normal to the null hypersurfaces. The **bicharacteristics** are the orbits of the contravariant vector field l^a, that is, they have equation

$$x^a = x^a(\rho) \tag{21.7}$$

for some parameter ρ, where

$$\frac{dx^a}{d\rho} = l^a = g^{ab} \partial_b u. \tag{21.8}$$

Then, taking the absolute derivative of (21.8), we get

$$\frac{D}{D\rho}\left(\frac{dx^a}{d\rho}\right) = \frac{D}{D\rho}(g^{ab} \partial_b u)$$

$$= \frac{dx^c}{d\rho} \nabla_c (g^{ab} \partial_b u)$$

$$= g^{ab} \frac{dx^c}{d\rho} (\nabla_c \partial_b u)$$

$$= g^{ab} \frac{dx^c}{d\rho} (\nabla_b \partial_c u)$$

$$= g^{ab} g^{cd} \partial_d u (\nabla_b \partial_c u)$$

$$= \tfrac{1}{2} g^{ab} \nabla_b (g^{cd} \partial_c u \partial_d u)$$

$$= 0, \tag{21.9}$$

using the symmetry of the connection in the fourth equality and (21.6) in the last. Hence, the bicharacteristics are **null geodesics** and ρ is an **affine parameter**. These null geodesics are often called **null rays**.

We choose as a second coordinate

$$x^1 = r, \tag{21.10}$$

where r is a radial parameter along the null rays, and we then use the remaining coordinates x^2 and x^3 to label the null rays. Assuming that space-time is asymptotically flat, that is,

$$\lim_{r \to \infty} g_{ab} = \eta_{ab}, \tag{21.11}$$

we can then take x^2 and x^3 to be the usual spherical polar angles

$$x^2 = \theta, \qquad x^3 = \phi \tag{21.12}$$

defined on each 2-sphere ($u = $ constant, $r = \infty$) at future null infinity $\mathscr{I}^+$. These coordinates are called **Bondi** or **radiation coordinates**. They are really only defined in a neighbourhood of $\mathscr{I}^+$ because if we follow the null rays back into the interior the gravitational field will cause them to focus and cross in general (Fig. 21.3). However, we shall ultimately be working asymptotically and so the coordinate system will be adequate for our needs.

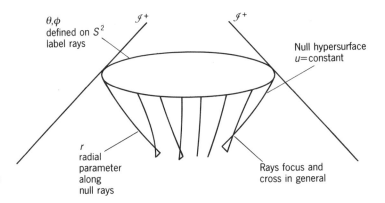

Fig. 21.3 Bondi's radiation coordinates (u, r, θ, ϕ).

21.4 Bondi's radiating metric

A null ray is one of the coordinate curves

$$u = u_0, \quad \theta = \theta_0, \quad \phi = \phi_0,$$

where u_0, θ_0, and ϕ_0 are constants, and r is varying. The tangent vector to this curve is

$$\frac{\mathrm{d}x^a}{\mathrm{d}r} = (0, 1, 0, 0) = \delta_1^a,$$

and so it must be parallel to l^a, that is, $l^a = \lambda \delta_1^a$ for some proportionality factor λ. But, by (21.8) and (21.5),

$$l^a = g^{ab} \partial_b u = g^{ab} \delta_b^0 = g^{0a},$$

for which we get

$$g^{00} = g^{02} = g^{03} = 0. \tag{21.13}$$

These conditions on the contravariant metric are equivalent to the conditions on the covariant metric (exercise)

$$g_{11} = g_{12} = g_{13} = 0. \tag{21.14}$$

Newman and Penrose, in work on gravitational radiation subsequent to Bondi, took x^1 to be an affine parameter, in which case (exercise) $\lambda = 1$ and

$$g_{01} = g^{01} = 1 \tag{21.15}$$

However, Bondi chose $x^1 = r$ to be a **luminosity distance parameter** defined by requiring

$$\begin{vmatrix} g_{22} & g_{23} \\ g_{23} & g_{33} \end{vmatrix} = r^4 \sin^2 \theta. \tag{21.16}$$

The significance of this choice is that the 2-surfaces ($u = $ constant, $r = $ constant) have the usual surface area of a 2-sphere, namely, $4\pi r^2$.

We next impose the symmetry assumptions of **axial symmetry** (21.1), which results in (exercise)

$$\frac{\partial g_{ab}}{\partial \phi} = 0, \tag{21.17}$$

and **azimuth reflection invariance** (21.2), which results in (exercise)

$$g_{03} = g_{13} = g_{23} = 0, \tag{21.18}$$

or equivalently

$$g^{03} = g^{13} = g^{23} = 0. \tag{21.19}$$

Putting all these assumptions together, we can write the metric in the particular form of **Bondi's radiating metric** (exercise)

$$ds^2 = \left(\frac{V}{r} e^{2\beta} - U^2 r^2 e^{2\gamma} \right) du^2 + 2e^{2\beta}\, du\, dr + 2Ur^2 e^{2\gamma}\, du\, d\theta$$
$$- r^2 (e^{2\gamma} d\theta^2 + e^{-2\gamma} \sin^2\theta\, d\phi^2), \tag{21.20}$$

where V, U, β, and γ are four arbitrary functions of the three coordinates u, r, and θ by (21.17), that is,

$$V = V(u, r, \theta), \quad U = U(u, r, \theta), \quad \beta = \beta(u, r, \theta), \quad \gamma = \gamma(u, r, \theta). \tag{21.21}$$

21.5 The characteristic initial value problem

We now consider the initial value problem for Bondi's radiating metric. The situation is different from the Cauchy problem because this time initial data is set on a characteristic or null hypersurface rather than on a spacelike hypersurface. As a consequence, it is called the **characteristic initial value problem**. Bondi showed that the ten vacuum field equations break up into four groups:

(1) three **symmetry conditions**

$$R_{03} = R_{13} = R_{23} \equiv 0; \tag{21.22}$$

(2) four **main equations**

$$R_{11} = R_{12} = R_{22} = R_{33} = 0; \tag{21.23}$$

(3) one **trivial equation**

$$R_{01} = 0; \tag{21.24}$$

(4) two **supplementary conditions**

$$R_{00} = R_{02} = 0. \tag{21.25}$$

The three components R_{03}, R_{13}, R_{23} vanish identically as a consequence of the symmetry assumptions. Recall that in the Cauchy problem we proved a result which states that if the dynamical equations hold everywhere and the constraint equations hold on an initial hypersurface then the contracted Bianchi identities ensure that the constraint equations hold everywhere. There is an analogous result for the characteristic initial value problem, except that in this case the 'constraint equations' consist of the trivial equation and the supplementary conditions, and the trivial equation is automatically satisfied as an algebraic consequence.

> **Lemma:** If the main equations hold everywhere, then the contracted Bianchi identities ensure that
> (a) the trivial equation holds as an algebraic consequence,
> (b) the supplementary conditions hold everywhere if they hold on a hypersurface $r = $ constant.

Hence, the initial value problem reduces to solving the main equations and satisfying the supplementary conditions for one value of r. The main equations break up further into the following:

(2a) one **dynamical equation**

$$R_{33} = 0; \tag{21.26}$$

(2b) three **hypersurface equations**

$$R_{11} = R_{12} = g^{22} R_{22} + g^{33} R_{33} = 0. \tag{21.27}$$

The dynamical equation is the only main equation which involves a term differentiated with respect to u and hence propagating into the future (that is, from one null hypersurface to the next). The hypersurface equations only involve differentiation within the hypersurface $u = $ constant.

If we assume that the solution is analytic everywhere, then a detailed analysis of the main equations leads to the following schema for integration. We first prescribe γ on $u = u_0$, that is, on some initial hypersurface N_0, say. The three hypersurface equations then determine β, U, and V on N_0. The dynamical equation serves to determine $\gamma_{,0}$ on N_0, which means that γ is determined on the 'next neighbouring' null hypersurface N_1, say. We then go through the whole cycle again on N_1 (Fig. 21.4). Proceeding in this way, we can generate a solution of the field equations in some region to the future of N_0. However, we have neglected functions of integration in the schema and it turns out that one of them, called the 'news' function, plays a key role in the analysis.

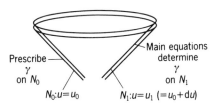

Prescribe γ on N_0 Main equations determine γ on N_1

$N_0{:}u{=}u_0$ $N_1{:}u{=}u_1 \; (=u_0{+}du)$

Fig. 21.4 An integration schema for Bondi's solution.

21.6 News and mass loss

In order to proceed further, we need to expand everything in inverse powers of the radial parameter r and carry out an **asymptotic analysis**. We shall outline the procedure. We start by strengthening the condition (21.11) and a detailed analysis reveals that the asymptotic behaviour of the metric is given by

$$
\left.
\begin{aligned}
g_{00} &= 1 + O(r^{-1}), \\
g_{01} &= 1 + O(r^{-1}), \\
g_{02} &= O(1), \\
g_{22} &= -r^2 + O(r), \\
g_{33} &= -r^2 \sin^2\theta + O(r).
\end{aligned}
\right\} \tag{21.28}
$$

We mention briefly that the coordinate transformations which preserve the form of the metric (21.20) together with the above asymptotic conditions form a group called the **Bondi–Metzner–Sachs**, or BMS, group. The BMS group is important because it plays the same role asymptotically for an isolated radiative system as the Poincaré group does in special relativity. Bondi adopts a final assumption, namely,

$$\lim_{r \to \infty} \left[\frac{\partial(r\gamma)}{\partial r} \right]_{u = \text{const}} = 0, \tag{21.29}$$

in an attempt to prevent radiation coming in from past null infinity and affecting the source. He chose this condition in analogy to the Sommerfield condition in electromagnetic theory which prevents incoming radiation, but it turns out that (21.29) is not strong enough to prevent the occurrence of sufficiently weak incoming radiation.

We now have sufficient starting assumptions to expand everything in inverse powers of r. It is only necessary to work to a certain limited order in inverse powers to obtain the mass-loss result. For example, to the required order, we get (changing the original notation slightly)

$$\gamma = \frac{n}{r} + \frac{q}{r^3} + O(r^{-4}),\qquad(21.30)$$

where $n = n(u, \theta)$ and $q = q(u, \theta)$ are arbitrary functions at this stage. The hypersurface conditions lead to

$$\left.\begin{array}{l}
\beta = -n^2/(4r^2) + O(r^{-3}),\\[4pt]
U = -(n_{,2} + 2n \cot \theta)/r^2 + (2d + 3n\,n_{,2} + 4n^2 \cot \theta)/r^3 + O(r^{-4})\\[4pt]
V = r - 2M + O(r^{-1}),
\end{array}\right\}\qquad(21.31)$$

where $d = d(u, \theta)$ and $M = M(u, \theta)$ are also arbitrary. The dynamical equation produces

$$4q_{,0} = 2Mn - d_{,2} + d\cot \theta.$$

The supplementary conditions lead to

$$M_{,0} = -n_{,0}^2 + \tfrac{1}{2}(n_{,22} + 3n_{,2}\cot \theta - 2n)_{,0},\qquad(21.32)$$

$$-3d_{,0} = M_{,2} + 3n\,n_{,02} + 4n\,n_{,0}\cot \theta + n_{,0}n_{,2},\qquad(21.33)$$

where these last two equations are **exact** results by the lemma, since this is the part of the equations which holds on $r =$ constant.

A detailed investigation reveals that, as before, our initial data involves prescribing one function of three variables, namely,

$$\gamma = \gamma(u, r, \theta),\qquad(21.34)$$

on N_0. However, in addition, we must prescribe one function of two variables, namely, the u derivative of n.

$$n_{,0} = n_{,0}(u, \theta),\qquad(21.35)$$

for any value of r. Since we are working asymptotically we shall prescribe $n_{,0}$ on $\mathscr{I}^+$. Finally, we must prescribe two functions of one variable, namely,

$$M = M(u_0, \theta),\qquad d = d(u_0, \theta),\qquad(21.36)$$

which we prescribe on the intersection of N_0 and $\mathscr{I}^+$ (see Fig. 21.5). With this initial data, **all** other quantities are determined. Clearly, it is the data $n_{,0}$ which determines the evolution of the source and, as a consequence, it is termed the **news** function. If the solution is static then the news function vanishes. If we restrict our attention to the periods when the source is static, then it is possible to find a coordinate transformation which relates the quantities M, d, and q to known physical parameters. It turns out that M is intimately connected to the mass and is termed the **mass aspect**. In fact, the quantity

$$m(u) = \tfrac{1}{2}\int_0^\pi M(u, \theta)\sin \theta \, d\theta\qquad(21.37)$$

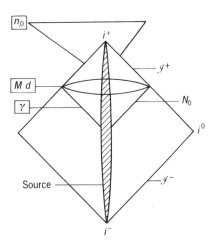

Fig. 21.5 Penrose diagram indicating initial data.

determines the mass of the system at $\mathscr{I}^+$ and is called the **Bondi mass**. (Similarly d is termed the **dipole aspect** and q the **quadrupole aspect**.) Multiplying (21.32) by $\sin\theta$, integrating with respect to θ from 0 to π, and using (21.37) together with some regularity conditions on the symmetry axis, we find (exercise)

$$m_{,0} = -\tfrac{1}{2} \int_0^\pi n^2_{,0} \sin\theta\, d\theta. \qquad (21.38)$$

The non-positive nature of the right hand side leads to the promised result.

Theorem: There is mass-loss if and only if there is news.

Thus, if a system remains quiescent, then there is no news and hence the Bondi mass remains constant. If, however, the system radiates, then there is news and the minus sign in (21.38) means there is a consequent mass **loss**. If a radiating system can become quiescent again, then this establishes the content of Fig. 21.1. The power of this result is that we have obtained it without having to assume that the gravitational field is weak everywhere and no linearization of the field is needed. We mention that, shortly after Bondi published his results, Sachs dropped the symmetry assumptions and obtained essentially the same result. The calculations are obviously longer, and, in general, it turns out that there are **two** news functions, corresponding to the two gravitational degrees of freedom, but otherwise the argument proceeds along similar lines.

21.7 The Petrov classification

The gravitational field is governed by the Riemann tensor. We can gain considerable insight into the possible types of gravitational field by considering the **algebraic** structure of the Riemann tensor. We restrict ourselves to the **vacuum** case, where the Riemann tensor coincides with the Weyl tensor, because in four dimensions, by (6.87),

$$C_{abcd} = R_{abcd} - g_{a[c}R_{d]b} + g_{b[c}R_{d]a} - \tfrac{1}{3}g_{a[d}g_{c]b}R, \qquad (21.39)$$

and in the vacuum case $R_{ab} = R = 0$.

The Weyl tensor has the same symmetries as the Riemann tensor and, in addition, possesses the trace-free property

$$C^a{}_{bac} \equiv 0. \qquad (21.40)$$

Since C_{abcd} is skew symmetric on each pair of indices and also symmetric under their interchange, we can start by thinking of it as a 6×6 symmetric matrix (exercise). We can then classify the Weyl tensor algebraically by classifying this 6×6 matrix in terms of its eigenvalues and eigenvectors. So, at first sight, we would expect this to involve classifying the possible roots of a sixth-order or sextic equation. However, the procedure is complicated by the additional symmetries (21.40) and $C_{a[bcd]} = 0$. We shall not pursue the details

Table 21.1

Petrov type:	I	II	D	III	N
Quartic roots:	all distinct	one double	two double	one triple	one four-fold
Distinct eigenvectors:	4	3	2	2	1

further, but it turns out that these symmetries reduce the problem to classifying the roots of a **quartic** equation. The resulting classification due to Petrov—and hence called the **Petrov** classification—itemizes the various possibilities of distinct eigenvalues and eigenvectors of the Weyl tensor at a point and gives them a name or **type** as shown in Table 21.1. If we add to this the completely degenerate case of conformally flat space-times in which C_{abcd} vanishes (called type 0), then there are six possibilities which can be conveniently arranged in a triangular hierarchy (Fig. 21.6), as suggested by Penrose. In the diagram, the arrows point in the direction of increasing specialization. The Petrov type of a given vacuum space-time is then defined as the type at those points which are highest up the hierarchy. Thus, a solution may be the same type everywhere, or may reduce to lower types at some points or region, but by definition the type cannot move up the hierarchy. A generic solution will be type I, which is called **algebraically general**, whereas all other types are called **algebraically special**.

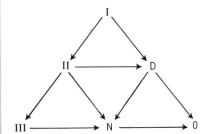

Fig. 21.6 The hierarchy of Petrov types.

A different but equivalent method, due to Debever, consists in classifying certain null vectors, called **principal null directions**, which have a special relationship to the Riemann tensor. The result rests on the following theorem.

> **Theorem:** Every vacuum space-time admits at least one and at most four null directions $l^a \neq 0$, $l^a l_a = 0$, which satisfy
> $$l_{[a} R_{b]ef[c} l_{d]} l^e l^f = 0$$

There is a corresponding result for non-vacuum space-times if we replace R_{abcd} by C_{abcd}. The Petrov type then relates to the coincidence of these null directions according to

Type:	I	II	D	III	N
Coincidence:	[1 1 1 1]	[2 1 1]	[2 2]	[3 1]	[4]

The coincidence also agrees with the coincidence of roots in the Petrov quartic equation.

The particular vacuum solutions of Schwarzschild, Reissner–Nordström, and Kerr are all algebraically special type D. Plane gravitational waves are type N, and hence the gravitational field from an isolated radiating source is expected to be asymptotically type N. However, any solution which is sufficiently complex to model a realistic solution will be type I. Bondi's radiating vacuum solution, namely (21.20), subject to (21.30) and (21.31), is type I, but asymptotically type N with

$$R_{abcd} \sim n_{,00}/r + O(r^{-2})$$

when $n_{,00} \neq 0$. We add that, in a non-vacuum space-time, the Petrov classification of the Weyl tensor is augmented by an analogous classification of the Ricci tensor called the **Plebanski type**. Moreover, the complete classification of the Weyl tensor and its covariant derivatives (in a canonically defined frame) leads to the **Karthede classification** mentioned in §13.7.

21.8 The peeling-off theorem

In the last section, we defined the possible algebraic types of the Riemann tensor in a vacuum space-time. In this section, we consider the physical significance of this classification. Sachs investigated the case of a retarded wave solution emanating from an isolated source in the linearized theory and was able to expand the Riemann tensor in terms of an affine parameter r along each outward null ray (null geodesic) producing the result

$$R = \frac{N_0}{r} + \frac{III_0}{r^2} + \frac{II_0}{r^3} + \frac{I_0}{r^4} + \frac{I_0'}{r^5} + O(r^{-6}) \tag{21.41}$$

where, for convenience, we have suppressed the indices. Thus, asymptotically, the leading order of the Riemann tensor in type N, then type III, type II and type I, respectively, at the subsequent orders. In the equation the 0 denotes a vanishing absolute derivative in the ray direction l^a. Unlike the other coefficients in (21.41), I_0' does not have a special relationship with l^a since its one principal null direction is not tangent to a null geodesic. Sachs also considered algebraically special fields and found that they do not have an expansion as general as (21.41), but, in generalizing the work of Bondi, he was able to show that the Riemann tensor for an asymptotically flat isolated radiative system has precisely the same form as (21.41). Indeed, starting in the **wave zone**, where the Riemann tensor is type N with a fourfold repeated ray direction l^a, the other principal null directions peel off as we move in towards the source, where terms of a less special nature predominate (Fig. 21.7). This is known as the **peeling-off theorem**.

Szekeres has investigated the properties of type N, III, and D fields by considering their effect on a cloud of test particles. An observer sets up an orthonormal triad $\{e_1{}^a, e_2{}^a, e_3{}^a\}$ of spacelike vectors adapted to the field in each case. For type N fields, the forces on the ring of particles results in the distortion shown in Fig. 21.8 (compare with Fig. 20.2). This clearly indicates the transverse character of such fields, since $e_1{}^a$ points in the direction of propagation of the field. Szekeres terms this a **pure transverse** gravitational wave. For type III fields, the effect on the particles is still planar, but in this case the plane contains the wave direction $e_1{}^a$ and the axis is tilted through

N r^{-1} III r^{-2} II r^{-3} I r^{-4}

Fig. 21.7 The peeling-off theorem.

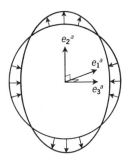

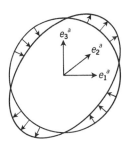

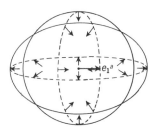

Fig. 21.8 The effects of a type N field on a ring of test particles.

Fig. 21.9 The effects of a type III field on a ring of test particles.

Fig. 21.10 The effects of a type D field on a sphere of test particles.

45° to the wave direction (Fig. 21.9). Szekeres terms this a **longitudinal wave** component. For type D fields, the effect ceases to be planar. In this case, a sphere of particles is distorted into an ellipsoid with major axis lying in the wave direction (Fig. 21.10). This is precisely the tidal force we discussed before in §16.10 for a radially infalling observer in the Schwarzschild field. Szekeres terms this a **Coulomb-type** field in analogy with electromagnetism. For type I and type II fields, nothing simple emerges.

21.9 The optical scalars

Consider a congruence of null geodesics with tangent vector field l^a. By a change of scale, it is always possible to obtain the geodesic equation in the simple form (exercise)

$$l^a{}_{;b}l^b = 0.$$

We assume this has been done and define three quantities called **optical scalars** determined by the congruence l^a as follows:

expansion (divergence): $\quad \theta = \tfrac{1}{2}l^a{}_{;a}$

twist (rotation): $\quad \omega = \{\tfrac{1}{2}l_{[a;b]}l^{a;b}\}^{\frac{1}{2}}$

shear (distortion): $\quad |\sigma| = \{\tfrac{1}{2}l_{(a;b)}l^{a;b} - \theta^2\}^{\frac{1}{2}}$

Their physical interpretation is embodied in the following result of Sachs. If a small object in a null geodesic congruence casts a shadow on a screen, then all portions hit it simultaneously. The shape, size, and orientation of the shadow depend only on the location of the screen and not on its velocity. If the screen is an infinitesimal distance dr from the object, then the shadow is expanded by $\theta\,dr$, rotated by $\omega\,dr$, and sheared by $|\sigma|\,dr$ (Fig. 21.11). The quantity shear

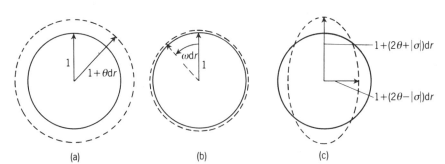

Fig. 21.11 The optical scalars: (a) expansion; (b) twist; (c) shear.

turns out to be the most important physically, as is evident from the following theorem.

> **Goldberg–Sachs theorem**: A vacuum solution is algebraically special if and only if it contains a shear-free null geodesic congruence.

In an isolated radiative system, the news function is also intimately connected to the shear.

The Petrov classification, optical scalars, and Killing vectors are three very important tools for classifying vacuum solutions in a coordinate-independent way. In particular, they have been used to find particular exact solutions of the field equations. Indeed, there are known vacuum solutions for each of the four classes of algebraically special Petrov type determined by the vanishing or otherwise of the expansion and twist. In some of these cases, all possible solutions are known. For example, all vacuum type D solutions have been found. There are also a large number of vacuum type I or algebraically general solutions known. However, few of these solutions are fully understood in the sense that we are able to understand their causal structure, geodesic structure, global structure, and singularity structure. Thus, there are relatively few solutions for which we can draw space-time, spatial, and Penrose diagrams. Moreover, there is evidence to suggest that many of them have a strange singularity structure and, as such, are pathological in nature and unlikely to approximate to any physically realistic solution.

Finally, we mention that, by using the Riemann identity on l^a

$$l^a{}_{;bc} - l^a{}_{;cb} = R^a{}_{dcb} l^d \tag{21.42}$$

and the definitions of optical scalars, it is a straightforward matter to derive propagation equations for the optical scalars. For example, setting

$$z = -\theta + i\omega, \tag{21.43}$$

we can deduce

$$\frac{Dz}{Dr} = z^2 + |\sigma|^2 + \tfrac{1}{2} R_{ab} l^a l^b. \tag{21.44}$$

This is a null version of the Raychauduri equation and equations such as these play a central role in the proof of the singularity theorems.

Exercises

21.1 (§21.1) Define cylindrical symmetry. What conditions does this impose on the metric coefficients in adapted coordinates $(x^a) = (x^0, x^1, \phi, z)$? Write down the metric of the 2-space ($\phi = $ constant, $z = $ constant). Use the result of Exercise 6.30 to deduce that there exist coordinates in which the line element can be written in the form

$$ds^2 = e^{2\alpha}(dt^2 - d\rho^2),$$

where α is a function of t and ρ only. What are the conditions for the (t, ρ) plane to be orthogonal to the (ϕ, z)-plane? Assuming these conditions, show that a cylindrically symmetric line element can be written in the canonical form

$$ds^2 = e^{2\gamma - 2\psi}(dt^2 - d\rho^2) - \rho^2 e^{-2\psi} d\phi^2$$
$$- e^{2\psi + 2\mu}(dz + \chi d\phi)^2,$$

where γ, ψ, μ, and χ are all functions of t and ρ only.

21.2 (§**21.1**) What is the condition for a cylindrically symmetric solution to be non-rotating? What effect does this have on the line element of Exercise 21.1?

21.3 (§**21.4**) Show that if a null ray is given by

$$u = u_0, \qquad \theta = \theta_0, \qquad \phi = \phi_0$$

then it leads to the conditions (21.13). Show that (21.13) is equivalent to (21.14). [Hint: consider inverting a general symmetric 4 × 4 matrix with zeros in the positions defined by (21.13).] Show that if x^1 is an affine parameter it leads to the conditions (21.15). Show that axial symmetry and azimuth reflection invariance leads to the conditions (21.18) or (21.19). Let $x^1 = r$ be a luminosity parameter defined by (21.16) and deduce Bondi's radiating metric (21.20) subject to (21.21). [Hint: show first that the conditions lead to a metric in which g_{00}, g_{01}, g_{02}, and g_{22} are four arbitrary functions of u, r, and θ; then the actual form of these coefficients are chosen to preserve the signature and for later convenience.]

21.4 (§**21.4**) Show that the surface area of the 2-surface $(u = u_0, r = r_0)$ is $4\pi r_0^2$.

21.5 (§**21.4**) Find the non-zero components of the metric connection Γ^a_{bc} of Bondi's radiating metric. [Hint: use the variational principle approach of §7.6.]

21.6 (§**21.5**) Use the results of Exercise 21.5 to establish the lemma of §21.5. [Hint: write out the contracted Bianchi identities in terms of Γ^a_{bc} and R_{ab}; do not insert the metric expressions for Γ^a_{bc} and R_{ab} in the identities, but merely consider which quantities are zero and which are not.]

21.7 (§**21.5**) Evaluate the components of the Ricci tensor which define the four main equations. [Hint: this is a long but straightforward calculation.] Use the results to confirm the integration schema for Bondi's solution.

21.8 (§**21.6**) The requirement that the Bondi metric remains regular on the symmetry axis $\theta = 0$, π leads to a number of conditions including $n(u, 0) = n(u, \pi) = 0$. Use these conditions together with (21.32) and (21.37) to deduce the mass-loss result.

21.9 (§**21.7**) Show that the symmetries

$$C_{abcd} = -C_{abdc} = -C_{bacd} = C_{cdab}$$

mean that we can treat C_{abcd} at a point as a symmetric 6 × 6 matrix.

21.10 (§**21.9**) Consider a congruence of null geodesics with tangent vector l^a. Write down the geodesic equation l^a satisfies in general. Show that if we rescale l^a so that

$$l^a \to \bar{l}^a = A l^a$$

then we can choose A so that the geodesic equation reduces to

$$\bar{l}^a_{;b} \bar{l}^b = 0.$$

21.11 (§**21.9**) Compute an expression for the expansion for Bondi's radiating line element (21.20) using $l^a \overset{*}{=} e^{-2\beta} \delta^a_1$.

F. Cosmology

Relativistic cosmology 22

22.1 Preview

Cosmology is the study of the dynamical structure of the universe as a whole. As in most modelling exercises, we shall start by trying to find a very simple model of the universe. This is done by smoothing out all the irregularities in space and in time and concentrating simply on the gross features of the universe. So, to start with, we ignore all details such as the solar system, our own galaxy (the Milky Way), the local cluster of galaxies and so on; the consideration of these details can then hopefully be introduced at a later stage to yield a more complete or better theory. We shall be concerning ourselves only with the very basics of cosmology, that is, the overall dynamics of the system. We shall see in this chapter that this is governed by a first-order ordinary differential equation called Friedmann's equation. The resulting solutions are the standard solutions of relativistic cosmology and are called the **Friedmann models**. We shall investigate some of these in the next chapter.

Cosmology as a separate scientific study really only came into existence with the advent of general relativity. It is possible to consider cosmology in a Newtonian framework, but this had not been seriously attempted prior to general relativity largely because, in as far as there was a generally accepted model of the universe in existence, it was considered devoid of dynamics; that is, the universe was considered **static**. Perhaps, surprisingly, it is possible to construct a 'Newtonian cosmology', based on Newtonian theory together with a number of **ad hoc** assumptions, which also results in Friedmann's equation. (However, the interpretation of some of the terms in the equation is different.) But it is emphasized that this Newtonian approach only came into existence after general relativity had first tackled the problem. We shall look at a discrete Newtonian model in §22.3. The starting point for both Newtonian and relativistic cosmology is a simplicity principle called the cosmological principle which states, essentially, that the universe is unchanging in space from point to point. This leads to the requirement that space is homogeneous and isotropic (the same in every direction) about each point.

In the early decades of cosmology, there were very few reliable observational results and, not surprisingly, different Friedmann models enjoyed periods of fashion, that is, periods when they were considered the best available model for our own universe. However, there was one school of thought that argued vociferously for a simple non-Friedmann model called the **steady-state solution** (§23.12), based on the perfect cosmological principle that the universe is unchanging in space **and** time. Most of these considerations are largely historical in nature, for, although the Friedmann models are still basic to much of cosmological thinking, the theory has progressed considerably beyond them. None the less, we shall largely content ourselves with obtaining and investigating the Friedmann models in this introduction to cosmology.

In more recent decades, one model has emerged as the best available, at least as far as the origins the universe are concerned, and that is the **hot big bang**. In this model it is assumed that there occurred a cataclysmic event (some 10^{10} years ago), called the big bang, when the universe sprang into existence and expanded away from a singular point. In the earliest phases, the universe consisted of radiation at incredibly high temperatures and densities. As the universe expanded, the temperature and density fell and protons, electrons, and neutrons emerged from the radiation bath. As the system cooled further, the simple atoms such as hydrogen and helium emerged first, followed later by the heavier elements. This phase can be treated mathematically and one of the great successes of this approach has been the agreement of the theoretical prediction of the abundancies of the heavy elements with the observed abundancies. As the system expands and cools yet further, then conditions become favourable for the condensation of the nebulae, that is, the stars and galaxies, from the primeval matter. The model then encompasses the dynamics of these nebulae up to the present epoch.

The development of the hot big bang model brings out an important point; namely, in modelling the universe in the large we have made use of our understanding of **local** physical laws. The justification for this is that we are more or less forced to do so — otherwise we would hardly be able to start — and yet it has proved extremely successful, to date, in providing insight into the structure of the universe. However, we cannot rule out the possibility that there exist additional interactions which only reveal themselves on a cosmological scale. One example of this is the cosmological term (Λg_{ab}) which Einstein incorporated into general relativity. Another important point relates to the fact that, in most branches of physics, it is possible to investigate phenomena by repeatedly carrying out experiments in the laboratory in controlled conditions where all but a small number of parameters are held fixed. No such possibility occurs in cosmology. Indeed, cosmology is unlike any other branch of physics in that the system we are studying is unique. Given this constraint, it is perhaps surprising that we are able to construct such apparently successful models. This success is so marked that, in some cosmological circles, the claim is that the universe is well understood after the first 10^{-43} seconds from its birth, so well in fact that the period after this time is referred to in these circles as the 'late universe'!

22.2 Olbers' paradox

The fact that, prior to general relativity, the universe was considered static is perhaps even more surprising when one is confronted by a paradox put forward by Olbers in 1826 which stems from the observation that the sky is dark at night. (In fact, others had considered similar ideas before, but Olbers gave a more precise statement of the paradox.) He assumed that space is Euclidean and infinite and that the average number of stars per unit volume and the average luminosity of each star is constant throughout space and time, provided these averages are taken over sufficiently large regions. He also assumed that the universe has been in existence for an infinite time and that, on the large scale, it is static. Now consider a shell of radius r and thickness dr, and let l denote the product of the average number of stars per unit volume and the average luminosity per star. The intensity at the centre of the shell will be given by the total luminosity produced by the shell divided by its area, that is, approximately,

$$\frac{(4\pi r^2 \, dr)l}{4\pi r^2} = l \, dr. \tag{22.1}$$

If we surround any point P by an infinite succession of shells, each of thickness dr, then clearly the intensity at P will be $\int_0^\infty l \, dr$, which is infinite! However, we have omitted to account for the possibility that light from a star may be intercepted on its way by another star (Fig. 22.1). When this is taken into account, it can be shown that the result is no longer infinite but equal to the average luminosity at the surface of a star. Since P is arbitrary, the result must hold everywhere. This leads to a paradox, because the sky is observed to be dark at night. The same conclusion may be reached by thermodynamic arguments. For, if the system is static and of infinite age, then it must have reached thermodynamic equilibrium, which means that each star must be absorbing as much radiation as it emits, and the result follows. Yet another argument is that, if one looks in any direction in an infinite universe in which the average number of stars per unit volume is finite, then the line of sight will eventually end on a star. Since the system is static, the light received from the star is not degraded, and the result again follows.

Fig. 22.1 Light intercepted by another star.

It is interesting to note that the bulk of this enormous amount of radiation arrives from very distant parts, half, in fact, from regions so distant that the light has only a 50% chance of arriving without being absorbed by other stars. An estimate from observations in our own neighbourhood suggests that half of this radiation should be due to stars more than 10^{20} light years distant.

Olbers tried to resolve this paradox by postulating the existence of a tenuous gas which would absorb the radiation in transit over long distances. This argument will not work, though, because the gas would be heated until it reaches a temperature at which it radiates as much as it receives, and hence it will not reduce the average density of radiation. The same paradox arises even if the assumption that the universe is Euclidean is dropped (exercise). Nor does it make any difference whether the universe is infinite (open) or bounded (closed).

As we look further out into space, we are looking further back in time. One resolution of the paradox rests on assuming that l is a function of time which is sufficiently small in the distant past that the distant regions do not contribute significantly to the radiation density. If it is assumed that the universe is static and that the stars do not start radiating until some finite period in the past, then it is possible to arrange for this period to be short enough to lead to the radiation density we observe today. However, some estimates would then suggest that the universe is younger than the age of the oldest stars. The accepted resolution rests on assuming that the universe is not static but rather undergoing large-scale expansion. Then, because of the Doppler shift, light received from receding stars will be shifted to the red and, if the recessional velocity is large enough, the loss of energy will be sufficient to reduce the radiation density to the observed level.

In summary, assuming that a dark night sky is not just a phenomenon of our current epoch, then Olbers' paradox requires that either the universe is young or it is expanding. In the latter case the question may be asked as to what happens to the 'lost' energy resulting from the Doppler shift. In fact, it is precisely this energy which is doing the work involved in the expansion of the universe.

22.3 Newtonian cosmology

In this section, we shall introduce Newtonian cosmology by investigating a simple discrete model in which it is assumed that the universe consists of a finite number of galaxies. Let the ith galaxy have mass m_i and position $r_i(t)$ as measured from a fixed origin O. We now impose the cosmological principle (see §22.4) in the form that the motion about O must be spherically symmetric, in which case the motion of the galaxies is purely radial, i.e.

$$r_i(t) = r_i(t)\hat{r}. \tag{22.2}$$

The kinetic energy T of the system is then

$$T = \tfrac{1}{2} \sum_{i=1}^{n} m_i \dot{r}_i^2.$$

The gravitational potential energy between a pair of galaxies m_i and m_j is given by $-Gm_i m_j / |r_i - r_j|$, and so the total potential energy V of the system is

$$V = -G \sum_{\substack{i,j=1 \\ (i<j)}}^{n} \frac{m_i m_j}{|r_i - r_j|}, \tag{22.3}$$

where the inequality in the double sum means that each pair of particles is only counted once. We also assume that there is a cosmological force acting on the ith galaxy of the form

$$F_i = \tfrac{1}{3}\Lambda m_i r_i, \tag{22.4}$$

where Λ is a constant called the cosmological constant. This yields an additional potential energy, called the cosmological potential energy V_c of the system, given by

$$V_c = -\tfrac{1}{6}\Lambda \sum_{i=1}^{n} m_i r_i^2. \tag{22.5}$$

The total energy E of the system is therefore

$$E = \tfrac{1}{2} \sum_{i=1}^{n} m_i \dot{r}_i^2 - G \sum_{\substack{i,j=1 \\ (i<j)}}^{n} \frac{m_i m_j}{|r_i - r_j|} - \tfrac{1}{6}\Lambda \sum_{i=1}^{n} m_i r_i^2. \tag{22.6}$$

Let us assume that the distribution and motion of the system is known at some fixed epoch t_0. Then the radial motion required by the cosmological principle implies that, at any time t,

$$r_i(t) = S(t) r_i(t_0), \tag{22.7}$$

where $S(t)$ is a universal function of time which is the same for all particles and is called the **scale factor**. This means that the only motions compatible with homogeneity and isotropy are those of uniform expansion or contraction, that is, a simple scaling up or down by a time-dependent scale factor.

The radial velocity of the ith galaxy is then

$$\dot{r}_i(t) = \dot{S}(t)r_i(t_0) = \frac{\dot{S}(t)}{S(t)}r_i(t) \tag{22.8}$$

by (22.7). We define a quantity called the **Hubble parameter** $H(t)$ by

$$H(t) = \dot{S}(t)/S(t), \tag{22.9}$$

and then (22.8) can be written as

$$\dot{r}_i(t) = H(t)r_i(t), \tag{22.10}$$

which is called **Hubble's law**. This states that, in an expanding universe, at any one epoch, the radial velocity of recession of a galaxy from a given point is proportional to the distance of the galaxy from the point. The value of the Hubble parameter at our epoch is known as the **Hubble constant**.

If we substitute (22.7) and (22.8) into (22.6), we find (exercise)

$$E = A[\dot{S}(t)]^2 - \frac{B}{S(t)} - D[S(t)]^2, \tag{22.11}$$

where the coefficients are positive constants defined by

$$A = \tfrac{1}{2} \sum_{i=1}^{n} m_i[r_i(t_0)]^2, \tag{22.12}$$

$$B = G \sum_{\substack{i,j=1 \\ (i<j)}}^{n} \frac{m_i m_j}{|r_i(t_0) - r_j(t_0)|}, \tag{22.13}$$

$$D = \tfrac{1}{6}\Lambda \sum_{i=1}^{n} m_i[r_i(t_0)]^2 = \tfrac{1}{3}\Lambda A. \tag{22.14}$$

This is one form of the **cosmological differential equation** for the scale factor $S(t)$. It has a simple interpretation. First of all, consider what happens when Λ vanishes, in which case we can neglect the last term. If the universe is expanding, then the second term on the right-hand side decreases and, since the total energy remains constant, it follows that the first term must decrease as well. Therefore the expansion must slow down. If Λ is positive, then all galaxies experience a cosmic repulsion, pushing them away from the origin out to infinity. In this case, the cosmological term contributes positively to the expansion. If Λ is negative, then the opposite happens and all galaxies experience a cosmic attraction towards the origin. In a later section, we shall go on to consider what solutions of the differential equation are possible for different values of the parameters occurring in them. In particular, we shall investigate whether it is possible for the expansion to slow down, stop, and reverse so that eventually the universe will collapse — the so-called 'big crunch'.

We finish this section by rewriting the differential equation in a form closer to the general relativistic equation. Solving (22.11) for $\dot{S}^2$, we find

$$\dot{S}^2 = \left(\frac{B}{A}\right)\frac{1}{S} + \frac{D}{A}S^2 + \frac{E}{A}$$

$$= \left(\frac{B}{A}\right)\frac{1}{S} + \tfrac{1}{3}\Lambda S^2 + \frac{E}{A} \tag{22.15}$$

by (22.14). We now rescale the scale factor $S(t)$ to obtain a new scale factor $R(t)$, where

$$R(t) = \mu S(t). \tag{22.16}$$

Then, multiplying (22.15) by μ^2, we can write it in the form

$$\dot{R}^2 = \frac{C}{R} + \tfrac{1}{3}\Lambda R^2 - k, \tag{22.17}$$

where the constants C and k are defined by

$$C = B\mu^3/A, \qquad k = -\mu^2 E/A.$$

If $E = 0$ we choose μ arbitrarily, but if $E \neq 0$ we choose it so that

$$\mu^2 = A/|E|. \tag{22.18}$$

This choice of rescaling means that k can only have the values $+1, 0$, or -1. In this case, (22.17) has exactly the same form as the Friedmann differential equation of relativistic cosmology. In a similar manner, it is possible to construct a finite **continuum** Newtonian model in which the universe is taken to be a perfect gas. With some additional assumptions, the model can then be extended to an infinite continuum one. We shall, however, proceed directly to the models of relativistic cosmology.

22.4 The cosmological principle

Cosmology is based on a principle of simplicity, namely, the cosmological principle. It is, in essence, a generalization of the Copernican principle that the Earth is not at the centre of the solar system. In the same spirit, we would not expect the Earth, or the solar system, or our galaxy, or our local group of galaxies either to occupy any specially favoured position in the universe. We state the principle in the following form.

The cosmological principle: At each epoch, the universe presents the same aspect from every point, except for local irregularities.

We need to make this statement mathematically precise. We assume that there is a cosmic time t and formulate the principle in each of the spacelike slices $t = $ constant. The statement that each slice has no privileged points means that it is **homogeneous**. Technically, a spacelike hypersurface is homogeneous if it admits a group of isometries which maps any point into any other point (Fig. 22.2). The principle requires that not only should a slice have no privileged points but it should have no privileged directions about any point either. A manifold which has no privileged directions about a point is called **isotropic** and it clearly must be spherically symmetric about that point. A manifold is globally isotropic if it is isotropic about every point. It can be shown that if a manifold is globally isotropic then it is necessarily homogeneous. Thus, the cosmological principle requires that space-time can be sliced up or 'foliated' into spacelike hypersurfaces which are spherically symmetric about any point in them. The homogeneity of the universe has to be understood in the same sense as the homogeneity of a gas: it does not apply to the universe in detail, but only to a 'smeared-out' universe averaged over cells of diameter 10^8 to 10^9 light years, which are large enough to include many clusters of galaxies.

Thus, the cosmological principle is a simplicity principle which leads to the requirement that the universe is both isotropic and homogeneous. What observational evidence is there for each? Observations from visible galaxies are not so precise but suggest that their distribution is isotropic to perhaps 30%. Associated measurements of the Hubble constant suggest it is isotropic to perhaps 25%. The observations of radio galaxies reveal them to be much more isotropic with a distribution which is isotropic to below 5%. The universe seems to be pervaded by cosmic X-rays and these again are isotropic to below 5%. But the greatest support for isotropy came in 1965 with the discovery of the cosmic microwave background by Penzias and Wilson. They discovered that the universe is currently pervaded by a bath of thermal radiation with a temperature of 2.7 K and, moreover, that this radiation is isotropic to fractions of a per cent. The generally accepted explanation is that this radiation is a thermal remnant of the hot big bang. Spatial homogeneity is also supported by the counts of galaxies and the linearity of the Hubble law.

Despite the high degree of isotropy and homogeneity which we observe now, some cosmologists have considered anisotropic and inhomogeneous models. There are basically three reasons for this. First of all, calculations of statistical fluctuations in Friedmann models suggest that they cannot collapse fast enough to form the observed galaxies. Secondly, although there are strong reasons to support a big bang, there is less reason to suppose that the original singularity has the simple spherically symmetric pointlike structure of a Friedmann singularity. Indeed, calculations by Belinski, Khalatnikov, and Lifschitz — the so-called BKL approach — suggest that a general cosmological singularity would have a quite different structure. Finally, there is the idea that the universe may have been anisotropic and inhomogeneous in the past, but that there is some mechanism by which these characteristics would be washed out in the subsequent evolution, regardless of the initial conditions.

Considerable work has been done on the theoretical side in investigating anisotropic and inhomogeneous solutions. One of the biggest group of such solutions is that of the **Bianchi** models, which are spatially homogeneous anisotropic models (technically they admit a three-dimensional group of transformations which map any point in a hypersurface of homogeneity into any other point). These are subdivided into classes and labelled I, II, III, IV, V, VI, VII, VIII, and IX. The field equations then reduce to ordinary differential equations with time as the independent variable. These equations

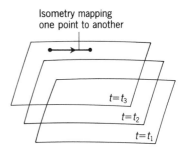

Fig. 22.2 Manifold sliced up into homogeneous 3-spaces.

can then be studied by either qualitative or numerical methods. These models, in general, have singularities. For example, the vacuum Bianchi I models are described by the **Kasner** solution

$$ds^2 = dt^2 - t^{2p_1}dx^2 - t^{2p_2}dy^2 - t^{2p_3}dz^2$$

where p_1, p_2, and p_3 are constants satisfying

$$p_1 + p_2 + p_3 = p_1^2 + p_2^2 + p_3^2 = 1,$$

which means that there is only one freely specifiable constant. In general, these solutions have a 'cigar'-like singularity when $t = 0$, that is, a small spatial region which is spherical at some time becomes infinitely long and thin as $t \to 0$. There is also a special case when the initial singularity is apparently of a 'pancake' type where the spherical region becomes an infinitely thin disc. Indeed, if we now include matter, it turns out that most of the Bianchi solutions have physical singularities, in the sense that the density becomes infinite, of these cigar or pancake types. Some special solutions give rise to weaker singularities called 'whimper' singularities which have the property that the Ricci components in an orthonormal frame parallely propagated along a curve hitting the singularity are unbounded, whereas the components in some other frame are bounded. However, the physical singularities are the generic ones. There is a fair amount known about the qualitative nature of the evolution of these models, but we will not consider them further.

A more radical notion is that there is no 'smeared-out' universe at all, but only clusters of galaxies, and clusters of clusters, and clusters of clusters of clusters, and so on, as in the hierarchical model proposed in 1908 by C. V. I. Charlier. There is in fact some observational evidence for super-clustering centred on the Virgo cluster, but the hierarchy appears to stop at cluster of clusters of galaxies, and shows no evidence of inhomogeneities on a larger scale.

We shall, from now on, adopt the cosmological principle. The real reason for this is not that it is definitely correct, but rather that it allows us to make use of the limited data provided to cosmology by observational astronomy. Any weaker assumptions, as in the anisotropic models or hierarchical models, would lead to metrics for which there would be insufficient data to determine the unknown functions occurring in them. By making such simplifying assumptions, we have a real chance of confronting theory with observation.

22.5 Weyl's postulate

In 1923, H. Weyl addressed the problem of how a theory like general relativity, based on general covariance, can be applied to a **unique** system like the universe. From one viewpoint, general relativity was specifically designed to deal with the equivalence of the observations of relatively accelerated observers. The universe consists of a single system which looks different to observers in different states of motion. Weyl argued that in attempting to understand the distant we must base ourselves, as far as possible, on the theories verified in our neighbourhood. General relativity offers the best available summary of local macroscopic physics and is accordingly a suitable theory. Other assumptions are needed such as the cosmological principle.

Weyl also added to this the assumption that there is a privileged class of observers in the universe, namely, those associated with the smeared-out motion of the galaxies. The fact that one can work with this smeared-out motion follows from the observation that the relative velocities of matter in each astronomical neighbourhood — each group of galaxies — are small. He then posits the introduction of a 'substratum' or fluid pervading space in which the galaxies move like 'fundamental particles' in the fluid, and assumes a special motion for these particles. This is contained in the following postulate.

Weyl's postulate: The particles of the substratum lie in space-time on a congruence of timelike geodesics diverging from a point in the finite or infinite past.

The postulate requires that the geodesics do not intersect except at a singular point in the past and possibly a similar singular point in the future. There is, therefore, one and only one geodesic passing through each point of space-time, and consequently the matter at any point possesses a unique velocity. This means that the substratum may be taken to be a **perfect fluid** and this is the essence of Weyl's postulate. Although the galaxies do not follow this motion exactly, the deviations from the general motion appear to be random and less than one-thousandth of the velocity of light. This is to be compared with the relative velocities of the galaxies due to the general motion which is comparable with the velocity of light. Accordingly, the random motion may be neglected in the first instance. Combined with the observation that the general motion is one of expansion, Weyl's postulate is seen to closely reflect the actual situation in the universe.

22.6 Relativistic cosmology

Relativistic cosmology is based on three assumptions, namely:

(1) the cosmological principle
(2) Weyl's postulate
(3) general relativity.

Weyl's postulate requires that the geodesics of the substratum are orthogonal to a family of spacelike hypersurfaces. We introduce coordinates (t, x^1, x^2, x^3) such that these spacelike hypersurfaces are given by $t = $ constant and the coordinates (x^1, x^2, x^3) are constant along the geodesics. This means that the spacelike coordinates of each particle are constant along its geodesic and, as a consequence, such coordinates are called **co-moving**. The orthogonality condition means that t can be chosen so that the line element is of the form

$$ds^2 = dt^2 - h_{\alpha\beta} \, dx^\alpha dx^\beta$$

where, as usual, Greek indices run from 1 to 3 and

$$h_{\alpha\beta} = h_{\alpha\beta}(t, x).$$

The coordinate t then plays the role of a **cosmic time** or **world time**.

The world time defines a concept of simultaneity. A **world map** is then the distribution of events on the surfaces of simultaneity (Fig. 22.3). The **world picture** is the aspect of the universe presented to an observer at any instant of world time, that is, it comprises the events seen looking along the observers past light cone (Fig. 22.4). Clearly, events from distant parts of the universe occur at earlier values of the world time than those nearby.

Consider a small triangle formed of three particles at some time t and also the triangle formed by these particles some time later. The second triangle will, in general, differ from the first in many respects. But, when we use the fact that the cosmological principle requires that the 3-spaces are isotropic and homogeneous, so that no point and no direction in the hypersurfaces may be preferential, then it follows that the second triangle must be geometrically similar to the first. Moreover, the magnification factor must be independent of the position of the triangle in the 3-space by similar arguments. It follows then that the time can enter $h_{\alpha\beta}$ only through a common factor in order that the ratios of the distances corresponding to the small displacements may be the same at all times. Hence, the time may only enter $h_{\alpha\beta}$ in the form

$$h_{\alpha\beta} = [S(t)]^2 g_{\alpha\beta}(x^\alpha). \tag{22.19}$$

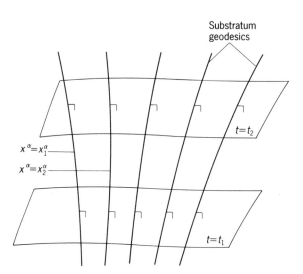

Fig. 22.3 Cosmic time surfaces and substratum geodesics.

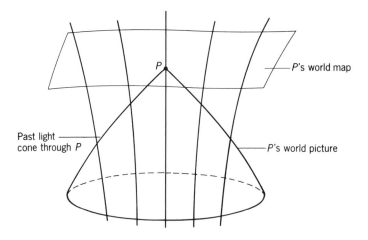

Fig. 22.4 The world map and world picture of an observer.

The ratio of the two values of $S(t)$ at two different times is the magnification factor and because of this it is called the **scale factor**. The scale factor $S(t)$ must be real, for otherwise the lapse of time could change a spacelike into a timelike interval. Next, we have to impose the condition that each slice is homogeneous and isotropic and also independent of time. This requires that the curvature at any point must be a constant, for otherwise all points would not be geometrically identical. Such a space is called a space of **constant curvature**, which we discuss next.

22.7 Spaces of constant curvature

Mathematically, a space of constant curvature is characterized by the equation

$$R_{abcd} = K(g_{ac}g_{bd} - g_{ad}g_{bc}),\tag{22.20}$$

where K is a constant called the **curvature**. As we shall see, the geometries of these spaces are qualitatively different depending on whether the curvature is positive, negative, or zero. In the case of a three-dimensional space, this becomes

$$R_{\alpha\beta\gamma\delta} = K(g_{\alpha\gamma}g_{\beta\delta} - g_{\alpha\delta}g_{\beta\gamma}).\tag{22.21}$$

Contracting with $g^{\alpha\gamma}$, we get

$$\begin{aligned}
g^{\alpha\gamma}R_{\alpha\beta\gamma\delta} &= R_{\beta\delta}\\
&= Kg^{\alpha\gamma}(g_{\alpha\gamma}g_{\beta\delta} - g_{\alpha\delta}g_{\beta\gamma})\\
&= K(3g_{\beta\delta} - g_{\beta\delta})\\
&= 2Kg_{\beta\delta}.
\end{aligned}\tag{22.22}$$

Now, since the 3-space is isotropic about every point, it must be **spherically symmetric** about every point. It follows that the line element will have the form (compare with (14.33))

$$d\sigma^2 = g_{\alpha\beta}dx^\alpha dx^\beta = e^\lambda dr^2 + r^2(d\theta^2 + \sin^2\theta d\phi^2),\tag{22.23}$$

where $\lambda = \lambda(r)$. The non-vanishing components of the Ricci tensor are

$$R_{11} = \lambda'/r, \quad R_{22} = \mathrm{cosec}^2\theta R_{33} = 1 + \tfrac{1}{2}re^{-\lambda}\lambda' - e^{-\lambda},\tag{22.24}$$

and the conditions for a space of constant curvature (22.22) reduce to the two equations

$$\lambda'/r = 2Ke^\lambda, \quad 1 + \tfrac{1}{2}re^{-\lambda}\lambda' - e^{-\lambda} = 2Kr^2.\tag{22.25}$$

The solution of these equations is

$$e^{-\lambda} = 1 - Kr^2.\tag{22.26}$$

We have shown that the metric for a 3-space of constant curvature is

$$d\sigma^2 = \frac{dr^2}{1 - Kr^2} + r^2(d\theta^2 + \sin^2\theta d\phi^2),\tag{22.27}$$

where K is positive, negative, or zero. We can introduce a new radial parameter $\bar{r}$ related to r by

$$r = \bar{r}/(1 + \tfrac{1}{4}K\bar{r}^2), \qquad (22.28)$$

in which case the metric takes on the conformally flat form (exercise)

$$d\sigma^2 = (1 + \tfrac{1}{4}K\bar{r}^2)^{-2}[d\bar{r}^2 + \bar{r}^2(d\theta^2 + \sin^2\theta\,d\phi^2)]. \qquad (22.29)$$

Combining this with the results of the last section, we obtain the line element for relativistic cosmology, namely,

$$ds^2 = dt^2 - [S(t)]^2\left(\frac{dr^2}{1 - Kr^2} + r^2(d\theta^2 + \sin^2\theta\,d\phi^2)\right), \qquad (22.30)$$

or, in terms of the barred radial coordinate,

$$ds^2 = dt^2 - [S(T)]^2 \frac{d\bar{r}^2 + \bar{r}^2(d\theta^2 + \sin^2\theta\,d\phi^2)}{(1 + \tfrac{1}{4}K\bar{r}^2)^2}. \qquad (22.31)$$

We prefer to write these line elements in an alternative form where the arbitrariness in the magnitude of K is absorbed into the radial coordinate and the scale factor. Assuming $K \neq 0$, we define k by $K = |K|k$, so that k is $+1$ or -1 depending on whether K is positive or negative, respectively. If we introduce a **rescaled radial coordinate**

$$r^* = |K|^{\frac{1}{2}}r, \qquad (22.32)$$

then (22.30) becomes (exercise)

$$ds^2 = dt^2 - \frac{[S(t)]^2}{|K|}\left(\frac{dr^{*2}}{1 - kr^{*2}} + r^{*2}(d\theta^2 + \sin^2\theta\,d\phi^2)\right). \qquad (22.33)$$

Finally, we define a rescaled scale function $R(t)$ by (see (22.16))

$$R(t) = S(t)/|K|^{\frac{1}{2}} \quad \text{if } K \neq 0,$$
$$R(t) = S(t) \qquad \text{if } K = 0.$$

Then, dropping the stars on the radial coordinate, we have shown that the line element of relativistic cosmology can be written in the alternative form

$$ds^2 = dt^2 - [R(t)]^2\left(\frac{dr^2}{1 - kr^2} + r^2(d\theta^2 + \sin^2\theta\,d\phi^2)\right), \qquad (22.34)$$

or, in terms of the barred radial coordinate,

$$ds^2 = dt^2 - [R(t)]^2 \frac{[d\bar{r}^2 + \bar{r}^2(d\theta^2 + \sin^2\theta\,d\phi^2)]}{[1 + \tfrac{1}{4}k\bar{r}^2]^2}, \qquad (22.35)$$

where k is now either $+1$, -1, or 0. This second form is called the **Robertson–Walker** line element after the first investigators to obtain it. At any epoch $t = t_0$, the geometry of the slice is given by

$$d\sigma^2 = R_0^2\left(\frac{dr^2}{1 - kr^2} + r^2(d\theta^2 + \sin^2\theta \, d\phi^2)\right), \qquad (22.36)$$

where the constant R_0 is given by $R_0 = R(t_0)$. In the next section, we shall investigate further the geometry of these 3-spaces of constant curvature for the three cases $k = +1, 0$, and -1.

22.8 The geometry of 3-spaces of constant curvature

Case 1: k=+1

Notice that in this case the coefficient of dr^2 becomes singular as $r \to 1$. We therefore introduce a new coordinate χ, where

$$r = \sin \chi, \qquad (22.37)$$

so that

$$dr = \cos \chi \, d\chi = (1 - r^2)^{\frac{1}{2}} d\chi$$

and (22.36) becomes

$$d\sigma^2 = R_0^2[d\chi^2 + \sin^2\chi(d\theta^2 + \sin^2\theta \, d\phi^2)]. \qquad (22.38)$$

We can now embed this 3-surface in a four-dimensional Euclidean space with coordinates (w, x, y, z), where

$$\left.\begin{array}{l} w = R_0 \cos \chi, \\ x = R_0 \sin \chi \sin \theta \cos \phi, \\ y = R_0 \sin \chi \sin \theta \sin \phi, \\ z = R_0 \sin \chi \cos \theta. \end{array}\right\} \qquad (22.39)$$

The embedding is possible because (exercise)

$$d\sigma^2 = dw^2 + dx^2 + dy^2 + dz^2 = R_0^2[d\chi^2 + \sin^2\chi(d\theta^2 + \sin^2\theta \, d\phi^2)]$$

in agreement with (22.38). Also, from (22.39), we get (exercise)

$$w^2 + x^2 + y^2 + z^2 = R_0^2, \qquad (22.40)$$

which shows that the surface can be regarded as a three-dimensional sphere in four-dimensional Euclidean space. This is depicted in Fig. 22.5, where one dimension ($y = 0$ or $\phi = 0$) is suppressed. The hypersurface is defined by the coordinate range

$$0 \leqslant \chi \leqslant \pi, \qquad 0 \leqslant \theta \leqslant \pi, \qquad 0 \leqslant \phi < 2\pi.$$

The 2-surfaces $\chi = $ constant, which appear as circles in the pictures, are 2-spheres of surface area (exercise)

$$A_\chi = \int_{\theta=0}^{\pi} \int_{\phi=0}^{2\pi} (R_0 \sin \chi \, d\theta)(R_0 \sin \chi \sin \theta \, d\phi) = 4\pi R_0^2 \sin^2\chi,$$

and (θ, ϕ) are the standard spherical polar coordinates of these 2-spheres.

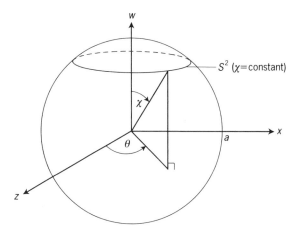

Fig. 22.5 A surface of constant positive curvature embedded in a four-dimensional Euclidean space ($\phi = 0$).

Thus, the area of these 2-spheres is zero at the North Pole, increases to a maximum at the equator, and decreases again to zero at the South Pole. The surface has 3-volume given by (exercise)

$$V = \int_{\chi=0}^{\pi} \int_{\theta=0}^{\pi} \int_{\phi=0}^{2\pi} (R_0 \, d\chi)(R_0 \sin\chi \, d\theta)(R_0 \sin\chi \sin\theta \, d\phi)$$
$$= 2\pi^2 R_0^3 = 2\pi^2 R^3(t_0), \tag{22.41}$$

which is why $R(t_0)$ is often referred to as the 'radius of the universe'.

This 3-space is clearly the generalization of an S^2, or 2-sphere, to a three-dimensional entity and is called an S^3, or **3-sphere**. The physical space should not really be thought of as embedded in anything else, since it is the totality of everything that exists at any one epoch. Thus, there are no physical points outside it nor does it have a boundary. It may be helpful to think of it as follows. If we introduce yet another radial-type coordinate r', where $r' = R_0 \chi$, then (22.38) becomes

$$d\sigma^2 = dr'^2 + R_0^2 \sin^2(r'/R_0)(d\theta^2 + \sin^2\theta \, d\phi^2),$$

and the surface area of the 2-spheres $\chi = $ constant is given by

$$A_\chi = 4\pi R_0^2 \sin^2(r'/R_0).$$

Notice that, for small r', $\sin r' \sim r'$, and so $A_\chi \sim 4\pi r'^2$. Now choose any point P and consider the surface area of a series of 2-surfaces centred on P of increasing radius r', all at one epoch t_0. For small values of the radius r' (compared with $R(t_0)$), the area is close to the Euclidean value $4\pi r'^2$. As r' increases, the area increases but becomes increasingly less than $4\pi r'^2$. The surface area reaches a maximum value when $r' = \frac{1}{2}\pi R_0$ and decreases from then on until it again becomes zero when $r' = \pi R_0$. In this space, any radial geodesic returns to its starting point. The topology of this space is variously called **closed**, **bounded**, or **compact**. The topology of the whole space-time is called **cylindrical**, since it is the product $\mathbb{R} \times S^3$, where $\mathbb{R}$ represents the one-dimensional cosmic time (Fig. 22.6).

Case 2: $k = 0$

If we set

$$x = R_0 r \sin\theta \cos\phi,$$
$$y = R_0 r \sin\theta \sin\phi,$$
$$z = R_0 r \cos\theta,$$

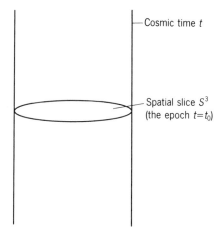

Fig. 22.6 The cylindrical topology $\mathbb{R} \times S^3$ of space-time when $k = +1$.

then (22.36) becomes

$$d\sigma^2 = dx^2 + dy^2 + dz^2,$$

which is clearly three-dimensional Euclidean space. The 3-space is covered by the usual coordinate range

$$0 \leqslant r < \infty, \qquad 0 \leqslant \theta \leqslant \pi, \qquad 0 \leqslant \phi < 2\pi.$$

The topology of the space-time is the same as that of four-dimensional Euclidean space, namely $\mathbb{R}^4$, and is called **open**.

Case 3: k=−1

If we introduce a new coordinate χ, where

$$r = \sinh \chi, \tag{22.42}$$

then

$$dr = \cosh \chi \, d\chi = (1 + r^2)^{\frac{1}{2}} dr,$$

and (22.36) becomes

$$d\sigma^2 = R_0^2 [d\chi^2 + \sinh^2 \chi (d\theta^2 + \sin^2 \theta \, d\phi^2)]. \tag{22.43}$$

We can no longer embed this 3-surface in a four-dimensional Euclidean space, but it can be embedded in a flat Minkowski space with signature $+2$ (exercise),

$$d\sigma^2 = -dw^2 + dx^2 + dy^2 + dz^2, \tag{22.44}$$

where

$$\left.\begin{aligned} w &= R_0 \cosh \chi, \\ x &= R_0 \sinh \chi \sin \theta \cos \phi, \\ y &= R_0 \sinh \chi \sin \theta \sin \phi, \\ z &= R_0 \sinh \chi \cos \theta. \end{aligned}\right\} \tag{22.45}$$

These equations imply that (exercise)

$$w^2 - x^2 - y^2 - z^2 = R_0^2, \tag{22.46}$$

so that the 3-surface is a three-dimensional hyperboloid in four-dimensional Minkowski space. This is depicted in Fig. 22.7 where one dimension ($y = 0$ or $\phi = 0$) is suppressed. The hypersurface is defined by the coordinate range

$$0 \leqslant \chi < \infty, \qquad 0 \leqslant \theta \leqslant \pi, \qquad 0 \leqslant \phi < 2\pi.$$

The 2-surfaces $\chi = $ constant, which appear as circles in the figure, are 2-spheres of surface area

$$A_\chi = 4\pi R_0^2 \sinh^2 \chi,$$

where (θ, ϕ) are the standard spherical polar coordinates on these 2-spheres. As χ ranges from 0 to ∞, the area of the successive 2-spheres increases from zero to infinity. For large χ, the surface area increases far more rapidly than it would if the hypersurface were flat. The 3-volume of the surface is infinite. The topology is again $\mathbb{R}^4$ and open. In each of the three cases, we have only specified the simplest topology possible; in fact, other topologies are possible by identifying points or regions, but we will not consider the issue further.

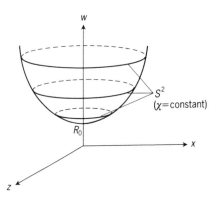

Fig. 22.7 Surface of constant negative curvature embedded in a four-dimensional Minkowski space ($\phi = 0$).

22.9 Friedmann's equation

Our three ingredients of relativistic cosmology are as follows.

(1) The cosmological principle, which leads to the Robertson–Walker line element, namely,

$$ds^2 = dt^2 - [R(t)]^2 \frac{[d\bar{r}^2 + \bar{r}^2(d\theta^2 + \sin^2\theta \, d\phi^2)]}{[1 + \frac{1}{4}k\bar{r}^2]^2}; \qquad (22.47)$$

(2) Weyl's postulate, which requires that the substratum is a perfect fluid, namely,

$$T_{ab} = (\rho + p)u_a u_b - p g_{ab}; \qquad (22.48)$$

(3) General relativity, with cosmological term, namely,

$$G_{ab} - \Lambda g_{ab} = 8\pi T_{ab}. \qquad (22.49)$$

Then using the fact that, in our preferred coordinate system

$$u^a \overset{*}{=} (1, 0, 0, 0),$$

the field equations lead to two independent equations (exercise)

$$3\frac{\dot{R}^2 + k}{R^2} - \Lambda = 8\pi\rho, \qquad (22.50)$$

$$\frac{2R\ddot{R} + \dot{R}^2 + k}{R^2} - \Lambda = -8\pi p, \qquad (22.51)$$

where we have used relativistic units and a dot denotes differentiation with respect to time. By homogeneity and isotropy, the density and pressure can only be functions of time t. Together with these equations, we have the requirements that the fluid is physically realistic, as expressed in the dominant energy conditions (12.56). Using our Newtonian analogue, (22.51) involves a second time-derivative of R and so may be thought of as an equation of motion, whereas (22.50) only involves a first time derivative of R and so may be considered an integral of the motion, that is an **energy** equation. If we differentiate (22.50) with respect to t, multiply through by $1/8\pi$, and add the result to (22.51) multiplied through by $-3\dot{R}/8\pi R$, we get

$$\dot{\rho} + 3p\frac{\dot{R}}{R} = -\frac{3}{8\pi}\frac{\dot{R}}{R}\left(\frac{3\dot{R}^2}{R^2} + \frac{3k}{R^2} - \Lambda\right) = -3\rho\frac{\dot{R}}{R},$$

again using (22.50). Multiplying through by R^3, we can rewrite this in the form

$$\frac{d}{dt}(\rho R^3) + p\frac{d}{dt}(R^3) = 0. \qquad (22.52)$$

Consider a set of particles in the substratum enclosing a volume V. Then, clearly, owing to the motion of the substratum, $V \sim R^3(t)$. If we now call the total mass-energy in the volume $E = \rho V$, then equation (22.52) can be written in the form

$$dE + p \, dV = 0. \qquad (22.53)$$

This is the first law of thermodynamics, or **conservation of energy**, and shows that the pressure does work in the expansion. This is exactly the same equation as results from the conservation equations (exercise)

$$T^{ab}_{;b} = 0. \qquad (22.54)$$

Thus, the field equations of the theory contain in them the equation for the conservation of energy. We have met this before in §13.4, and it arises from the fact that the field equations (22.49) satisfy the contracted Bianchi identities

$$(G^{ab} - \Lambda g^{ab})_{;b} = 0,$$

which in turn leads to the conservation equations (22.54).

The pressure p includes all types of pressure, such as that due to the random motion of the stars and galaxies, that due to heat motion of molecules, radiation pressure, and so forth. However, observation reveals that at the present epoch the pressure is far smaller than the energy density ρ due to matter. The ratio of the two quantities is about 10^{-5} or 10^{-6}. Accordingly, as long as only states of the universe differing not too widely from the present one are considered, we may take

$$p = 0, \qquad (22.55)$$

and so the substratum is comprised of dust. Then (22.51) integrates immediately to give

$$R(\dot{R}^2 + k) - \tfrac{1}{3}\Lambda R^3 = C, \qquad (22.56)$$

where C is a constant of integration, and, using (22.50), we find

$$C = \tfrac{8}{3}\pi R^3 \rho. \qquad (22.57)$$

Apart from a numerical factor, this is the energy content E of a volume V of the substratum and is constant immediately by (22.53), which becomes a conservation of mass equation when p vanishes. The value can be remembered as twice the mass of a spherical volume of a Euclidean universe of radius R and density ρ. If we now use (22.57) to eliminate ρ in (22.50), the result can be written in the form

$$\dot{R}^2 = \frac{C}{R} + \tfrac{1}{3}\Lambda R^2 - k. \qquad (22.58)$$

This is **Friedmann's equation** for the time variation of the scale factor in the absence of pressure. Note that it is identical with the Newtonian analogue (22.17). We shall consider the solutions of this equation — the **Friedmann models** — in the next chapter. Some authors refer to these as the **FRW models**, short for Friedmann–Robertson–Walker models. Recall that, in obtaining the Newtonian analogue (22.17), we imposed the assumption (22.7), which is essentially equivalent to Hubble's law (22.10) (see Exercise 22.4). In the rest of this chapter, we shall consider light propagation and distance in relativistic cosmology in order to **deduce** Hubble's law from the premises of the theory.

22.10 Propagation of light

We assume that light propagates in relativistic cosmology in the same way as it does in general relativity. Let us consider how an observer O receives light from a receding galaxy. We use the unbarred form of the Robertson–Walker line element (22.34). Since we assume that the time slices are homogeneous 3-spaces, we can, without loss of generality, take O to be at the origin of coordinates $r = 0$. Inserting the conditions for a radial null geodesic, namely,

$$ds^2 = d\theta = d\phi = 0,$$

into (22.34), we find

$$\frac{dt}{R(t)} = \pm \frac{dr}{(1 - kr^2)^{\frac{1}{2}}}, \tag{22.59}$$

where the $+$ sign corresponds to a receding light ray and the minus sign to an approaching light ray. Consider a light ray emanating from a galaxy P with world-line $r = r_1$, at coordinate time t_1, and received by O at coordinate time t_0 (Fig. 22.8). Using (22.59), we get (exercise)

$$\int_{t_1}^{t_0} \frac{dt}{R(t)} = -\int_{r_1}^{0} \frac{dr}{(1 - kr^2)^{\frac{1}{2}}} = f(r_1), \tag{22.60}$$

where

$$f(r_1) = \begin{cases} \sin^{-1} r_1 & \text{if } k = +1, \\ r_1 & \text{if } k = 0, \\ \sinh^{-1} r_1 & \text{if } k = -1. \end{cases} \tag{22.61}$$

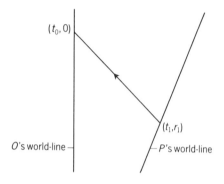

Fig. 22.8 Light ray from galaxy P to observer O.

Next, consider two successive light rays emanating from P at times t_1 and $t_1 + dt$, and received by O at times t_0 and $t_0 + dt_0$, respectively (Fig. 22.9). Then, from (22.60),

$$\int_{t_1 + dt_1}^{t_0 + dt_0} \frac{dt}{R(t)} = \int_{t_1}^{t_0} \frac{dt}{R(t)},$$

since each side is equal to the same function $f(r_1)$. Therefore (check),

$$\int_{t_1 + dt_1}^{t_0 + dt_0} \frac{dt}{R(t)} - \int_{t_1}^{t_0} \frac{dt}{R(t)} = \int_{t_0}^{t_0 + dt_0} \frac{dt}{R(t)} - \int_{t_1}^{t_1 + dt_1} \frac{dt}{R(t)} = 0,$$

and, assuming that $R(t)$ does not vary greatly over the intervals dt_1 and dt_0, we can take it outside the integral in the last equation and deduce that

$$\frac{dt_0}{R(t_0)} = \frac{dt_1}{R(t_1)}. \tag{22.62}$$

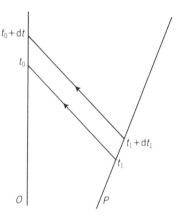

Fig. 22.9 Successive light rays from galaxy P to observer O.

All fundamental particles (galaxies) of the substratum have world-lines on which the coordinates r, θ, ϕ are constant and hence, from (22.34), $ds^2 = dt^2$. It follows that t measures the **proper time** along the substratum world-lines. The intervals dt_1 and dt_2 are the proper time intervals between the rays as measured at the source and observer, respectively. Hence, from (22.62), the interval, as measured by O, is $R(t_0)/R(t_1)$ times the interval measured by P. In an expanding universe,

$$t_0 > t_1 \quad \Rightarrow \quad R(t_0) > R(t_1),$$

and it follows that the observer O will experience a **red shift** z given by

$$1 + z = v_1/v_0 = R(t_0)/R(t_1), \qquad (22.63)$$

where v_1 and v_0 are the frequencies measured by the emitter and receiver, respectively. This red shift is also called a **Doppler shift**, but is not to be confused with the special relativistic Doppler shift. Clearly, in a contracting universe, O will detect a corresponding blue shift.

If, roughly speaking, P is 'near' to O, then the cosmic times of emission and reception differ only by a small amount, dt say, that is, $t_0 = t_1 + dt$, and so (22.63) produces

$$1 + z = \frac{R(t_0)}{R(t_0 - dt)} \simeq \frac{R(t_0)}{R(t_0) - \dot{R}(t_0)dt} \simeq 1 + \frac{\dot{R}(t_0)}{R(t_0)}dt \qquad (22.64)$$

to first order in dt. In addition,

$$\int_{t_1}^{t_0} \frac{dt}{R(t)} = \int_{t_1}^{t_1 + dt} \frac{dt}{R(t)} \simeq \frac{dt}{R(t_1)} = \frac{dt}{R(t_0 - dt)} \simeq \frac{dt}{R(t_0)}.$$

But for small r, using (22.61),

$$\int_{t_1}^{t_0} \frac{dt}{R(t)} = f(r_1) \simeq r_1,$$

and so

$$\frac{dt}{R(t_0)} \simeq r_1.$$

Combining this with (22.64), we get the result

$$z \simeq \dot{R}(t_0)r_1. \qquad (22.65)$$

Thus, at any one epoch, the **red shift z is proportional to the distance r_1**. Interpreting z as a velocity of recession, we have obtained a velocity–distance relation similar to Hubble's law. To make this more precise, we need to consider how distance is measured, as least theoretically, on a cosmologically interesting scale.

22.11 A cosmological definition of distance

Because we have a world time, it is mathematically easy to define an **absolute** distance between fundamental particles by considering them at the same value of world time and then measuring the geodesic distance between them in the slice (Fig. 22.10). If we set $dt = d\theta = d\phi = 0$ in (22.34), then the absolute distance d_A between O and P at time t is

$$d_A = R(t) \int_0^{r_1} \frac{dr}{(1 - kr^2)^{\frac{1}{2}}}. \qquad (22.66)$$

This is of no practical use and so we try another tack. If we know the actual size of a distant nebulae, then we can define an observational distance by

$$d_O = \alpha/\beta, \qquad (22.67)$$

where α is the **actual** diameter of the nebulae and β is the **observed** angular diameter. Such a definition would be satisfactory if some means of determining α were known. Since this is not known, we instead use a definition based

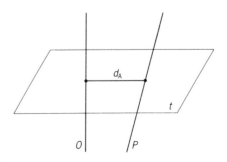

Fig. 22.10 An absolute distance between fundamental particles.

on the **apparent** luminosity of a nebula. Let E be the energy radiated per unit time by the distant nebula and let I be the intensity of the radiation received per unit area per unit time. Then, assuming the energy is distributed uniformly on a sphere in a Euclidean space and neglecting the red shift, the distance can be defined as $(E/4\pi I)^{\frac{1}{2}}$. But, in an expanding universe, the interval of time during which a certain amount of energy is received is longer than the interval of emission by virtue of the Doppler shift, and hence the number of photons received per unit time is reduced by the factor $1 + z$. In addition, the energy of each photon of light is reduced by the same factor (because energy is the time component of a 4-vector and so the transformation from one observer to another introduces the factor $1 + z$). These considerations lead to the definition of a **luminosity distance** d_{L}, where

$$d_{\text{L}}^2 = \frac{E}{4\pi I(1 + z)^2}. \tag{22.68}$$

The luminosity distance is in essence the distance used by astronomers. However, the detailed way in which astronomical distances are measured is quite complicated and beyond the scope of this book (for further details, see Weinberg (1972)), although we mention, without being precise, that one unit of measurement is called **apparent magnitude** m. It is related to the energy received E_R by the relation

$$m = \text{constant} - 0.4 \log_{10} E_R.$$

Moreover, there is a problem with the definition (22.68), because it involves E, the **absolute** luminosity of the source, which is not observationally measurable. This definition thus appears to suffer from the same defects as (22.67). However, the distances to nearby galaxies may be determined by other means and hence their absolute luminosities may be calculated, and it appears that all galaxies have roughly the same absolute luminosity. So a first approximation is to take all galaxies as having the same absolute luminosity. This assumption is almost certainly wrong, though, because if we live in an evolutionary universe then the mean age of the more distant galaxies is much less than the mean age of nearby galaxies and so there is no reason to believe that they will have the same mean luminosity. We will not pursue the matter further, but simply employ (22.68).

22.12 Hubble's law in relativistic cosmology

We start by finding an expression for the luminosity distance in terms of the coordinates of the Robertson–Walker line element in the unbarred form (22.34). Consider light emanating from galaxy P at time t_1, and observed by us 'now' at O at a time $t = t_0$ $(t_1 < t_0)$ (Fig. 22.11). The light will have spread out over the surface of a sphere with centre at the event P_0 $(t = t_0, r = r_1)$ and passing through the event O_0 $(t = t_0, r = 0)$. The surface area of this sphere is the same as that of the sphere centred on O_0 passing through P_0 (dotted line in Fig. 22.11), owing to the homogeneity of the 3-sphere. The line element for this sphere $(t = t_0, r = r_1)$ is, from (22.34),

$$\text{d}s^2 = -[R(t_0)r_1]^2 (\text{d}\theta^2 + \sin^2\theta \, \text{d}\phi^2).$$

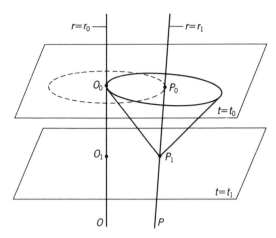

This is the usual line element for a sphere of radius $R(t_0)r_1$, and so the sphere has surface area $4\pi R^2(t_0)r_1^2$. Hence, the observed intensity is given by

$$I = \frac{E}{4\pi r_1^2 R^2(t_0)(1 + z)^2},$$

taking into account the double Doppler shift factor. Comparing this with (22.68), we obtain an expression for the luminosity distance in terms of the scale factor, namely,

$$d_L = r_1 R(t_0). \tag{22.69}$$

If we define the **Hubble parameter** by (see (22.9))

$$H(t) = \dot{R}(t)/R(t), \tag{22.70}$$

then (22.65) and (22.69) give

$$z \simeq H(t_0)d_L, \tag{22.71}$$

where $H(t_0)$ is the value of the Hubble parameter at the current epoch and is called **Hubble's constant**. This is the famous **Hubble law** in relativistic cosmology. It states that for 'nearby' nebulae the radial velocity of recession as measured by the red shift z is proportional to its distance. The dimension of $H(t)$ is that of inverse time, and so if we define $T = 1/H(t)$ then T has the dimension of time. Current observations give the value

$$T_0 \approx 10^{10} \text{ years}, \tag{22.72}$$

which is believed correct to within a factor of 2. We stress that Hubble's law is an approximate one in relativistic cosmology. We define the **deceleration parameter** q by

$$q(t) = -R\ddot{R}/\dot{R}^2. \tag{22.73}$$

Then, since $R > 0$ and $\dot{R}^2 > 0$, it follows that

$$\ddot{R} < 0 \quad \Rightarrow \quad q > 0,$$

and so a positive q measures the rate at which the expansion of the universe is slowing down. The current value of the deceleration parameter q_0 is uncertain, but most measurements make it positive and close to 1 with a typical range

$$q_0 = 1 \pm 0.5. \tag{22.74}$$

Then, taking the second-order term into account in (22.64), we find the relationship (exercise)

$$d_L = zT_0[1 - \tfrac{1}{2}(1 + q_0)z + \cdots]. \tag{22.75}$$

For objects too close to the observer, the random motions which we have excluded from our model do in fact obscure the general motion. But there is a good range of nebulae satisfying the velocity–distance law (out to about the 18th magnitude) from which a good determination of T can be made. For more distant observations, the relationship (22.75) must be used, which is crucially dependent on the value of q_0. In Fig. 22.12, we present some data given in a 1970 review by Sandage. It is remarkable to note Hubble first proposed his law in 1929 on the basis of observations of only 18 nearby galaxies, and this data corresponds to a tiny part of the graph in Fig. 22.12.

Differentiating Friedmann's equation (22.58), we get

$$2\dot{R}\ddot{R} = -\frac{C}{R^2}\dot{R} + \tfrac{2}{3}\Lambda R\dot{R},$$

and multiplying by $-R/2\dot{R}^3$ gives

$$-\frac{R\ddot{R}}{\dot{R}^2} = \frac{C}{2R\dot{R}^2} - \tfrac{1}{3}\Lambda\frac{R^2}{\dot{R}^2}.$$

Then, using (22.73), (22.57), and (22.70), we can write this in the form

$$q = (\tfrac{4}{3}\pi\rho - \tfrac{1}{3}\Lambda)/H^2. \tag{22.76}$$

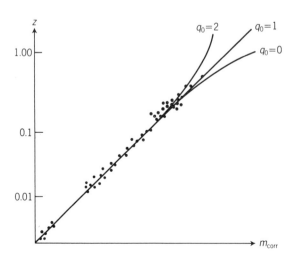

Fig. 22.12 Red shift versus corrected apparent magnitudes (Sandage 1970).

This shows that there is an intimate connection between the deceleration parameter q, the Hubble parameter H, and ρ, the mean density of the universe.

There is one other important observable and that is N, the number of nebulae in a given volume. We assume that there are $n(t)$ nebulae per unit volume. Using (22.34), the volume enclosed by a sphere centred at O bounded by $r = r_1$ at time t_0 is (exercise)

$$V = 4\pi R^3(t_0) \int_0^{r_1} \frac{r^2 \, dr}{(1 - kr^2)^{\frac{1}{2}}}. \tag{22.77}$$

The number of nebulae in this volume is then

$$N = Vn(t_0). \tag{22.78}$$

In relativistic cosmology, the assumption is usually made that all nebulae existing at $t = t_0$ have also existed at the time of emission of light ($t = t_1$ in the case of P in Fig. 22.8), which is a somewhat doubtful assumption. Mathematically, this amounts to putting

$$R^3(t)n(t) = \text{constant.} \tag{22.79}$$

In order to replace this assumption by something else, we would need to know more about nebular evolution. In the next chapter, we shall see that the observational parameters H, q, ρ, and N play a crucial role in discriminating between different cosmological models.

Exercises

22.1 (§22.2) Show that Olbers's paradox remains if we assume space is non-Euclidean but still homogeneous by the cosmological principle.

22.2 (§22.3) Write down an expression for the cosmological potential energy V_{c_i} of the ith particle such that $F_i = -\text{grad } V_{c_i}$.

22.3 (§22.3) Substitute (22.7) into (22.6) and establish (22.11)–(22.14). Identify A physically. Show that if $E \neq 0$ the choice (22.18) of μ in (22.16) allows (22.11) to be written in the standard form (22.17) with $k = \pm 1$.

22.4 (§22.3) Integrate Hubble's law (22.10) and deduce the scale factor law (22.7), identifying the function $S(t)$ in terms of $H(t)$.

22.5 (§22.7) If three-dimensional space is isotropic (i.e. it has no preferred directions, in which case every direction is an eigendirection), then deduce that $R^\alpha_{\ \beta} = c\delta^\alpha_{\ \beta}$ at each point. If, in addition, the space is homogeneous, what can we conclude about c? Show that $c = \frac{1}{3}R$. Use the definition of the Weyl tensor in §6.13 together with the fact that it vanishes in three dimensions to show that the Riemann tensor can be expressed in the form (22.21) and relate K to c.

22.6 (§22.7) Work out the non-vanishing components (22.24) of the Ricci tensor for the line element (22.23). [Hint: use the results of Exercise 6.31(iv).] Confirm that if this line element is a space of constant curvature then λ is given by (22.26).

22.7 (§22.7) Confirm the results of the following transformations:

(i) (22.28) transforms (22.27) into (22.29);

(ii) (22.32) transforms (22.30) into (22.33).

22.8 (§22.8) Confirm the results of the following transformations:

(i) (22.37) transforms (22.36) into (22.38);

(ii) (22.42) transforms (22.36) into (22.43).

22.9 (§22.8)
(i) Show that (22.39) is a parametric form of the surface (22.40) in Euclidean 4-space. Confirm that its line element reduces to (22.38).
(ii) Show that (22.45) is a parametric form of the surface (22.46) in Minkowski space with line element (22.44). Confirm that its line element reduces to (22.43).

22.10 (§**22.8**) Write down the line elements for the 2-spheres $\chi = $ constant in the cases $k = +1$ and $k = -1$. By comparing them with the standard line element for a sphere of radius a, namely,

$$ds^2 = a^2(d\theta^2 + \sin^2\theta\,d\phi^2),$$

confirm the formula for the surface area A_χ in each case. Confirm (22.41) and show that the volume of the 3-surface in the case $k = -1$ is infinite.

22.11 (§**22.9**) Establish the field equations of relativistic cosmology (22.50) and (22.51). [Hint: this involves working out the Ricci and Einstein tensors for the line element (22.47).]

22.12 (§**22.9**) Use (22.50) and (22.51) to establish the result (22.53). Confirm that the same equation results from the conservation law (22.54)

(i) by direct computation
(ii) without utilizing expressions for the connection. [Hint: take the covariant derivative of every term in (22.48) and use the results that u^a is a tangent to a geodesic, $u^a \overset{*}{=} \delta^a_0$ and $u^a_{\ ;a} = \frac{1}{2}(d/dt)(\ln g)$ — why?]

22.13 (§**22.9**) Use (22.50) and (22.51) to obtain (22.58) subject to (22.57) in the case (22.55).

22.14 (§**22.10**) Confirm the result (22.60) subject to (22.61). Deduce (22.62) and (22.65).

22.15 (§**22.12**) Confirm Hubble's law in the form
(i) (22.71) to first order
(ii) (22.75) to second order.

22.16 (§**22.12**) Confirm (22.77).

Cosmological models **23**

23.1 The flat space models

Our considerations of the last chapter led to Friedmann's equation

$$\dot{R}^2 = \frac{C}{R} + \tfrac{1}{3}\Lambda R^2 - k \qquad (23.1)$$

governing the scale factor in the pressure-free epochs of relativistic cosmology (and in Newtonian cosmology suitably interpreted). The task is to solve this non-linear first-order ordinary differential equation for different values of the parameters occurring in it. Recall that the values of these parameters are governed by the requirements

$$C > 0, \qquad -\infty < \Lambda < +\infty, \qquad k = -1, 0, +1. \qquad (23.2)$$

There are a number of ways of proceeding. The equation can be solved, in general, by using elliptic functions, or resort can be made to computer plots of numerically generated solutions. However, many of the sub-cases can be integrated directly using elementary functions or, failing that, elementary functions can be used to investigate their qualitative features. We shall not give an exhaustive account of this approach here (but see Landsberg and Evans (1977) for details). Instead, we shall restrict our attention to the important cases of flat space ($k = 0$) and vanishing cosmological constant ($\Lambda = 0$). The techniques employed may be applied to the other cases.

In the flat space case, (23.1) reduces to

$$\dot{R}^2 = C/R + \tfrac{1}{3}\Lambda R^2. \qquad (23.3)$$

We first assume $\Lambda > 0$ and introduce a new variable

$$u = \frac{2\Lambda}{3C} R^3.$$

Differentiating, we get

$$\dot{u} = \frac{2\Lambda}{C} R^2 \dot{R},$$

and, substituting in (23.3), we find

$$\dot{u}^2 = \frac{4\Lambda^2}{C^2} R^4 \left(\frac{C}{R} + \tfrac{1}{3}\Lambda R^2 \right)$$

$$= \frac{4\Lambda^2}{C} R^3 + \frac{4\Lambda^3}{3C^2} R^6$$

$$= 6\Lambda u + 3\Lambda u^2$$

$$= 3\Lambda(2u + u^2). \tag{23.4}$$

Taking the positive square root, we have

$$\dot{u} = (3\Lambda)^{\frac{1}{2}} (2u + u^2)^{\frac{1}{2}},$$

which can be integrated by parts. If we assume a big bang model, namely, $R = 0$ when $t = 0$, then $u = 0$ initially, and so integrating gives

$$\int_0^u \frac{du}{(2u + u^2)^{\frac{1}{2}}} = \int_0^t (3\Lambda)^{\frac{1}{2}} \, dt = (3\Lambda)^{\frac{1}{2}} t.$$

If we complete the square in the u-integral and set $v = u + 1$ and $\cosh w = v$, then we get

$$\int_0^u \frac{du}{[(u + 1)^2 - 1]^{\frac{1}{2}}} = \int_1^v \frac{dv}{(v^2 - 1)^{\frac{1}{2}}} = \int_0^w \frac{\sinh w \, dw}{(\cosh^2 w - 1)^{\frac{1}{2}}} = \int_0^w dw = w.$$

In terms of R, the solution becomes

$$R^3 = \frac{3C}{2\Lambda} [\cosh (3\Lambda)^{\frac{1}{2}} t - 1]. \tag{23.5}$$

If $\Lambda < 0$, we introduce a new variable

$$u = -\frac{2\Lambda}{3C} R^3 \tag{23.6}$$

and then, proceeding as before, we obtain the solution (exercise)

$$R^3 = \frac{3C}{2(-\Lambda)} \{1 - \cos [3(-\Lambda)]^{\frac{1}{2}} t\}. \tag{23.7}$$

The case $\Lambda = 0$ may be obtained by taking the series for cosh, namely,

$$\cosh x = 1 + \frac{x^2}{2!} + \frac{x^4}{4!} + \cdots .$$

Then (23.5) gives

$$R^3 = \frac{3C}{2\Lambda} \left[\left(1 + \frac{3\Lambda t^2}{2} + \frac{3\Lambda^2 t^4}{8} + \cdots \right) - 1 \right],$$

and so, in the limit as $\Lambda \to 0$, we have $R^3 = \frac{9}{4}Ct^2$, or equivalently

$$R = (\tfrac{9}{4}Ct^2)^{\frac{1}{3}}. \tag{23.8}$$

This is called the **Einstein–de Sitter model**. Alternatively, we can obtain this result directly from (23.3) which becomes $\dot{R}^2 = C/R$. Taking square roots, the equation is immediately separable, producing

$$R^{\frac{1}{2}}\,dR = C^{\frac{1}{2}}\,dt.$$

Integrating, using $R = 0$ when $t = 0$, we get

$$\tfrac{2}{3}R^{\frac{3}{2}} = C^{\frac{1}{2}}t,$$

so again

$$R = (\tfrac{9}{4}Ct^2)^{\frac{1}{3}},$$

as in (23.8). The Hubble parameter $H(t)$ and the deceleration parameter $q(t)$ can be easily computed from (23.5), (23.7), or (23.8). For example, in the Einstein–de Sitter case (exercise),

$$H(t) = \dot{R}/R = 2/(3t) \tag{23.9}$$

and

$$q(t) = -R\ddot{R}/\dot{R}^2 = \tfrac{1}{2}. \tag{23.10}$$

In the initial stages of a big bang universe, R is small and so the term C/R dominates over $\frac{1}{3}\Lambda R^2$ in (23.1). Hence, for small t,

$$\dot{R}^2 \sim C/R, \tag{23.11}$$

and, integrating, we obtain, as in (23.8),

$$R \sim (\tfrac{9}{4}Ct^2)^{\frac{1}{3}}. \tag{23.12}$$

So, in the early stages, all big bang models behave like the Einstein–de Sitter model, namely, they expand at the rate $t^{\frac{2}{3}}$. If we write (23.3) in the form

$$\dot{R}^2 = F(R), \tag{23.13}$$

where

$$F(R) = C/R + \tfrac{1}{3}\Lambda R^2, \tag{23.14}$$

then much of the qualitative behaviour of R can be inferred from the behaviour of $F(R)$. For example,

$$\Lambda < 0 \quad \Rightarrow \quad F(R) = 0 \quad \text{when } R = R_{\mathrm{m}} = [3C/(-\Lambda)]^{\frac{1}{3}},$$

so that $\dot{R}$ vanishes at R_{m}, which is a local minimum (exercise). Conversely, if $\Lambda \geqslant 0$, the solution grows without bound. In the case when $\Lambda > 0$, then, for large t, the second term on the right dominates in (23.3), and so

$$\dot{R}^2 \sim \tfrac{1}{3}\Lambda R^2, \tag{23.15}$$

and, integrating, we find (exercise)

$$R \sim \exp[(\tfrac{1}{3}\Lambda)^{\frac{1}{2}}t]. \tag{23.16}$$

We now have enough information to sketch the graphs of the three models. We postpone this to §23.3.

23.2 Models with vanishing cosmological constant

In this section, we consider the case when Λ vanishes. Friedmann's equation then becomes

$$\dot{R}^2 = C/R - k. \tag{23.17}$$

To solve this, we need to consider separately the cases $k = +1$ and $k = -1$. In the former case, (23.17) becomes $\dot{R}^2 = C/R - 1$. This time we start with a change of variable given by

$$u^2 = R/C. \tag{23.18}$$

Then $2u\dot{u} = \dot{R}/C$, and, substituting in (23.17), we find

$$\dot{u}^2 = \frac{\dot{R}^2}{4C^2 u^2} = \frac{1}{4C^2 u^2}\left(\frac{C}{R} - 1\right) = \frac{1}{4C^2 u^2}\left(\frac{1}{u^2} - 1\right).$$

Taking positive square roots, the equation is separable, and, integrating with big bang initial conditions, we get

$$2\int_0^u \frac{u^2}{(1-u^2)^{\frac{1}{2}}}\,du = \frac{1}{C}\int_0^t dt = \frac{t}{C}.$$

To evaluate the u-integral, we set $u = \sin\theta$. Then

$$2\int_0^u \frac{u^2}{(1-u^2)^{\frac{1}{2}}}\,du = 2\int_0^\theta \frac{\sin^2\theta\cos\theta\,d\theta}{(1-\sin^2\theta)^{\frac{1}{2}}}$$

$$= 2\int_0^\theta \sin^2\theta\,d\theta$$

$$= \int_0^\theta (1 - \cos 2\theta)\,d\theta$$

$$= \theta - \tfrac{1}{2}\sin 2\theta$$

$$= \theta - \sin\theta\cos\theta$$

$$= \sin^{-1}u - u(1-u^2)^{\frac{1}{2}}.$$

Writing the solution in terms of R, we obtain the result

$$C[\sin^{-1}(R/C)^{\frac{1}{2}} - (R/C)^{\frac{1}{2}}(1 - R/C)^{\frac{1}{2}}] = t. \tag{23.19}$$

Similarly, in the case $\Lambda = 0$, $k = -1$, the solution becomes (exercise)

$$C[(R/C)^{\frac{1}{2}}(1 + R/C)^{\frac{1}{2}} - \sinh^{-1}(R/C)^{\frac{1}{2}}] = t. \tag{23.20}$$

The case $\Lambda = 0$, $k = 0$ is the Einstein–de Sitter model and has already been dealt with in §23.1. Again, the Hubble parameter and deceleration parameter can be computed directly from (23.19) or (23.20). For example, when $k = +1$,

$$H = C^{-1}(R/C)^{-\frac{3}{2}}(1 - R/C)^{\frac{1}{2}} \tag{23.21}$$

and

$$q = \tfrac{1}{2}(1 - R/C)^{-1} \qquad (23.22)$$

with R determined implicitly in terms of t by (23.19).

As in the last section, if we write (23.17) in the form

$$\dot{R}^2 = G(R), \qquad (23.23)$$

where

$$G(R) = C/R - k, \qquad (23.24)$$

then we find that the model for which $k = +1$ has a local minimum, whereas the other two models grow without bound. When $k = -1$, for large t, we have $\dot{R}^2 \sim 1$, and so $R \sim t$. We again have enough information to sketch the graphs of the models.

23.3 Classification of Friedmann models

In Fig. 23.1, we collect together the graphs of all the various possibilities. They are divided up into three major cases, namely, $k = -1, 0$, or $+1$, and subdivided into 3, 3, and 8 sub-cases, respectively, depending on the sign or value of Λ. We describe the sub-cases briefly.

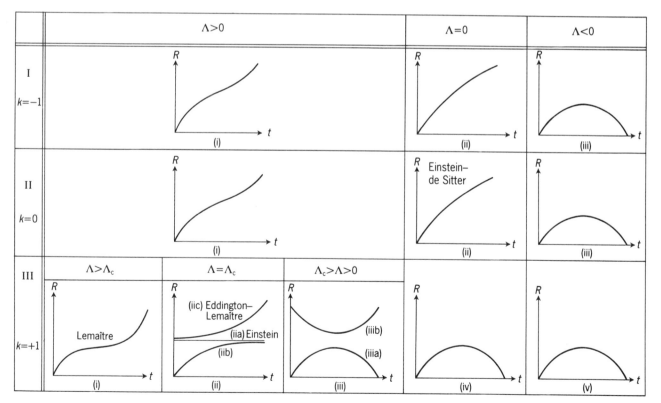

Fig. 23.1 Classification of Friedmann models.

Case I: k = −1

All of these models have open topology.

(i) $\Lambda > 0$. This is an indefinitely expanding model, but it possesses a 'kink' in it where the rate of expansion slows down for a period before picking up again, and asymptotically it approaches $\exp[(\frac{1}{3}\Lambda)^{\frac{1}{2}}t]$. Initially, like **all** big bang models, the rate of expansion goes like that of the Einstein–de Sitter model, namely, like $t^{\frac{2}{3}}$.

(ii) $\Lambda = 0$. An indefinitely expanding model without a kink and which goes like t asymptotically.

(iii) $\Lambda < 0$. In this case, the cosmological force is attractive and eventually halts the expansion and forces the model to collapse ending in an apocalyptic event called the **big crunch**. It is usually referred to as an **oscillating model**. There is also the possibility that the model is indefinitely oscillating with each cycle followed by another, as in Fig. 23.2. All models for which $\Lambda < 0$ are oscillating models.

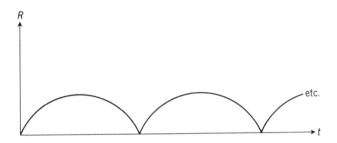

Fig. 23.2 An indefinitely oscillating model.

Case II: k = 0

All of these models have open topology.

(i) $\Lambda > 0$. This is identical in character to the sub-case I(i) above, again possessing a kink and asymptotically approaching $\exp[(\frac{1}{3}\Lambda)^{\frac{1}{2}}t]$.

(ii) $\Lambda = 0$. The Einstein–de Sitter model where $R \sim t^{\frac{2}{3}}$.

(iii) $\Lambda < 0$. An oscillating model.

Case III: k = +1

All of these models have closed topology.
In this case, there are more possibilities since there is a positive **critical value** of the cosmological constant Λ_c given by

$$\Lambda_c = 4/(9C^2) \tag{23.25}$$

and an associated critical value of the scale factor R_c given by

$$R_c = \tfrac{3}{2}C. \tag{23.26}$$

(i) $\Lambda > \Lambda_c$. This is called **Lemaître's model** and is again similar to the indefinitely expanding models I(i) and II(i). However, the closer Λ is to Λ_c, the more pronounced the kink is and the closer the expansion is brought to a halt in this period.

(ii) $\Lambda = \Lambda_c$.   There are three possibilities in this sub-case, which depend on the value of a constant of integration.

(a) This is the **Einstein static model** in which the gravitational attraction is exactly counterbalanced by the cosmic repulsion. The scale factor then has the constant value R_c.

(b) This is a big bang model which asymptotically approaches the Einstein static model.

(c) This is the **Eddington–Lemaître model** in which if time is run backwards it asymptotically approaches the Einstein static model. In forward time, it is an ever-expanding model asymptotically approaching $\exp[(\frac{1}{3}\Lambda)^{\frac{1}{2}}t]$.

(iii) $\Lambda_c > \Lambda > 0$.   There are again two possibilities depending on a constant of integration.

(a) An oscillating model.
(b) This is a model which has a contracting phase followed by an expanding phase in which the scale factor always remains positive. It is symmetric about its point of minimum radius with $R \sim \exp[(\frac{1}{3}\Lambda)^{\frac{1}{2}}t]$ as $t \to \infty$ and $R \sim \exp[(\frac{1}{3}\Lambda)^{\frac{1}{2}}(-t)]$ as $t \to -\infty$.
(iv) $\Lambda = 0$.   An oscillating model.
(v) $\Lambda < 0$.   An oscillating model.

23.4 The de Sitter model

This is not a model of relativistic cosmology because it is devoid of matter. However, as we shall see, it is important historically. It is obtained by setting $p = \rho = k = 0$ in (22.50) and (22.51). Then (22.50) gives

$$3\dot{R}^2/R^2 - \Lambda = 0,$$

or

$$\dot{R}/R = (\tfrac{1}{3}\Lambda)^{\frac{1}{2}}, \qquad (23.27)$$

which on integration becomes

$$R = A\exp[(\tfrac{1}{3}\Lambda)^{\frac{1}{2}}t],$$

where A is a constant of integration. Since the origin of this curve is arbitrary, let us choose $R = 1$ when $t = 0$, in which case $A = 1$. Alternatively, we can rescale r and absorb the factor A into it. This leads to the **de Sitter model**, for which

$$R = \exp[(\tfrac{1}{3}\Lambda)^{\frac{1}{2}}t]. \qquad (23.28)$$

The graph of the scale factor is shown in Fig. 23.3. This solution is the common limiting case to which all the models I(i), II(i), III(i), III(iic), and III(iiib) tend as $t \to \infty$.

From (22.34), (23.28), and the requirement that k vanishes, the line element becomes

$$ds^2 = dt^2 - [\exp 2(\tfrac{1}{3}\Lambda)^{\frac{1}{2}}t][dr^2 + r^2(d\theta^2 + \sin^2\theta\, d\phi^2)],$$

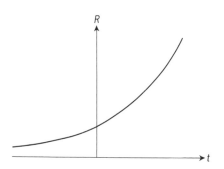

Fig. 23.3 The de Sitter model.

or, in Cartesian coordinates, the standard form

$$ds^2 = dt^2 - [\exp(2t/\alpha)][dx^2 + dy^2 + dz^2], \qquad (23.29)$$

where

$$\alpha = (3/\Lambda)^{\frac{1}{2}}. \qquad (23.30)$$

This line element is invariant under a shift in t and a simultaneous change of scale in the space coordinates (exercise). Note that the metric is completely specified by α. The coordinate range of t is from $-\infty$ to $+\infty$ with the zero of t being conventional. This is because the exponential curve is 'self-similar', that is, one cannot tell where one is along it by intrinsic measurements; it has no natural origin. If we introduce new coordinates $(\bar{t}, \bar{x}, \bar{y}, \bar{z})$, where

$$\left. \begin{aligned} \bar{t} &= t - \tfrac{1}{2}\alpha \ln[1 - \alpha^{-2}(x^2 + y^2 + z^2)\exp(2t/\alpha)], \\ \bar{x} &= x\exp(t/\alpha), \\ \bar{y} &= y\exp(t/\alpha), \\ \bar{z} &= z\exp(t/\alpha), \end{aligned} \right\} \qquad (23.31)$$

then (23.29) becomes, dropping bars, (exercise)

$$ds^2 = [1 - \alpha^{-2}(x^2 + y^2 + z^2)]dt^2 - dx^2 - dy^2 - dz^2$$
$$- \frac{\alpha^{-2}(x\,dx + y\,dy + z\,dz)^2}{1 - \alpha^{-2}(x^2 + y^2 + z^2)}, \qquad (23.32)$$

which is visibly **stationary** (why?). We shall return to this solution in §23.16.

We are now in a position to give a semi-historical account of the models which have been considered at one time or another as models of our universe.

23.5 The first models

The Einstein static model (III(iia))

This was the first relativistic cosmological model ever to be considered. As we have said before, it was constructed by Einstein in an attempt to incorporate Mach's principle into general relativity and also to overcome the boundary conditions of the theory. It was discarded as soon as it became clear that the matter of the universe is not at rest on average, but is undergoing a large-scale expansion. In addition, Eddington has shown that it is unstable under small perturbations. This is fairly apparent because, if we consider a universe in an Einstein state which for some reason suffers a slight expansion, then, since an expansion would decrease the gravitational attraction **and** increase the cosmic repulsion, the system would continue to expand indefinitely.

The de Sitter model ($p=\rho=k=0$, $\Lambda > 0$)

This solution was discovered in 1917 and provided an example of an empty space solution which satisfies the Einstein equations with cosmological term.

The de Sitter model is expanding and yet contains no matter and so is clearly in violation with Mach's principle. Since it was the only model in existence at the time which could accommodate expansion, it was seized upon as a possible model for our universe. To explain the emptiness of the model, it was argued that the density of the matter in the universe was in any case low, though the meaning of the remark was not discussed seriously until much later. However, it was thought that there might well be solutions of the Einstein equations with Λ intermediate between Einstein's 'matter with motion' and de Sitter's 'motion without matter', and some solution not far from de Sitter's might well represent the actual universe. The work of Freidmann largely solved the problem.

23.6 The time-scale problem

Before 1952, the reciprocal of the Hubble constant was estimated to be $T_0 \approx 1.8 \times 10^9$ years. On the other hand, the age of the earth was thought to be at least 3×10^9 years and many stars were thought to have existed for 5×10^9 years. (Modern estimates are considerably longer.) Hence, it was argued, if the universe has only existed for a finite time, then its age must be at least 5×10^9 years. Now consider the graph of R against t for any big bang model (Fig. 23.4). If $t = t_0$ represents 'now', then the tangent to the curve at $t = t_0$ cuts the t-axis at a time $t = t_1$, say, and so

$$t_0 - t_1 = R(t_0)/\dot{R}(t_0) = T_0. \tag{23.33}$$

This rests on the assumption that $\ddot{R} \leqslant 0$ for all t, or equivalently $q > 0$, that is, the rate of expansion has been slowing down since the big bang. It follows that the big bang must have occurred at a time **less than** T_0. So, with the old value of T_0, there would not have been time for the stars to develop, and hence all such models were rejected. This is known as the **time-scale** problem, and it was thought that the only way to overcome it was to consider models possessing periods for which $\ddot{R} > 0$. However, if we multiply (22.51) by -3 and add it to (22.50), we obtain

$$8\pi(\rho + 3p) = 2\Lambda - 6\ddot{R}/R. \tag{23.34}$$

The left-hand side is always positive and, if $\Lambda < 0$ or $\Lambda = 0$, then $\ddot{R}$ would have to be negative for all time. Therefore big bang models were thought to require $\Lambda > 0$ to overcome the time-scale problem.

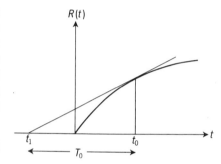

Fig. 23.4 The time-scale problem for big bang models.

23.7 Later models

The Eddington–Lemaître solution (III(iic))

The model was discovered by Lemaître in 1925, and was put forward forcefully by Eddington. It is not a big bang model, but was chosen because it overcame the time-scale problem. It is best pictured by reference to the static Einstein universe in which gravitational attraction and cosmic repulsion are just in equilibrium. As discussed earlier, the model is unstable and a small expansion would result in continuing expansion, as in III(iic), and similarly a small contraction would result in a time reversal of III(iib). This model therefore has an infinite past which was spent in the Einstein state. Thus it appeared to surmount the time-scale problem since it permits an arbitrarily long time scale of evolution. The picture of the history of the universe derived from this model was that, for an infinite period in the distant past, there was a completely homogeneous distribution of matter in equilibrium in the Einstein state until some event started off the expansion. This expansion has been going at an increasing pace ever since, which requires a negative value for q. The condensation of the stars and galaxies from the primeval matter took place at the time the expansion began, but this development was stopped later by the decrease of average density due to the progress of expansion.

One objection to this model is that the initial condensation and its neighbourhood would differ from the rest of the universe. No such distinction is known from observation and, anyway, its existence would be incompatible with the cosmological principle. Again, a condensation must have been unlikely at the temperature and density assumed to exist in an Einstein state, for otherwise the life of that state would be very short. But, once the expansion starts, circumstances become still less favourable owing to the decrease in density. The requirement that q is negative also seems to contradict observations. These and other objections remain unanswered, and so the model eventually fell from favour.

Lemaître's model (III(i))

In 1935, there was a swing away from the Eddington–Lemaître model when the question of the generation of the elements — **nucleogenesis** — came under active consideration. The universe is believed to consist very largely of hydrogen, the simplest atom, and it was considered desirable by many to account for the generation of the heavy elements from hydrogen. Nuclear physics showed that the building of heavy elements requires conditions of extremely high density and temperature. The point source models appear to offer such a possibility, since, as R tends to zero, the density tends to infinity as R^{-3} and it can be shown that the temperature goes as R^{-1}. Such models are consequently called **hot big bang** models. The purely gravitational effects will not dominate in such conditions and so we cannot follow the model up to its origin using solely the equations of general relativity.

Lemaître investigated the problem and concluded that three distinct periods in the evolution of the universe should be distinguished:

(1) a period of explosion from a point source, during which the elements were formed;

(2) a period of very much reduced expansion, during which conditions were favourable for the formation of condensations leading to the nebulae;

(3) a final period of renewed expansion, during which the recession of the nebulae is accelerating and the formation of new condensations is made unlikely by the diminishing density.

The three models I(i), II(i), and III(i) each are point source models with a period of decreasing rate of expansion followed by an increasing rate of expansion. However, only in the last model is the expansion brought almost to a standstill, thus allowing for a period of the form (2) above. If Λ is very slightly in excess of Λ_c, then this model starts off like III(iib) and slowly approaches the Einstein state, but, since gravitation can never completely counteract the cosmic repulsion, the development then essentially follows the

Eddington–Lemaître solution. This model, like all in which $\Lambda > 0$, requires that the formation of new galaxies is impossible since the cosmic repulsion is, and has been for some time, more powerful than the gravitational attraction of the smoothed-out universe. This means that hardly any nebulae can be younger than a certain time, in fact of the order of $\frac{2}{3}T$, which is something that can be directly tested.

23.8 The missing matter problem

Many of the considerations discussed so far became profoundly altered when in 1952 Baade and Sandage looked again at the observations leading to a value for Hubble's constant, and came up with a revised value of $T_0 \approx 1.3 \times 10^{10}$ years. Thus, at one stroke, the time-scale problem for big bang models was largely resolved. In addition, there are known alternative mechanisms for nucleogenesis. For example, it is thought that supernovae explode from a stellar condition capable of nucleogenesis and result in the heavy elements being distributed over a wide area. Again, it is thought that the neutrons in the nuclei of red giants would rapidly transmute light elements into heavy ones. One of the most remarkable pieces of work was that of Hoyle, Burbidge, and Fowler, who, using these processes, were able to account for the observed abundancies of elements with remarkable accuracy. (Indeed, Hoyle was forced to predict correctly a previously ill-determined nuclear energy level in ^{12}C in the process).

The resolution of the time-scale problem means it is no longer necessary to take $\Lambda > 0$ in order to consider big bang models. Indeed, there is a strong body of opinion that says that we should take $\Lambda = 0$. After all, Einstein introduced it on grounds which turned out to be erroneous, so on simplicity grounds, if on no other, it should be dropped from consideration. The three models with $\Lambda = 0$ are called the **standard models** and are the ones to which most attention is given today. We shall discuss these in the next section. Setting $\Lambda = 0$ in (22.58) and using (22.57), we find (exercise)

$$k = -\dot{R}^2 + \tfrac{8}{3}\pi R^2 \rho(t) = [\rho(t) - 3H^2/8\pi]\tfrac{8}{3}\pi R^2. \qquad (23.35)$$

We denote the current value of H by H_0 and ρ by ρ_0, i.e.

$$H_0 = H(t_0), \qquad \rho_0 = \rho(t_0)$$

and define a density, called the **critical density** ρ_c, by

$$\rho_c = 3H_0{}^2/8\pi. \qquad (23.36)$$

Then it follows immediately from (23.35) that ρ_0 discriminates between the three standard models as follows:

$$
\left.
\begin{aligned}
\rho_0 > \rho_c &\Rightarrow k = +1 \quad \text{(oscillating model)}, \\
\rho_0 = \rho_c &\Rightarrow k = \;\;\,0 \quad \text{(Einstein–de Sitter model)}, \\
\rho_0 < \rho_c &\Rightarrow k = -1 \quad \text{(ever-expanding model)}.
\end{aligned}
\right\} \qquad (23.37)
$$

In addition, setting $\Lambda = 0$ in (22.76) gives

$$q = \tfrac{4}{3}\pi\rho/H^2, \tag{23.38}$$

which determines q_0 in terms of ρ_0 and H_0. If we use the estimate $T_0 \approx 10^{10}$ years, then we find (exercise)

$$\rho_c = 2 \times 10^{-29} \, \mathrm{g\,cm^{-3}}. \tag{23.39}$$

We prefer the historic c.g.s. units because much of the literature on cosmology uses them. The smoothed out density of luminous matter in galaxies, which we denote ρ_l, is believed to be rather less than this, in fact

$$\rho_l \lesssim 10^{-30} \, \mathrm{g\,cm^{-3}}$$

according to present-day estimates. This discrepancy is known as the **missing matter problem** and has led many to speculate about what other forms of matter (other than the luminous matter in galaxies) may exist. One possibility is that intergalactic space contains a gas density of $\sim 2 \times 10^{-29} \, \mathrm{g\,cm^{-3}}$. However, although some evidence exists for intergalactic hydrogen (the most invisible of all elements) and even ionized hydrogen, it would appear again to be less than $10^{-30} \, \mathrm{g\,cm^{-3}}$. There is also the possibility that a large amount of matter is hidden inside black holes. Indeed, current observations suggest that there may well be massive black holes situated at the centre of every galaxy, gobbling up stars in the galactic nucleus. Even so, the question of the actual density of the universe is still an open one, but it would seem that it is most likely a good deal less than the critical value ρ_c.

23.9 The standard models

The Einstein–de Sitter model (II(ii))

In this model there is an exact expression for T, namely, from (23.9), $t = \tfrac{2}{3}T$. Using the value $T_0 \approx 1.3 \times 10^{10}$ years, it follows that the age of the universe $t_0 \approx 8.6 \times 10$ years, which with present-day estimates is less than the age of the oldest stars and close to the age of the Sun. In fact, radioactive dating and the theory of stellar evolution gives uncertainties in the range 0.7×10^{10} to 1.6×10^{10} years, but even so it is clear that, unless T_0 is significantly greater, this model possesses a time-scale problem. The model predicts a fixed value for q, namely, from (23.10), $q = \tfrac{1}{2}$, which, compared with the currently observed value, is low but within the observed range (22.74). Substituting $q_0 = \tfrac{1}{2}$ in (23.35) reveals immediately that $\rho_0 = \rho_c$, in agreement with (23.37). Thus, the model also suffers from the missing matter problem discussed in the last section.

The oscillating model (III(iv))

In this case $\rho_0 > \rho_c$, by (23.37), so that the missing matter problem is made worse. Keeping H_0 fixed in (23.38), we then find

$$\rho_0 > \rho_c \quad \Rightarrow \quad q_0 > \tfrac{1}{2},$$

which is reasonable. Unfortunately, the time-scale problem is worse since $t_0 < \frac{2}{3} T_0$.

Indefinitely expanding model (I(i))

This model requires $\rho_0 < \rho_c$, which is good, but as a consequence $0 < q_0 < \frac{1}{2}$, which is not so good. Indeed, to obtain a value of ρ_0 consistent with ρ_l, we would need $q_0 \simeq \frac{1}{40}$, which seems unlikely. As we lower the value of ρ_0, so $t_0 \to T_0$, since the effect of gravity is being lowered. Thus, this model would appear to cope best with the time-scale problem.

In short, all three standard models have problems in accommodating the presently observed values of the three observational parameters. There is the possibility that future observations could lead to one or more of these parameters changing radically, but it seems unlikely. The two most favoured models are the oscillating and indefinitely expanding ones and they each have their advocates. Of course, they are markedly different models, since one requires the universe to be **closed** and the other requires it to be **open**. It is perhaps surprising that, even though cosmology as a serious science began with Einstein as long ago as 1915, the basic question as to whether we live in a finite or infinite universe is still **open**.

23.10 Early epochs of the universe

In constructing the simplest possible model of the universe, we have neglected pressure. However, in the early epochs of the universe, one would expect the radiation to dominate completely over matter as a source of gravitation. Let us look briefly at a simple model which includes pressure at the **extreme relativistic condition**

$$3p = \rho, \qquad (23.40)$$

which is the equation of state for radiation. Then, taking $\Lambda = 0$ in (22.50) and (22.51), the condition (23.40) requires that (exercise)

$$\frac{\ddot{R}}{R} + \frac{\dot{R}^2}{R^2} + \frac{k}{R^2} = 0. \qquad (23.41)$$

In the earliest phases, the first two terms will dominate, and so, neglecting the last term in (23.41), we find that, for small t (exercise), $R \sim t^{\frac{1}{2}}$. Comparing this with the small-time behaviour we had previously, namely, $R \sim t^{\frac{2}{3}}$, we see that this corresponds to a more rapid expansion. The effect of the pressure of radiation is that it exerts its own gravitational field, thereby increasing the amount of gravity acting. This increases the rate of expansion, as is clear if we reverse the sense of time and consider the resulting rate of collapse.

23.11 Cosmological coincidences

There are a number of startling numerical coincidences which have led some authors to try and construct cosmological models which incorporate these coincidences. They are best stated in terms of dimensionless ratios. One such is (in non-relativistic units)

$$\frac{\text{radius of universe}}{\text{classical radius of electron}} = \frac{c\,T_0}{e^2/m_e c^2} \approx 10^{40},$$

where e and m_e are the charge and mass of an electron. Another ratio is

$$\frac{\text{electric force between electron and proton}}{\text{gravitational force between electron and proton}} \approx 2 \times 10^{39},$$

which is more or less equal to the first ratio. The same number is approximately equal to $N^{\frac{1}{2}}$, where N is the 'number of particles in the universe', that is, the number of hydrogen atoms within a sphere of radius $c\,T$, as derived from observations of the mean density ρ of the universe (which as we have seen is uncertain by one or two orders of magnitude). It follows directly from these coincidences that

$$G\rho T^2 \approx 1, \qquad (23.42)$$

a result which crops up a great deal in cosmology (see, for example, (23.38)). It is possible to use Mach's principle to give an explanation of this result. As we stated in the consideration of Mach's principle, the Machian interaction depends on the value of the gravitational constant G and the amount of matter in the universe. This latter is given by the density multiplied by the volume of the universe, namely, $\sim \rho(c\,T)^2$. The problem of calculating the total effect of all matter in the universe is rather similar to the problem of calculating the background radiation due to a uniform distribution of sources and involves the introduction of a cut-off distance, in the former case of the order of $c\,T$. The condition that there is just enough matter in the universe to induce the observed amount of inertia into a local body can then be shown to be precisely of the form of (23.42). It is also possible to give plausible arguments for the other coincidences and thereby circumvent the need for the introduction of a new theory, but we shall not pursue the matter further.

23.12 The steady-state theory

In 1948, Bondi and Gold and, independently in the same year, Hoyle produced a cosmological theory which was in many ways radically different from the models of relativistic cosmology. It is a theory of charming simplicity, but, unfortunately, one which involves a modification of the law of conservation of energy — a law close to the hearts of many physicists. The theory provides definite answers to cosmological questions and so is more amenable to direct tests. Put another way, since it makes unique predictions, it is easier to disprove. Unfortunately, the theory seems to be at variance with much of present-day observations, and hence many consider it to be of historic interest only. Nonetheless, it has made something of a comeback from time to time and it does raise a number of important questions about the interpretation of observations which were perhaps glossed over too easily in the past. We shall summarize the original formulation here, and not enter into the more recent reinterpretations of the theory.

The fundamental assumption of the theory as derived by Bondi and Gold is the following principle.

Perfect cosmological principle: The universe presents an unchanging aspect on the large scale.

It follows immediately from this principle that the universe must be expanding on thermodynamic grounds. For, if the universe were static and unchanging, a state of thermodynamic equilibrium would exist in which there would be no time — the so-called 'heat death' of the universe. In a contracting universe, the Doppler shift leads to a disequilibrium in which radiation preponderates over matter; whereas, in an expanding universe, the opposite is true. The observational fact that matter predominates over radiation in the current universe establishes its expansion.

The perfect cosmological principle also requires that the average density of matter must not change in time. There is only one way in which a constant density can be compatible with a continual expansion and that is for there to exist a **continuous creation of matter**. It can be shown that this is, on average, given by $3\rho T^{-1} \sim 10^{-46} \mathrm{g\,cm^{-3}\,s^{-1}}$. This represents a creation rate of one proton per litre every 5×10^{11} years. It is clear that it is utterly impossible to observe directly such a rate and so in this sense it does not contradict our local conservation of matter law. However, the fact that the matter is created out of **nothing** is a rather startling thought and as such has been the basis of much controversy.

The **steady-state theory** derives from the three assumptions:

(1) the perfect cosmological principle;

(2) Weyl's postulate;

(3) the general relativistic properties of light propagation.

As we have seen in the last chapter, assumption (1), in the weaker form of the ordinary cosmological principle, and assumption (2) lead to the Robertson–Walker line element. We next use the requirement of **stationarity** contained in (1). Since the universe is expanding, $R(t)$ must be an increasing function of time. But the curvature of a 3-space of constant curvature in a Robertson–Walker space-time goes like kR^{-2} (exercise), and this is an observable quantity (affecting, for example, the rate of increase of the number of nebulae with distance). The fact that it is observable means that it must be constant by (1), and since R varies with time we must conclude that $k = 0$. The function $R(t)$ is not directly observable, but the Hubble parameter $\dot{R}(t)/R(t)$ is and so again must remain constant. Thus, $\dot{R}/R = 1/T_0$, where T_0 is a constant and, proceeding as in (23.27), we get

$$R(t) = \exp(t/T_0) \qquad (23.43)$$

and the line element becomes

$$ds^2 = dt^2 - \exp(2t/T_0)[dr^2 + r^2(d\theta^2 + \sin^2\theta)d\phi^2]. \qquad (23.44)$$

This is the same as the line element of the **de Sitter model**, which we considered in relativistic cosmology but discarded because it led to an empty universe. This difficulty does not arise in the steady-state theory because the field equations of general relativity no longer hold. Note that (23.44) is completely specified by the scale factor T_0. We leave the question of coordinate range until §23.16.

Assumption (3) means that we can consider light propagation in the same way as we did in §22.10. Then (22.59) becomes in this case

$$\frac{dt}{\exp(t/T_0)} = \pm dr. \tag{23.45}$$

For an incoming ray reaching $r = 0$ at $t = t_0$, we get

$$r = T_0(e^{-t/T_0} - e^{-t_0/T_0}). \tag{23.46}$$

The luminosity distance d_L is given by (22.69), which in this case becomes

$$d_L = r_1 e^{t_0/T_0}, \tag{23.47}$$

so that the coordinate r is proportional to the luminosity distance. Then, using (22.63), we have

$$1 + z = R(t_0)/R(t_1) = e^{(t_0 - t_1)/T_0} = 1 + r_1 e^{t_0/T_0}/T_0 \tag{23.48}$$

by (23.46). Combining this result with (23.47), we find

$$d_L = z T_0. \tag{23.49}$$

Thus, in the steady-state theory, Hubble's law is **exact**. It follows from (23.43) that (exercise)

$$q = -1, \tag{23.50}$$

that is, the universe is continuously expanding at an ever-increasing rate.

The number n of nebulae per unit volume is observable and so must be constant. It follows from the line element (23.44) and (23.45) that the number of nebulae with a radial coordinate between r and $r + dr$ from which light reaches $r = 0$ at $t = t_0$ is

$$4\pi r^2 n dr(e^{-t_0/T_0} + r/T_0)^{-3}. \tag{23.51}$$

In relativistic cosmology, it is the number of nebulae per unit coordinate volume, $r^2 \sin\theta \, dr \, d\theta \, d\phi$, which is taken to be constant, and so the last factor in (23.51) does not occur. Hence, owing to the expansion of the co-moving system of coordinates, relativistic cosmology requires a **higher** nebulae density per unit volume in the past than now. Thus, as we look further out, we are looking further back in time and we should see an increase in nebulae density. The steady-state theory, in contradistinction, requires that this density should remain a constant. The theory as presented so far does not predict the mean average density of matter in space, so we outline a formulation of the theory which does.

At the same time that Bondi and Gold put forward the steady-state theory on the basis of the perfect cosmological principle, Hoyle proposed the theory as a set of local physical laws contained in a set of field equations. He took general relativity without a cosmological term as his starting point and modified the theory in two distinct ways. First, he changed the conservation of matter property of general relativity. Secondly, he changed the tensor character of the theory by introducing a privileged class of observers much along the lines of the fundamental particles of Weyl's postulate. As in Weyl's postulate, we start with a congruence of timelike geodesics diverging from a point O in the past. Then, through any point P, there will, in general, be a unique geodesic emanating from O. Let $C(x^a)$ be a scalar function defined to be proportional to the geodesic length OP (the definition needs modification if O is an infinite distance from P). The first derivative $C_{,a}$ defines a field of vectors tangent to the geodesic congruence and of constant length, and the second derivative $C_{,ab}$ defines a symmetric tensor field. Hoyle then takes as the modified field equations

$$G_{ab} = 8\pi T_{ab} + C_{,ab}. \tag{23.52}$$

It can be shown that this equation possesses the de Sitter metric as a solution if the universal length of $C_{,a}$ is $3/T_0$. Moreover, the de Sitter solution is stable and any other solution tends to it asymptotically. The density of matter is given by

$$\tfrac{8}{3}\pi\rho T_0^2 = 1 \tag{23.53}$$

(compare with (23.38), taking $q = \tfrac{1}{2}$), which, as in the Einstein–de Sitter model, leads to a mean density of ρ_c. The term $C_{,ab}$ is of the same order as the cosmological term in general relativity and hence has no detectable effect on local physical laws. The energy–momentum tensor is not conserved since the C field has negative energy density and

$$8\pi T^{ab}{}_{;b} = -C^{,ab}{}_{;b} \neq 0,$$

as is required by the continuous creation of matter.

The steady-state theory would not appear to stand up too well to observation. It requires Hubble's law to be exact, whereas observation suggests that on a large scale the linear relation ceases to hold. The deceleration parameter $q = -1$, whereas observation indicates that q is non-negative. There is no time-scale problem in the steady-state theory, but the problem of the missing matter is just as apparent as in the Einstein–de Sitter universe. The theory requires that there are the same number of galaxies per unit volume in the remote past as there are in the present, whereas evolutionary theories require that, the further back in the past we look, the more galaxies per unit volume we should see. Optical observations are inconclusive in this respect, but observations of radio galaxies tend to support the evolutionary theories. In addition, the red shift of quasars tell heavily against the theory. Perhaps it was the discovery of the 3 K cosmic background microwave radiation and its generally held interpretation as the remnants of the hot phase of a big bang universe which finally made the steady-state theory seem untenable.

23.13 The event horizon of the de Sitter universe

As we have seen in (23.46), the equation of the past light cone through the point $r = 0$, $t = t_0$ has the equation

$$r = T_0(e^{-t/T_0} - e^{-t_0/T_0}). \tag{23.54}$$

At any particular time, information can be received only from events inside an observer's past light cone. If we take the limit in (23.54) as $t_0 \to \infty$, then it follows that an observer whose world-line is $r = 0$ can **never** receive any information from events occurring **outside** the hypersurface

$$r = T_0 e^{-t/T_0}. \tag{23.55}$$

In other words, this hypersurface is an **event horizon** for O and has a similar character to the event horizon of black holes. Let us consider what an observer would see while observing a particle of the substratum P with world-line $r = r_1$ (Fig. 23.5). If we set $r = r_1$ in (23.55), then P crosses O's event horizon at the event P_1 at time

$$t_1 = T_0 \ln(T_0/r_1). \tag{23.56}$$

Observer O can only receive signals from P at events of P's world-line for which $t < t_1$. These signals travel on null geodesics (the dotted line in Fig. 23.5), which reach O at time (exercise)

$$\tau = -T_0 \ln(e^{-t/T_0} - e^{-t_1/T_0}). \tag{23.57}$$

So, by (22.69), O ascribes to P the luminosity distance

$$d_L = r_1 e^{\tau/T_0} = r_1(e^{-t/T_0} - e^{-t_1/T_0})^{-1} \tag{23.58}$$

and a red shift, by (23.49), of

$$z = d_L/T_0. \tag{23.59}$$

Therefore, as $t \to t_1$, it follows that $\tau \to \infty$ and the light takes longer and

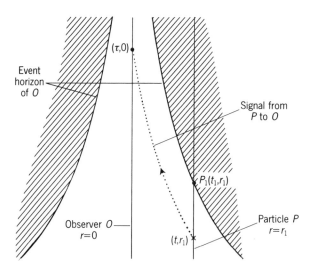

Fig. 23.5 The event horizon of an observer in the de Sitter universe.

longer to reach O from P. In addition, both $d_L \to \infty$ and $z \to \infty$ as P disappears over O's horizon and this happens in a finite proper time as measured by P. At time t, the geodesic distance l from O to P is, by (22.66)

$$l = r_1 e^{t/T_0}, \tag{23.60}$$

which is still finite at the event P_1. The velocity of recession is

$$\frac{dl}{dt} = \frac{r_1}{T_0} e^{t/T_0} = e^{(t-t_1)T_0} \tag{23.61}$$

by (23.56), and this tends to 1 as $t \to t_1$. Thus, the geodesically measured velocity of recession tends to the velocity of light as the particle approaches the event horizon. So far, we have only considered an observer at $r = 0$, but, by homogeneity of the de Sitter solution, the above conclusions apply to any observer moving with the substratum.

The event horizon is rather like a curtain behind which one can see nothing. However, the curtain can be drawn, but at a price. Consider the world-line of an explorer E who is sent out into space by O and is asked to send back reports on all that E sees (Fig. 23.6). The explorer E will be able to see past O's horizon, but not until passing the event E_1 on this horizon, after which E can never return home to O nor send any information back to O. So we see that O can never receive information about events beyond O's horizon. However, their existence cannot be neglected, since, by travelling around, O's horizon can be changed and some of the forbidden knowledge can be found out—but no return home is then possible. We have met similar event horizons in Minkowski space-time (Fig. 3.8). In suitably chosen coordinates, the world-line of a uniformly accelerated observer travelling in the x-direction has equation $x^2 - t^2 = $ constant, $y = z = 0$. It is clear from Fig. 23.7 that light emitted from events in the shaded region will never reach the observer P, who therefore has an event horizon. No such horizon exists for inertial observers of course.

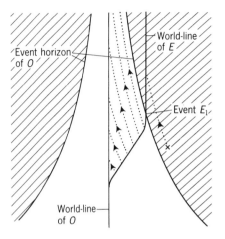

Fig. 23.6 Explorer E draws the curtain.

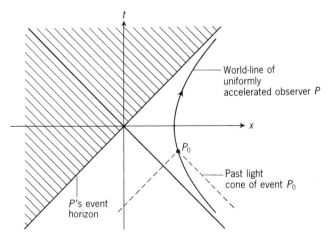

Fig. 23.7 Event horizons in Minkowski space-time.

23.14 Particle and event horizons

We can, in fact, distinguish between different sorts of horizons. Consider the world-line of an observer O moving on a timelike geodesic in a space-time in

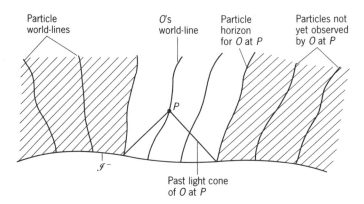

Fig. 23.8 Particle horizons of an observer ($\mathscr{I}^-$ spacelike).

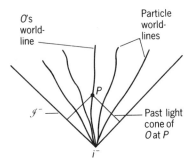

Fig. 23.9 The case when $\mathscr{I}^-$ is null.

which $\mathscr{I}^-$ is spacelike (Fig. 23.8). Then, at any point P on O's world-line, the past light cone at P is the set of events in space-time which can be observed by O at that time. The division of particles into those seen by O at P and those not seen by O at P gives rise to the **particle horizon** of O at P. It represents the history of those particles lying at the limits of O's vision. Of course, if $\mathscr{I}^-$ is null (for example, as in Minkowski space-time), then all particles are seen by O at P (Fig. 23.9). Now consider a space-time in which both $\mathscr{I}^-$ and $\mathscr{I}^+$ are spacelike (Fig. 23.10). If we consider the whole history of the observer O, then the past light cone of O at P on $\mathscr{I}^+$ is called the **future event horizon** of O. Events outside this horizon will never be seen by O. Next, consider the case when $\mathscr{I}^+$ is null (Minkowski space-time, for example). If O moves on a timelike geodesic, then O does not possess an event horizon. However, if observer $\bar{O}$ moves with uniform acceleration, then, asymptotically, the speed of the observer approaches the speed of light — which means that the world-line ends up on $\mathscr{I}^+$ — and then $\bar{O}$ possesses a future event horizon (Fig. 23.11). Notice that these event horizons are **observer-dependent**. This is to be contrasted with the event horizons of black holes which are more accurately termed **absolute event horizons** because they are **observer-independent**.

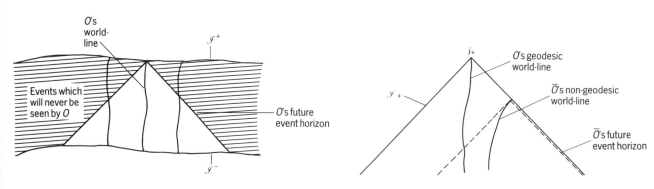

Fig. 23.10 The case when $\mathscr{I}^+$ and $\mathscr{I}^-$ are spacelike.

Fig. 23.11 The case when $\mathscr{I}^+$ is null.

23.15 Conformal structure of Robertson–Walker space-times

We proceed as we did in §22.8 and introduce a new radial coordinate χ so that the Robertson–Walker line element takes the form

$$ds^2 = dt^2 - [R(t)]^2[d\chi^2 + f^2(\chi)(d\theta^2 + \sin^2\theta\, d\phi^2)], \qquad (23.62)$$

where

$$\left.\begin{array}{ll} k = 0, & r = \chi = f(\chi), \\ k = +1, & r = \sin\chi = f(\chi), \\ k = -1, & r = \sinh\chi = f(\chi). \end{array}\right\} \qquad (23.63)$$

The coordinate χ runs from 0 to ∞ when $k = 0$ or -1, and from 0 to 2π when $k = +1$. Next, we introduce a new time coordinate τ defined by

$$d\tau = R^{-1}(t)dt,$$

so that (23.62) becomes

$$ds^2 = R^2(\tau)d\bar{s}^2 \qquad (23.64)$$

where

$$d\bar{s}^2 = d\tau^2 - d\chi^2 - f^2(\chi)(d\theta^2 + \sin^2\theta\, d\phi^2). \qquad (23.65)$$

Let us restrict our attention to the standard models $\Lambda = 0$, in which case $R(\tau)$ has one of the forms (23.8), (23.19), or (23.20). When $k = +1$, the unphysical line element (23.65) is precisely the Einstein static space (17.25). Indeed, all three models can be mapped on to different portions of the Einstein static space depending on the values taken by τ. In the case $k = 0$, the procedure is exactly the same as that employed in obtaining the conformal structure of Minkowski space-time (§17.4) except that now $0 < \tau < \infty$. The solution is therefore conformal to the half $t' > 0$ in Fig. 17.7. When $k = +1$, τ lies in the range $0 < \tau < \pi$. When $k = -1$, it can be shown that the space is conformal

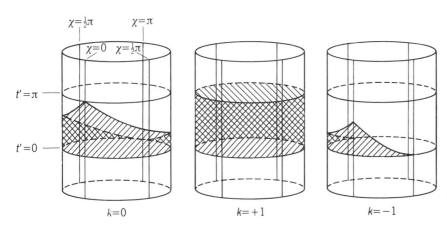

Fig. 23.12 Conformal Robertson–Walker space-times ($\Lambda = 0$).

to the region

$$-\tfrac{1}{2}\pi \leqslant t' + r' \leqslant \tfrac{1}{2}\pi,$$
$$-\tfrac{1}{2}\pi \leqslant t' - r' \leqslant \tfrac{1}{2}\pi,$$
$$t' \geqslant 0.$$

The various regions of the Einstein static cylinder for each case are depicted in Fig. 23.12.

These conformal diagrams are somewhat different from the others we have met so far, in that part of the boundary is not 'infinity' in the sense it was previously, but represents the initial singularity when $R = 0$. In fact, this makes little difference to the conformal diagrams. The Penrose diagram for the ever-expanding cases $k = 0$ and -1 is given in Fig. 23.13 (two dimensions suppressed). The initial singularity—the big bang—is a spacelike surface. The Penrose diagram for the oscillating universe $k = +1$ is given in Fig. 23.14 (two dimensions suppressed). In this case, both the initial and the final singularity—the big crunch—are spacelike surfaces. It can be shown that matter-filled Robertson–Walker universes are, in fact, inextendible.

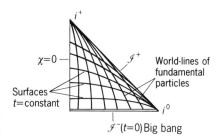

Fig. 23.13 Penrose diagram for $k = 0$ and -1 ($\Lambda = 0$).

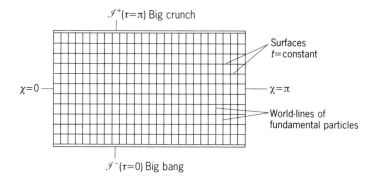

Fig. 23.14 Penrose diagram for $k = +1$ ($\Lambda = 0$).

23.16 Conformal structure of de Sitter space-time

De Sitter space-time is most easily visualized as the hyperboloid

$$-\hat{v}^2 + \hat{w}^2 + \hat{x}^2 + \hat{y}^2 + \hat{z}^2 = T_0^2 \tag{23.66}$$

embedded in flat **five-dimensional** Euclidean space with a Minkowski-type line element

$$ds^2 = d\hat{v}^2 - d\hat{w}^2 - d\hat{x}^2 - d\hat{y}^2 - d\hat{z}^2. \tag{23.67}$$

One can introduce coordinates $(\hat{t}, \chi, \theta, \phi)$ on the hyperboloid by the relations

$$\begin{aligned}
\hat{v} &= T_0 \sinh(\hat{t}/T_0),\\
\hat{w} &= T_0 \cosh(\hat{t}/T_0)\cos\chi,\\
\hat{x} &= T_0 \cosh(\hat{t}/T_0)\sin\chi\cos\theta,\\
\hat{y} &= T_0 \cosh(\hat{t}/T_0)\sin\chi\sin\theta\cos\phi,\\
\hat{z} &= T_0 \cosh(\hat{t}/T_0)\sin\chi\sin\theta\sin\phi,
\end{aligned} \tag{23.68}$$

in which case the line element has the form

$$ds^2 = d\hat{t}^2 - T_0^2\cosh^2(\hat{t}/T_0)[d\chi^2 + \sin^2\chi(d\theta^2 + \sin^2\theta\, d\phi^2)]. \tag{23.69}$$

Apart from coordinate singularities at $\chi = 0, \pi$ and $\theta = 0, \pi$, the hyperboloid is covered by the coordinate range

$$-\infty < \hat{t} < +\infty,$$
$$0 \leqslant \chi \leqslant \pi,$$
$$0 \leqslant \theta \leqslant \pi,$$
$$0 \leqslant \phi < 2\pi.$$

The surfaces $\hat{t} = $ constant are 3-spheres of constant positive curvature, the particles of the substratum travel on timelike geodesics normal to these surfaces, and the overall topology is cylindrical, being $\mathbb{R} \times S^3$ (Fig. 23.15). If we then introduce coordinates

$$
\left.
\begin{aligned}
t &= T_0 \ln \left[(\hat{w} + \hat{v})/T_0 \right], \\
x &= T_0 \hat{x}/(\hat{w} + \hat{v}), \\
y &= T_0 \hat{y}/(\hat{w} + \hat{v}), \\
z &= T_0 \hat{z}/(\hat{w} + \hat{v}),
\end{aligned}
\right\}
\tag{23.70}
$$

then the line element (23.67) reduces to the form (23.29) in Cartesian coordinates with $\alpha = T_0$, or (23.44) in the corresponding spherical polar coordinates on the hyperboloid. However, the coordinates (t, x, y, z) only cover half the hyperboloid since t is not defined for $\hat{w} + \hat{v} \leqslant 0$ (Fig. 23.16). In these coordinates, the surfaces $t = $ constant are flat 3-spaces, and the particles of the substratum are geodesics normal to these 3-spaces diverging from a point in the infinite past. Thus, the region of de Sitter space-time corresponding to $\hat{w} + \hat{v} > 0$ forms the space-time for the steady-state model.

We can obtain the conformal structure by defining a new time coordinate

$$t' = 2 \tan^{-1} \left[\exp (\hat{t}/T_0) \right] - \tfrac{1}{2}\pi,$$

where

$$-\tfrac{1}{2}\pi < t' < \tfrac{1}{2}\pi.$$

Then

$$\mathrm{d}s^2 = T_0^2 \cosh^2(t'/T_0)\, \mathrm{d}\bar{s}^2,$$

where $\mathrm{d}\bar{s}^2$ is the Einstein static line element (17.25) on identifying $r' = \chi$. The

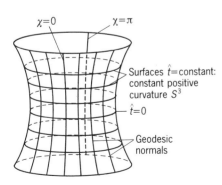

Fig. 23.15 de Sitter space-time embedded in five-dimensional Minkowski space-time.

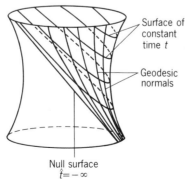

Fig. 23.16 de Sitter space-time in (t, x, y, z)-coordinates.

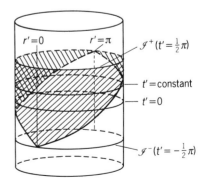

Fig. 23.17 Conformal de Sitter space-time.

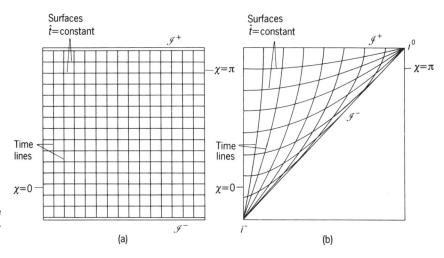

Fig. 23.18 Penrose diagram of (a) de Sitter space-time and (b) the steady-state model.

region to which de Sitter space is conformal is shown in Fig. 23.17. The Penrose diagrams of de Sitter space-time and the steady-state universe are shown in Fig. 23.18. It is clear that the steady-state theory suffers, at least aesthetically, from being geodesically incomplete in the past.

23.17 Inflation

No discussion of modern cosmology would be complete without some mention of inflation, since it has attracted so much attention in recent years. This is the idea that some 10^{-30} seconds after the big bang, there existed a phase in which the universe expands much faster (in fact exponentially) than the rate given by the standard scenario. The mechanism driving this expansion arises from modern physical theories called GUTs (grand unified theories), which attempt to unify three of the four fundamental forces, namely, the strong nuclear, weak nuclear, and electromagnetic forces. In particular, it relies on the existence of a scalar field introduced to break the symmetry between these forces. As such, these ideas lie beyond the scope of this book. None the less, we can get the gist of the ideas involved in inflation if we simply use the result that in this phase the energy density ρ is dominated by the vacuum energy density V_0 of the scalar field, i.e. $\rho \simeq V_0$. In the standard models ($\Lambda = 0$), this requires, from (22.57) and (22.58), that

$$\dot{R}^2 = \tfrac{8}{3}\pi R^2 V_0 - k. \tag{23.71}$$

Moreover, physical arguments reveal that, at the onset of this phase, the size of the scale factor is such that the term involving R^2 dominates over k in (23.71), so that we can neglect k. Then, dividing (23.71) by R^2, we get

$$\dot{R}^2/R^2 = \tfrac{8}{3}\pi V_0, \tag{23.72}$$

which is the square of the Hubble parameter, so that

$$H = (\tfrac{8}{3}\pi V_0)^{\frac{1}{2}}. \tag{23.73}$$

Taking the positive square root of (23.72) and integrating, we find that

$$R = R_0 \exp(Ht), \tag{23.74}$$

where R_0 is the value of the scalar factor at the start of the phase (compare with the de Sitter model (23.28)). Thus, we have obtained an exponential rate of expansion for this early phase, whereas previously, in the standard scenario, we had only obtained power laws.

The idea of inflation has attracted attention because it seems to answer a number of fundamental problems arising in the standard cosmological models. Two of these are the flatness and horizon problems, which we now briefly discuss. If we define

$$\Omega(t) = \rho(t)/\rho_c,$$

then present-day estimates give bounds on Ω of

$$0.01 < \Omega(t_0) < 10.$$

As we saw in (23.37), the universe is ever expanding, flat, or closed depending on whether $\Omega < 1$, $\Omega = 1$, or $\Omega > 1$, respectively. A major difficulty with standard cosmology is that it requires very fine tuning of the initial parameters to result in a universe consistent with present-day observations. In particular, it requires that Ω is very close to unity in the early universe. A typical estimate is that $|\Omega - 1| < 10^{-57}$ at the Planck time of some 10^{-43} seconds after the big bang. This necessity for fine tuning is called the **flatness problem**. Inflation helps to overcome this problem because the exponential rate of expansion makes the universe very flat. This is analogous to imagining the universe as a balloon, with the curved surface of the balloon representing curved space. If the balloon is not blown up very much, then the curvature is high. However, if the balloon is 'inflated' by a large amount, its surface becomes very flat.

The observable universe is highly homogeneous and isotropic on the large scale. Detailed investigations reveal that this would only appear to be possible if the universe was also highly homogeneous and isotropic in the earliest epochs. Moreover, these investigations lead to the conclusion that the region which evolved into the observable universe would have been too large for all points in it to be causally connected (i.e. there would not have been sufficient time for influences travelling with the speed of light to connect all points in the region). Thus, one is unable to use physical forces to account for the homogeneity and isotropy of the early universe, unless one uses forces capable of violating causality. This is the **horizon problem**. Inflation helps

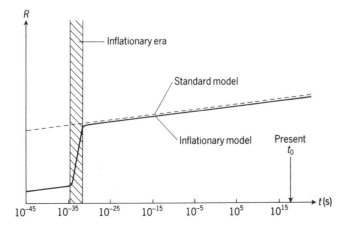

Fig. 23.19 Scale factor for inflationary and standard models.

overcome the problem because it allows the initial region of space which evolved into our present-day observed universe to be much smaller than the horizon distance. Mechanisms can then be discussed which account for the homogeneity and isotropy then required.

There is certainly much more to inflation than this brief introduction suggests, but we shall content ourselves by ending this section with a diagram (Fig. 23.19) illustrating how the behaviour of the scale factor differs qualitatively in an inflationary model. The need for an inflationary model is motivated in part by ideas originating in discussions of the anthropic principle, which is where we finally turn.

23.18 The anthropic principle

Cosmology is based on the cosmological principle which is a simplicity principle leading to a smoothed out universe, namely one which is homogeneous and isotropic. But why is the universe so smooth? One answer is that if **we** are to exist then it could hardly be otherwise. Put another way, a non-smooth universe would not have allowed us humans to have developed. This is an example of the **anthropic principle** which in simple terms, states the following.

> **The anthropic principle:** we see the universe the way it is because we exist.

The principle comes in two versions, the weak and the strong, which we consider in turn.

> **The weak anthropic principle:** the conditions for the development of life are only met in certain regions of the universe.

This form of the principle can be used to 'explain' why the bing bang occurred some ten thousand million years ago; namely, because it takes that long for sentient beings to emerge. More precisely, this is the time needed for all the intervening processes, such as the condensation of the galaxies from the primeval matter, the subsequent formation of the heavy elements (in supernovae), the eventual birth of our own galaxy, the formation of the solar system, the cooling of the Earth and the slow process of evolution up to the present day.

The earliest epochs of the universe really involve quantum ideas, and this leads to the area of **quantum cosmology**. As is well known, quantum theory involves deep problems of interpretation (see Penrose (1989) for an intriguing discussion). One interpretation leads to the 'many worlds' of Everett and Wheeler, in which the universe is bifurcating from one instant to the next into many (indeed infinite) disjoint new universes (Fig. 23.20). Or again, the universe may consist of many different regions, each with its own initial configuration and perhaps with its own set of laws of science. If we consider

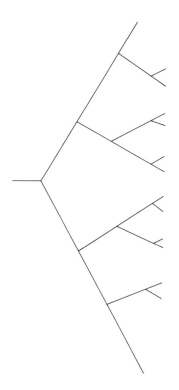

Fig. 23.20 The 'many worlds' interpretation of Everett and Wheeler.

these disjoint regions as different universes then the strong version of the anthropic principle can be stated as follows.

> **The strong anthropic principle**: the conditions for the development of life are only met in a few universes.

The laws of science involve a number of fundamental constants (such as the charge on the electron) which, at present, cannot be predicted from theoretical considerations, but can only be found by observation. Moreover, their actual values seem to be very finely adjusted. The slightest alteration of these values would lead to very different universes, most of which could not support life. One can interpret them in two ways: as evidence of a divine purpose or Creator (the argument from design in theology) and with it the choice of a particular set of laws of science, or as support for the strong anthropic principle. Although it is not clear the extent to which Einstein believed in a personal God, it is worth remarking that Einstein believed profoundly in the argument of a divine purpose. He considered that God could not have created the universe in any other way.

There are a number of objections to the strong formulation of the principle. If all these universes are really separate from us, in what sense can they be said to exist? If what happens in another universe has no observable consequences on ours then, on simplicity grounds alone, we can ignore them. If, on the other hand, they are different but accessible regions of our universe, then they are just the result of different initial configurations and so the strong anthropic principle would reduce to the weak one. Another objection is that the principle runs counter to the way that ideas have developed throughout the history of science, which has continuously demoted the importance of humankind in the scheme of things. For example, the cosmological principle leads us to believe that we live in a typical part of the universe, attached to a typical star, in a typical galaxy, belonging to a typical cluster, and so on. Yet the strong form of the anthropic principle turns this on its head and says that the whole giant structure exists simply for our sake.

The attempt to find a model of the universe in which many different initial configurations could have evolved into something like the present universe led to the idea of inflation. So inflation, together with the weak form of the anthropic principle, may be used to explain why the universe looks the way it does now.

The anthropic principle can also be used to throw light on whether the three **arrows of time** agree or not. These are the thermodynamic arrow (as expressed in the idea that disorder or entropy is always increasing), psychological time (as perceived by humans) and cosmological time (world time). For further development of these ideas see Hawking (1988) (on which this account is based), and for a more technical account see Barrow and Tippler (1986). It seems appropriate to end with a reference to Hawking, given that one of the goals of the book is to make contact with Hawking and Ellis (1973). It also seems appropriate to finish with an amusing representation of the development of life subsequent to the big bang in the universe (the big U) by Wheeler (Fig. 23.21), because this is reminiscent of the surrealistic pictures by Hugh Lieber in Lillian Lieber's book — which is where I came in.

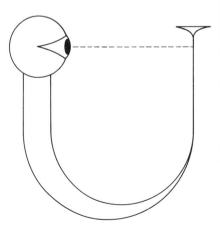

Fig. 23.21 Wheeler's 'big U' from the big bang (upper right) to the development of the human eye.

23.19 Conclusion

We have seen in this chapter that de Sitter space-time possesses event horizons and that most of the Friedmann models possess singularities: both phenomena we have met before in our considerations of classical black holes in Part D. The big bang singularity is a very drastic one, in which both the density and temperature increase without bound as $R \to 0$, and indeed space-time itself becomes singular at $R = 0$, where it is squeezed out of existence. However, the results have been deduced from the assumptions of exact spatial homogeneity and spherical symmetry. Although these assumptions may be reasonable on the large scale, they certainly do not hold locally. One might expect that, if one traced the evolution of the universe back in time, the local irregularities would grow and could prevent the occurrence of a singularity — causing the universe to 'bounce' instead. Yet, once again, the singularity theorems of Hawking and Penrose reveal that the occurrence of singularities is generic and, as a consequence, there is good evidence to believe that the physical universe was singular in the past.

There is another difference about the initial singularity of cosmology, compared with the black hole singularities, in that the big bang singularity is in principle **observable**. And it is observation that is the linchpin of cosmology. It is perhaps disappointing that the observations of cosmology are not sufficiently precise to yet determine whether we live in an open or closed universe. But there is good reason to believe that there will be a considerable increase in our observational knowledge in the not-too-distant future. For example, orbiting space stations should improve much of the astronomical data, X-ray astronomy being just one example of an area which will benefit significantly. Again, it is to be hoped that gravitational astronomy, that is, the detection of gravitational waves, will not be long in the offing. This would open up a whole new window on the world and, in all likelihood, allow us to put our cosmological theories more rigorously to the test.

It is a natural consequence of our inquisitive nature that we should wish to understand our own origins and that of the universe we inhabit. The hot big bang theory would appear to be a great stride forward in our search for this understanding. Whether or not the universe had this singular origin is perhaps the central question of cosmology. The mathematical basis of this question and the attempt to answer it is the principal problem dealt with in the book of Hawking and Ellis (1973). In turn, it has been one of the main objectives in writing this book to make their book, or at least parts of it, more accessible and the hope is that some readers may make this their next port of call.

And so, we end our considerations of cosmology and with it we end the book. There are many topics in general relativity which have not been mentioned, and even those that we have met have been covered in a largely introductory manner. None the less, we have acquainted ourselves with the essential components of the precursor to the general theory, namely, special relativity, we have looked carefully at the principles behind general relativity and investigated both the formulation of the theory and its principal consequences. In particular, we have reached the three endpoints we had promised ourselves, namely, classical black holes, gravitational waves, and cosmology. In the process, it is hoped that some of the richness and beauty of the theory and some of its absorbing and bizarre consequences have been revealed. At the start of this book, we set out on a long journey of discovery. It would seem that we have come a long way, but the journey is really only just begun.

Exercises

23.1 (§**23.1**) Show that taking the negative square root in (23.4) leads to the same result (23.5).

23.2 (§**23.1**) Use the substitution (23.6) to establish the solution (23.7) in the case $k = 0$, $\Lambda < 0$.

23.3 (§**23.1**) Confirm (23.9) and (23.10) for the Einstein–de Sitter model. Show that (23.15) leads to (23.16).

23.4 (§**23.2**) Use the substitution (23.18) to establish the solution (23.20) in the case $\Lambda = 0$, $k = -1$.

23.5 (§**23.2**) Confirm (23.21) and (23.22) for the model with solution (23.19).

23.6 (§**23.3**) Show that the general differential equation for $\ddot{R}$ and $\dot{R}$ can be written in the form

$$\ddot{R} = \frac{2}{9C}(-1/x^2 + \lambda x),$$

$$\dot{R}^2 = \tfrac{1}{3}(2/x + \lambda x^2 - 3k),$$

where $R_c = 3C/2$, $\Lambda_c = 4/(9C^2)$, $x = R/R_c$ and $\lambda = \Lambda/\Lambda_c$. [Hint: Let $R = R_c$, $\Lambda = \Lambda_c$ when $\ddot{R} = \dot{R} = 0$.]
 (i) Deduce that if $\ddot{R} = \dot{R} = 0$ at some time then $R = R_c$, $\Lambda = \Lambda_c$, $k = 1$, and $\ddot{R} = \dot{R} = 0$ for all times.
 (ii) If at some finite time t_0, $\dot{R}(t_0) = 0$ and $\ddot{R}(t_0) \neq 0$, then what type of cosmological model results?
 (iii) Show that
  (a) If $\Lambda < 0$, then all models are oscillating;
  (b) If $\Lambda > 0$, then oscillating models require $k = 1$ and $\Lambda < \Lambda_c$.
[Hint: consider the equations for x and λ in turn when x is small and large and λ is positive and negative.]

23.7 (§**23.3**) A straight channel contains a fixed particle of mass M at its origin O, while another particle P of mass m moves under gravitational attraction. Let OP be denoted by x, and take the time to be zero when the particle starts off from O in the positive x-direction. If the particle has velocity v_0 at x_0, then show there exists a value of x, $x = x_1$ say (positive, negative, or infinite), at which the velocity vanishes and find it in terms of x_0 and v_0. Show that the energy equation can be written in the form

$$\dot{x}^2 = 2GM/x - 2GM/x_1.$$

Compare this with Friedmann's equation and hence interpret the types of motion possible for various values of x_1.

23.8 (§**23.4**) Show that the line element (23.29) is invariant under a shift in t and a simultaneous change of scale in the space coordinates. Confirm that (23.29) is transformed into (23.32) under the transformation (23.31).

23.9 (§**23.8**) Establish (23.35) for the standard models and deduce (23.37), where ρ_c is defined by (23.36). Use the estimate $T_0 \approx 10^{10}$ years to obtain the value (23.39) for ρ_c. [Hint: in non-relativistic units (23.36) is $\rho_c = 3H_0^2/8\pi G$.]

23.10 (§**23.10**) Show that if $\Lambda = 0$ then (22.50) and (22.51), subject to (23.40), lead to (23.41). Neglecting the last term, deduce that $R \sim t^{\frac{1}{2}}$.

23.11 (§**23.12**) Use the Robertson–Walker line element in the form (22.34) to show that the three-dimensional Ricci scalar curvature of a 3-space $t = t_0$ is $6k[R(t_0)]^{-2}$.

23.12 (§**23.12**) Confirm the results (23.49), (23.50), and (23.51) for the steady-state theory.

23.13 (§**23.13**) Confirm the results (23.57) and (23.60).

23.14 (§**23.16**) Check that (23.68) satisfies (23.66). Show that the line element (23.67) reduces to the forms (23.69) and (23.29) on the hypersurface. [Hint: use (23.68) and (23.70), respectively.]

Answers to exercises

2.1

$$x = x' + v_1 t$$
$$y = y'$$
$$z = z'$$
$$t = t'.$$
$$x' = x - v_1 t$$
$$y' = y$$
$$z' = z$$
$$t' = t.$$

Interchange primes and unprimes and replace v_1 by $-v_1$.

$$x'' = x' + v_2 t$$
$$y'' = y'$$
$$z'' = z'$$
$$t'' = t'.$$
$$x'' = x + (v_1 + v_2)t$$
$$y'' = y$$
$$z'' = z$$
$$t'' = t.$$

2.2

(i)

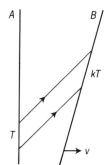
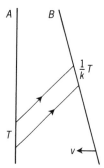

(ii)

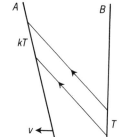
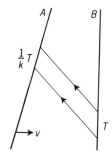

2.3 Blue shift.

2.6 Draw circle centre O, radius OG and two light rays entering and leaving G which cut the circle at points P and Q, as shown.

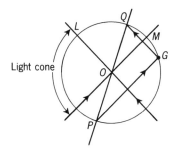

Then POQ is the world-line of an inertial observer who considers O and G to be simultaneous (since $PO = OQ$). Observers whose world-lines through O intersect LQ consider that G occurs later than O, and observers whose world-lines intersect QM consider that G occurs before O.

2.7

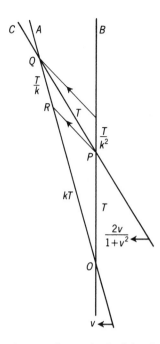

According to B, the coordinates (t, x) of the three events are

O $(0, 0)$

R $(\frac{1}{2}(k^2 + 1)T, \frac{1}{2}(k^2 - 1)T)$

Q $(\frac{1}{2}(k^2 + 1)(1 + 1/k^2)T, \frac{1}{2}(k^2 - 1)(1 + 1/k^2)T)$

Thus, whereas A's clock has elapsed by $(k + 1/k)T$ between events O and Q, the time lapse of B's clock is $\frac{1}{2}(1 + k^2)(1 + 1/k^2)T$ (which for $k > 1$ is greater than A's time lapse).

2.9 $v = \pm T(x^2 + T^2)^{-\frac{1}{2}}$.

2.10 $s^2 = -(x_1 - x_2)^2 - (y_1 - y_2)^2 - (z_1 - z_2)^2 = -\sigma^2$.

3.1 $(2/3)^{\frac{1}{2}}$.

3.3 $\theta = i\phi$.

3.6 Take the room to be in the frame S' moving along the x-axis of the rest frame of the pole with speed $-v$, as shown (not to scale):

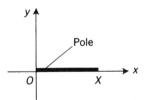

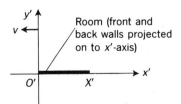

Then in S's frame

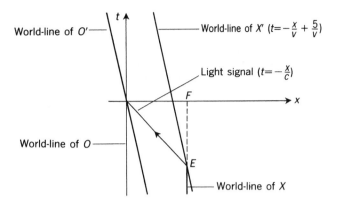

When O and O' coincide, S will 'see' X at F, as a result of a light signal from event E.

3.9 (a) 7.5×10^{-5} s.

(b) 17 min.

3.10 3.4×10^9 light years.

940 years.

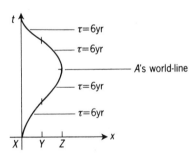

3.11

$$v = [(1 - u/c)/(1 + u/c)]^{\frac{1}{2}},$$

$$v = [(1 + u/c)/(1 - u/c)]^{\frac{1}{2}}.$$

3.12 0.32c.

4.1 One possibility is to define a **unit of force** F^1 as that which results in a standard mass m_S undergoing an acceleration g_L, that is

$$F^1 = m_S g_L, \tag{1}$$

where g_L is the acceleration due to gravity at a given latitude. We can then use Newton's second law to compare any other force F by measuring the acceleration a_S^F this produces when applied to the standard mass, that is

$$\frac{F}{F^1} = \frac{m_S a_S^F}{m_S g_L} = \frac{a_S^F}{g_L}. \tag{2}$$

We could then define **unit mass** m^1 as that mass which, when acted on by a unit force F^1, suffers a unit acceleration 1. Other masses could then be defined by either (i) measuring the acceleration a that a mass experiences under the influence of the unit force, that is

$$\frac{F^1}{F^1} = \frac{ma}{m^1 1},$$

or (ii) using (2) to measure a force F and then applying this force to a mass m and measuring the resulting acceleration a, so that

$$m = F/a.$$

4.2 The kinetic energy of the initial particle in motion.

4.3 $(\bar{m}_0^2 \gamma^4 + 2m_0 \bar{m}_0 \gamma + m_0^2)^{\frac{1}{2}}$ where $\gamma = (1 - u^2/c^2)^{-\frac{1}{2}}$,

4.8 $P = 2Mp_0/(m_0 + M)$, $p = (m_0 - M)p_0/(m_0 + M)$.

4.9

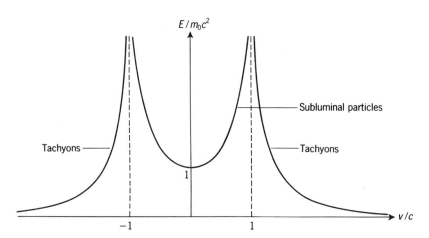

4.10 $-c\cos\theta$.

4.11 $chv/(hv + m_0 c^2)$,
$(m_0^2 + 2hvm_0/c^2)^{\frac{1}{2}}$.

5.1 (i) (a) $x = a\cos\phi$, $y = a\sin\phi$, $z = 0$ $(0 \leqslant \phi \leqslant 2\pi)$.
(b) $x^2 + y^2 - a^2 = 0$, $z = 0$.
(ii) $x = a\sin\theta\cos\phi$, $y = a\sin\theta\sin\phi$, $z = a\cos\theta$
$(0 \leqslant \theta \leqslant \pi, 0 \leqslant \phi \leqslant 2\pi)$.
(b) $x^2 + y^2 + z^2 - a^2 = 0$.

5.2

$$(x^a) = (x^1, x^2, x^3) = (x, y, z)$$
$$(x'^a) = (x'^1, x'^2, x'^3) = (r, \theta, \phi)$$
$$(x^a) \to (x'^a):$$
$$r = (x^2 + y^2 + z^2)^{\frac{1}{2}}$$
$$\theta = \tan^{-1}[(x^2 + y^2)^{\frac{1}{2}}/z]$$
$$\phi = \tan^{-1}(y/x)$$
$$(x'^a) \to (x^a):$$
$$x = r\sin\theta\cos\phi$$
$$y = r\sin\theta\sin\phi$$

$$z = r\cos\theta$$

$$\left(\frac{\partial x^a}{\partial x'^b}\right) = \begin{pmatrix} \sin\theta\cos\phi & r\cos\theta\cos\phi & -r\sin\theta\sin\phi \\ \sin\theta\sin\phi & r\cos\theta\cos\phi & r\sin\theta\cos\phi \\ \cos\theta & -r\sin\theta & 0 \end{pmatrix}$$

$$\left(\frac{\partial x'^a}{\partial x^b}\right) = \begin{pmatrix} \sin\theta\cos\phi & \sin\theta\sin\phi & \cos\theta \\ \cos\theta\cos\phi/r & \cos\theta\sin\phi/r & -\sin\theta/r \\ -\sin\phi/(r\sin\theta) & \cos\phi/(r\sin\theta) & 0 \end{pmatrix}$$

$$J = J'^{-1} = r^2\sin\theta.$$

$J' \to 0$ when $r \to \infty$,
$J' \to \infty$ when $r = 0$ and $\theta = 0, \pi$.

5.6

$$(x^a) \to (x'^a):$$
$$R = (x^2 + y^2)^{\frac{1}{2}}$$
$$\phi = \tan^{-1}(y/x)$$
$$\left(\frac{\partial x'^a}{\partial x^b}\right) = \begin{pmatrix} \cos\phi & \sin\phi \\ -\sin\phi/R & \cos\phi/R \end{pmatrix},$$
$$X^a = \frac{dx^a}{d\phi} = (-a\sin\phi, a\cos\phi),$$
$$X'^a = (0, 1).$$

5.7

$$X_c'^{ab} = \frac{\partial x'^a}{\partial x^d}\frac{\partial x'^b}{\partial x^e}\frac{\partial x^f}{\partial x'^c}X_f^{de}.$$

5.14 $\delta_a^a = \delta_b^a\delta_a^b = n$.

5.16 (i) $X'^a = (\cos\phi, -\sin\phi/R)$.
(ii)

$$\frac{\partial}{\partial x} = \cos\phi\frac{\partial}{\partial R} - \frac{\sin\phi}{R}\frac{\partial}{\partial\phi},$$

In the figure: $E/m_0 c^2$ axis (vertical), v/c axis (horizontal). Labels: Subluminal particles, Tachyons, Tachyons, with values -1, 1 on the horizontal axis and 1 on the vertical.

$$\frac{\partial}{\partial y} = \sin\phi\,\frac{\partial}{\partial R} + \frac{\cos\phi}{R}\,\frac{\partial}{\partial\phi},$$

$$\frac{\partial}{\partial R} = \frac{x}{(x^2+y^2)^{\frac{1}{2}}}\,\frac{\partial}{\partial x} + \frac{y}{(x^2+y^2)^{\frac{1}{2}}}\,\frac{\partial}{\partial y},$$

$$\frac{\partial}{\partial\phi} = -y\frac{\partial}{\partial x} + x\frac{\partial}{\partial y}.$$

(iii)

$$X^a\partial_a = \frac{\partial}{\partial x},$$

$$X'^a\partial'_a = \cos\phi\,\frac{\partial}{\partial R} - \frac{\sin\phi}{R}\,\frac{\partial}{\partial\phi}.$$

(iv)

$$Y'^a = (\sin\phi,\ \cos\phi/R),$$

$$Z'^a = (0, 1),$$

$$Y = \frac{\partial}{\partial y} = \sin\phi\,\frac{\partial}{\partial R} + \frac{\cos\phi}{R}\,\frac{\partial}{\partial\phi},$$

$$Z = -y\frac{\partial}{\partial x} + x\frac{\partial}{\partial y} = \frac{\partial}{\partial\phi}.$$

(v) The Lie brackets are given in the table below (the vector in the column being the first entry):

	X	Y	Z
X	0	0	Y
Y	0	0	$-X$
Z	$-Y$	X	0

6.2

$$L_X Z_{bc} = Z_{bc,d}X^d + Z_{dc}X^d_{,b} + Z_{bd}X^d_{,c}$$

$$L_X(Y^a Z_{bc}) = (Y^a Z_{bc})_{,d}X^d - Y^d Z_{bc}X^a_{,d}$$
$$+\ Y^a Z_{dc}X^d_{,b} + Y^a Z_{bd}X^d_{,c}$$

6.14

$$g_{ab} = \mathrm{diag}(1, 1, 1),\ g^{ab} = \mathrm{diag}(1, 1, 1),\ g = 1.$$

$$g_{ab} = \mathrm{diag}(1, R^2, 1),\ g^{ab} = \mathrm{diag}(1, R^{-2}, 1),\ g = R^2.$$

$$g_{ab} = \mathrm{diag}(1, r^2, r^2\sin^2\theta),\ g^{ab} = \mathrm{diag}(1, r^{-2},\ r^{-2}\sin^{-2}\theta),$$
$$g = r^4\sin^2\theta.$$

6.15 $T_{ab} = g_{ac}g_{bd}T^{cd}.$

6.16

$$g'_{ab} = \frac{\partial x^c}{\partial x'^a}\frac{\partial x^d}{\partial x'^b}g_{cd}.$$

6.17

$$\frac{\mathrm{d}^2 R}{\mathrm{d}u^2} - R\left(\frac{\mathrm{d}\phi}{\mathrm{d}u}\right)^2 = 0,$$

$$\frac{\mathrm{d}^2\phi}{\mathrm{d}u^2} + \frac{2}{R}\frac{\mathrm{d}R}{\mathrm{d}u}\frac{\mathrm{d}\phi}{\mathrm{d}u} = 0,$$

$$\frac{\mathrm{d}^2 z}{\mathrm{d}u^2} = 0.$$

6.21 (i) -2.

(ii) Yes.

(iii) Yes.

6.22 (i) $(x^1, x^2, x^3) = (r, \theta, \phi).$

(ii) Yes.

6.31 (i)

$$g_{ab} = \mathrm{diag}(e^\nu,\ -e^\lambda,\ -r^2,\ -r^2\sin^2\theta),$$

$$g = -e^{\nu+\lambda}r^4\sin^2\theta,$$

$$g^{ab} = \mathrm{diag}(e^{-\nu},\ -e^{-\lambda},\ -r^{-2},\ -r^{-2}\sin^{-2}\theta).$$

(ii) Non-zero independent components:

$$\Gamma^0_{00} = -\tfrac{1}{2}\dot\nu,\quad \Gamma^0_{01} = \tfrac{1}{2}\nu',\quad \Gamma^0_{11} = \tfrac{1}{2}e^{\lambda-\nu}\dot\lambda,$$

$$\Gamma^1_{00} = \tfrac{1}{2}e^{\nu-\lambda}\nu',\quad \Gamma^1_{01} = \tfrac{1}{2}\dot\lambda,\quad \Gamma^1_{11} = \tfrac{1}{2}\lambda',$$

$$\Gamma^1_{22} = -re^{-\lambda},\quad \Gamma^1_{33} = -re^{-\lambda}\sin^2\theta,$$

$$\Gamma^2_{12} = r^{-1},\quad \Gamma^2_{33} = -\sin\theta\cos\theta,$$

$$\Gamma^3_{13} = r^{-1},\quad \Gamma^3_{23} = \cot\theta.$$

(iii) Non-zero independent components:

$$R^0{}_{101} = -\tfrac{1}{2}\nu'' + \tfrac{1}{4}e^{\lambda-\nu}\dot\lambda^2 - \tfrac{1}{4}e^{\lambda-\nu}\dot\nu\dot\lambda$$
$$+ \tfrac{1}{2}e^{\lambda-\nu}\ddot\lambda - \tfrac{1}{4}\nu'^2 + \tfrac{1}{4}\nu'\lambda',$$

$$R^0{}_{202} = -\tfrac{1}{2}re^{-\lambda}\nu',$$

$$R^0{}_{212} = -\tfrac{1}{2}re^{-\nu}\dot\lambda,$$

$$R^0{}_{303} = -\tfrac{1}{2}re^{-\lambda}\sin^2\theta\,\nu',$$

$$R^0{}_{313} = -\tfrac{1}{2}re^{-\nu}\sin^2\theta\,\dot\lambda,$$

$$R^1{}_{212} = \tfrac{1}{2}re^{-\lambda}\lambda',$$

$$R^1{}_{313} = \tfrac{1}{2}re^{-\lambda}\sin^2\theta\,\lambda',$$

$$R^2{}_{323} = -e^{-\lambda}\sin^2\theta + \sin^2\theta.$$

(iv) Non-zero independent components:

$$R_{00} = \tfrac{1}{2}e^{\nu-\lambda}\nu'' - \tfrac{1}{4}\dot\lambda^2 + \tfrac{1}{4}\dot\nu\dot\lambda - \tfrac{1}{2}\ddot\lambda$$
$$+ \tfrac{1}{4}e^{\nu-\lambda}\nu'^2 - \tfrac{1}{4}e^{\nu-\lambda}\nu'\lambda' + r^{-1}e^{\nu-\lambda}\nu',$$

$$R_{01} = r^{-1}\dot\lambda,$$

$$R_{11} = -\tfrac{1}{2}\nu'' + \tfrac{1}{4}e^{\lambda-\nu}\dot\lambda^2 - \tfrac{1}{4}e^{\lambda-\nu}\dot\nu\dot\lambda$$
$$+ \tfrac{1}{2}e^{\lambda-\nu}\ddot\lambda - \tfrac{1}{4}\nu'^2 + \tfrac{1}{4}\nu'\lambda' + r^{-1}\lambda',$$

$$R_{22} = -\tfrac{1}{2}re^{-\lambda}\nu' + \tfrac{1}{2}re^{-\lambda}\lambda' - e^{-\lambda} + 1,$$

$$R_{33} = \sin^2\theta\,R_{22},$$

$$R = e^{-\lambda}\nu'' - \tfrac{1}{2}e^{-\nu}\dot\lambda^2 + \tfrac{1}{2}e^{-\nu}\dot\nu\dot\lambda$$
$$- e^{-\nu}\ddot\lambda + \tfrac{1}{2}e^{-\lambda}\nu'^2 - \tfrac{1}{2}e^{-\lambda}\nu'\lambda'$$
$$+ 2r^{-1}e^{-\lambda}\nu' - 2r^{-1}e^{-\lambda}\lambda' + 2r^{-2}e^{-\lambda} - 2r^{-2}.$$

$$G_{00} = r^{-1}e^{\nu - \lambda}\lambda' - r^{-2}e^{\nu - \lambda} + r^{-2}e^{\nu},$$

$$G_{01} = r^{-1}\dot{\lambda},$$

$$G_{11} = r^{-1}\nu' - r^{-2}e^{\lambda} + r^{-2},$$

$$G_{22} = \tfrac{1}{2}re^{-\lambda}\nu' - \tfrac{1}{2}re^{-\lambda}\lambda' + \tfrac{1}{2}r^2e^{-\lambda}\nu''$$
$$- \tfrac{1}{4}r^2e^{-\nu}\dot{\lambda}^2 + \tfrac{1}{4}r^2e^{-\nu}\dot{\nu}\dot{\lambda} - \tfrac{1}{2}r^2e^{-\nu}\ddot{\lambda}$$
$$+ \tfrac{1}{4}r^2e^{-\lambda}\nu'^2 - \tfrac{1}{4}r^2e^{-\lambda}\nu'\lambda',$$

$$G_{33} = \sin^2\theta\, G_{22}.$$

(v) Non-zero components:

$$G^0{}_0 = r^{-1}e^{-\lambda}\lambda' - r^{-2}e^{-\lambda} + r^{-2},$$

$$G^0{}_1 = r^{-1}e^{-\nu}\dot{\lambda},$$

$$G^1{}_0 = r^{-1}e^{-\lambda}\dot{\lambda},$$

$$G^1{}_1 = r^{-1}e^{-\lambda}\nu' - r^{-2}e^{-\lambda} + r^{-2},$$

$$G^2{}_2 = \tfrac{1}{2}r^{-1}e^{-\lambda}\lambda' - \tfrac{1}{2}r^{-1}e^{-\lambda}\nu'$$
$$- \tfrac{1}{2}e^{-\lambda}\nu'' + \tfrac{1}{4}e^{-\nu}\dot{\lambda}^2 - \tfrac{1}{4}e^{-\nu}\dot{\nu}\dot{\lambda}$$
$$+ \tfrac{1}{2}e^{-\nu}\ddot{\lambda} - \tfrac{1}{4}e^{-\lambda}\nu'^2 + \tfrac{1}{4}e^{-\lambda}\nu'\lambda',$$

$$G^3{}_3 = G^2{}_2.$$

7.1 $\Phi_{;a} = \Phi_{,a} - \Phi\Gamma^b{}_{ba}.$

7.5 (i) $y'' - y = 0.$

(ii) $2y_1y_1'' + y_1'^2 - y_2'^2 - 3xy_1^2 - y_2 = 0,$
$$2y_1y_2'' + 2y_1'y_2' - y_1 = 0.$$

7.8 $\dfrac{\partial}{\partial y}.$

7.10 $X^1 = \dfrac{\partial}{\partial x}, \quad X^2 = \dfrac{\partial}{\partial y}, \quad X^3 = \dfrac{\partial}{\partial z},$

$$X^4 = y\dfrac{\partial}{\partial z} - z\dfrac{\partial}{\partial y}, \quad X^5 = z\dfrac{\partial}{\partial x} - x\dfrac{\partial}{\partial z},$$

$$X^6 = x\dfrac{\partial}{\partial y} - y\dfrac{\partial}{\partial x}$$

	X^1	X^2	X^3	X^4	X^5	X^6
X^1	0	0	0	0	$-X^3$	X^2
X^2	0	0	0	X^3	0	$-X^1$
X^3	0	0	0	$-X^2$	X^1	0
X^4	0	$-X^3$	X^2	0	$-X^6$	X^5
X^5	X^3	0	$-X^1$	X^6	0	$-X^4$
X^6	$-X^2$	X^1	0	$-X^5$	X^4	0

7.13 $(\nabla_c\nabla_b - \nabla_b\nabla_c)X_a = R_{adcb}X^d.$

8.5 (a) $\tfrac{1}{2}n(n-1)\,\omega_{ab}, n\,t_a.$
(b) $6\omega_{ab}$ – 3 spatial rotations, 3 boosts.
$4t_a$ – 3 spatial translations, 1 time translation.

8.8
$$\ddot{x} = \frac{(m_1 - m_2)}{(m_1 + m_2)}g,$$

$$p = (m_1 + m_2)\dot{x},$$

$$H(p, x) = \frac{p^2}{2(m_1 + m_2)} - m_1gx - m_2g(l - x).$$

8.10 Zeroth component gives the rate of work done by force **F**, viz.
$$\frac{dE}{dt} = \mathbf{F}.\mathbf{u}.$$

8.11 (i) $u'_x = \dfrac{u_x - v}{1 - u_xv}, \quad u'_y = \dfrac{u_y}{\beta(1 - u_xv)},$

$$u'_z = \dfrac{u_z}{\beta(1 - u_xv)}.$$

(ii) $E' = \beta(E - vp_x), \quad p'_x = \beta(p_x - vE/c^2),$
$$p'_y = p_y, \quad p'_z = p_z.$$

(iii) $F'_x = \dfrac{F_x - v\mathbf{F}.\mathbf{u}}{1 - u_xv}, \quad F'_y = \dfrac{F_y}{\beta(1 - u_xv)},$

$$F'_z = \dfrac{F_z}{\beta(1 - u_xv)}.$$

Yes.

(iv) $\mathbf{F}' = (F'_x, F'_y, F'_z) = (F, 0, 0) = \mathbf{F}.$

9.1 (i) $\tan^{-1}(a/g).$
The inertial observer will see the mass accelerate in the direction of motion due to a tension in the rod whose horizontal component produces the acceleration and whose vertical component counterbalances the weight. A non-inertial observer in the car will consider that the pendulum mass experiences two forces, the weight mg down and an inertial force ma in the opposite direction to motion.
(ii) $\tan^{-1}(a/g).$
(iii) 0.

9.2 (ii) Four inertial forces, namely:-
(a) a linear accelerative force as discussed in Ex. 9.1
(b) a velocity-dependent Coriolis force
(c) a centrifugal force
(d) a non-uniform rotational force (analogous tangentially to (a)).

9.5 (1) The released body undergoes uniform motion (due to Newton's first law) and the rocket accelerates with acceleration g, so the inertial observer sees the floor of the rocket ship come up and hit the body with relative acceleration g.
(2) The rocket and the body have no forces acting on them and therefore, by Newton's first law, both undergo uniform motion i.e. travel with the same constant velocity.
(4) The body and the lift both fall under gravity with the same acceleration.

9.6 Ellipsoid (see §**16.10**).

9.7

$$\frac{d^2 R}{du^2} - R\left(\frac{d\phi}{du}\right)^2 = 0,$$

$$\frac{d^2 \phi}{du^2} + \frac{2}{R}\frac{dR}{du}\frac{d\phi}{du} = 0,$$

$$\frac{d^2 z}{du^2} = 0.$$

Inertial force (centrifugal and Coriolis components).

9.8 (i) Straight line.
(ii) Parabola (projectile motion).

9.9 One example:

$$\nabla_b T^{ab} + R^a{}_{mnc} R^{emn}{}_d \nabla_e T^{cd} = 0$$

9.10 $\nabla_{[a} F_{bc]} = 0$

9.11

In the limit, the null comes degenerate into planes of simultaneity. That is, all observers, irrespective of their motion, agree that events occurring in one of these planes do so simultaneously.

10.1

$$f(\mathbf{x} + \mathbf{h}) = f(\mathbf{x}) + \left(h_1 \frac{\partial}{\partial x} + h_2 \frac{\partial}{\partial y} + h_3 \frac{\partial}{\partial z}\right) f(\mathbf{x})$$

$$+ \frac{1}{2}\left(h_1^2 \frac{\partial^2}{\partial x^2} + 2h_1 h_2 \frac{\partial^2}{\partial x \partial y}\right.$$

$$+ 2h_1 h_3 \frac{\partial^2}{\partial x \partial z} + h_2^2 \frac{\partial^2}{\partial y^2}$$

$$\left.+ 2h_2 h_3 \frac{\partial^2}{\partial y \partial z} + h_3^2 \frac{\partial^2}{\partial z^2}\right) f(\mathbf{x}) + \cdots.$$

10.7

$$g^{ab} = \text{diag}(A^{-1}, -A, -r^{-2}, -r^{-2}\sin^{-2}\theta),$$

$$g_{ab} = \text{diag}(A, -A^{-1}, -r^2, -r^2\sin^2\theta),$$

$$ds^2 = A\,dt^2 - A^{-1}\,dr^2 - r^2\,d\theta^2 - r^2\sin^2\theta\,d\phi^2.$$

10.9 (i) Principle of equivalence
(ii) Principle of equivalence
(iii) Correspondence principle
(iv) Correspondence and covariance principles
(v) Principle of equivalence
(vi) Correspondence principle.

10.10 Principle of equivalence, principle of minimal gravitational coupling, Mach's principle (?) and correspondence principle.

11.7 (i) $\nabla^a((-g)^{\frac{1}{2}} G_{ab}) = 0$

(ii) $-\nabla_a((-g)^{\frac{1}{2}} G^{ab}) = 0$

(iii) $\nabla^a((-g)^{\frac{1}{2}} G_{ab}) = 0.$

11.12 $\mathscr{L}_{uv} = (-g)^{\frac{1}{2}}[g_{ud} R^{abcd} R_{abcv} + g_{vd} R^{abcd} R_{abcu}$

$$-\tfrac{1}{2} g_{uv} R^{abcd} R_{abcd}].$$

11.13 $T_{ab} = (-g)^{\frac{1}{2}}(\phi_{,a}\phi_{,b} - \tfrac{1}{2} g_{ab}\phi_{,c}\phi_{,d} g^{cd} - \tfrac{1}{2} g_{ab} m_0^2 \phi^2).$

12.4

$$E'_x = E_x,$$

$$E'_y = \beta(E_y - vB_z),$$

$$E'_z = \beta(E_z + vB_y),$$

$$B'_x = B_x,$$

$$B'_y = \beta(B_y + vE_z),$$

$$B'_z = \beta(B_z - vE_y).$$

$$\rho' = \beta(\rho - vj_x), \quad j'_x = \beta(j_x - v\rho), \quad j'_y = j_y,$$

$$j'_z = j_z.$$

12.6 $\phi \to \bar{\phi}_a = \phi_a + \partial_a \psi$

where ψ must be a solution of $\Box \psi = 0$.

13.3 $\mu = -\tfrac{1}{4}.$

$$R_{ab} - \tfrac{1}{4} g_{ab} R = 2(F_{ac} F^c{}_b + \tfrac{1}{4} g_{ab} F_{cd} F^{cd}).$$

14.1 One example: a particle falls in from infinity at time $t = -\infty$ reaching the origin at time $t = 0$ whereupon the motion is reversed.

14.5 (a) $ds = a\,d\theta$.

(b) $ds = a\sin\theta\,d\phi$.

14.10 $[G] = M^{-1}L^3T^{-2}$.

14.11

$$R_{0101} = 2mr$$

$$R_{0202} = -\left(1 - \frac{2m}{r}\right)\frac{m}{r}$$

$$R_{0303} = -\left(1 - \frac{2m}{r}\right)\frac{m}{r}\sin^2\theta$$

$$R_{1212} = \left(1 - \frac{2m}{r}\right)^{-1}\frac{m}{r}$$

$$R_{1313} = \left(1 - \frac{2m}{r}\right)^{-1}\frac{m}{r}\sin^2\theta$$

$$R_{2323} = -2mr\sin^2\theta$$

14.14 (i), (ii), (iii), (iv), (v), and (vi).

15.1 Motion in a straight line.

15.6 The laws become modified to:

K1′: Each planet moves in an ellipse about the centre-of-mass as one of the foci.

K2′: The radius vector from the centre-of-mass to the planet sweeps out equal area in equal times.

$$\text{K3′: } \tau = \frac{2\pi}{(G(m_{\text{planet}} + m_{\text{sun}}))^{\frac{1}{2}}}a^{\frac{3}{2}}.$$

15.7 $\left(1 - \frac{2m}{r}\right)^{-1}\ddot{r} - \left(1 - \frac{2m}{r}\right)^{-2}\frac{m}{r^2}\dot{r}^2 + \frac{m}{r^2}\dot{t}^2 - r\dot{\theta}^2$

$- r\sin^2\theta\dot{\phi}^2 = 0.$

15.8

$$\left(\frac{du}{d\phi}\right)^2 + u^2 = \frac{k^2 - 1}{h^2},$$

$$\frac{d^2u}{d\phi^2} + u = 0.$$

15.14

$$\pm t \approx (r^2 - D^2)^{\frac{1}{2}} + 2m\cosh^{-1}\left(\frac{r}{D}\right)$$

$$- m\frac{(r^2 - D^2)^{\frac{1}{2}}}{r} + \text{constant}$$

$$= (r^2 - D^2)^{\frac{1}{2}} + 2m\log\left(\frac{r + (r^2 - D^2)^{\frac{1}{2}}}{D}\right)$$

$$- m\frac{(r^2 - D^2)^{\frac{1}{2}}}{r} + \text{constant}.$$

16.1 This is the Schwarzschild solution under the renaming of the coordinates:

$$(\theta, \phi, t, r) \to (t, r, \theta, \phi).$$

16.2 $\dfrac{4\bar{t}^2\bar{r}^3\cos\bar{\theta}}{\bar{r}\cos\bar{\theta} + 2m} - (\bar{r}\cos\bar{\theta} + 2m)^2\bar{r}^2\bar{\theta}^2\bar{\phi}^2\sin^2(\bar{\phi}\bar{t}).$

16.3 (i) t-timelike; ρ, z, ϕ-spacelike.

(ii) u-null; r, x, y-spacelike.

16.4 $ds^2 = A(t)\,dt^2 - B(t)\,dx^2 - C(t)\,dy^2 - D(t)\,dz^2.$

16.5

$$ds^2 = \left(1 - \frac{2m}{r}\right)dt^2 - \left[1 + \frac{2mx^2}{r^2(r - 2m)}\right]dx^2$$

$$- \frac{4mxy}{r^2(r - 2m)}dx\,dy$$

$$- \frac{4mxz}{r^2(r - 2m)}dx\,dz - \left[1 + \frac{2my^2}{r^2(r - 2m)}\right]dy^2$$

$$- \frac{4myz}{r^2(r - 2m)}dy\,dz - \left[1 + \frac{2mz^2}{r^2(r - 2m)}\right]dz^2$$

where $r = (x^2 + y^2 + z^2)^{\frac{1}{2}}$.

16.6

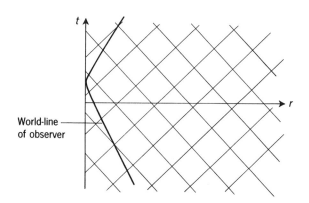

16.7 $r = \pm ku + c,$ c constant.

16.8 -1.

16.14 (Roughly)

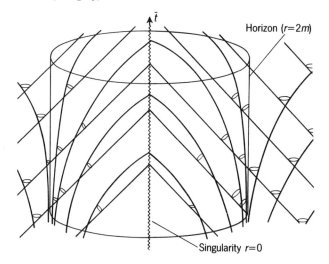

16.16 A non-rotating white hole consists of a visible singularity situated at the origin of coordinates, which suddenly erupts into a star whose radius increases inexorably through its Schwarzschild radius.

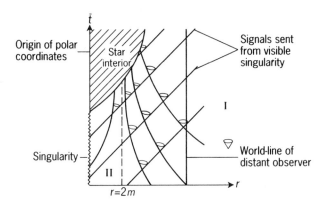

17.4 Region II.
They cannot escape from region II, but are ultimately crushed out of existence by the singularity.

17.7

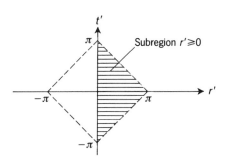

17.8
$$t' = \tan^{-1}(t + r) + \tan^{-1}(t - r),$$
$$r' = \tan^{-1}(t + r) - \tan^{-1}(t - r).$$
$$t = t_0 \leftrightarrow \tan\tfrac{1}{2}(t' + r') + \tan\tfrac{1}{2}(t' - r') = 2t_0,$$
$$r = r_0 \leftrightarrow \tan\tfrac{1}{2}(t' + r') - \tan\tfrac{1}{2}(t' - r') = 2r_0.$$

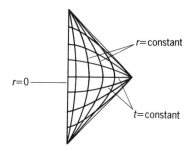

17.9

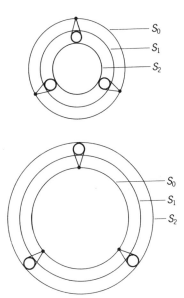

17.10 Figure 17.10 is the Penrose diagram of the Schwarzschild solution in the absence of a source. The introduction of a source suppresses regions I′ and II′.

18.4
$$T_{ab} = \frac{\varepsilon^2}{8\pi r^4} \operatorname{diag}\left[\left(1 - \frac{m}{r} + \frac{\varepsilon^2}{r^2}\right),\right.$$
$$\left. -\left(1 - \frac{m}{r} + \frac{\varepsilon^2}{r^2}\right)^{-1}, r^2, r^2 \sin^2\theta\right].$$

18.6 I: t timelike, r spacelike.
II: t spacelike, r timelike.
III: t timelike, r spacelike.
$$r = r_{\pm} = m \pm (m^2 - \varepsilon^2)^{\frac{1}{2}}.$$

18.7

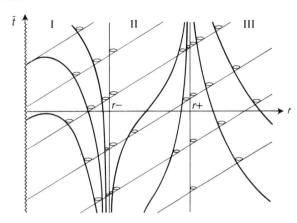

$$g^{ab} = \begin{pmatrix} 0 & -1 & 0 & 0 \\ -1 & -(1 - 2m/r) & 0 & 0 \\ 0 & 0 & -r^{-2} & 0 \\ 0 & 0 & 0 & -r^{-2}\sin^2\theta \end{pmatrix}.$$

19.11

$$g^{11} = -\frac{(r^2 - 2mr + a^2)}{(r^2 + a^2\cos^2\theta)}.$$

19.12 The other condition is identical, except that the sign of l is reversed and both signs of l are considered in (19.56) and the sequel.

18.8 $t + r + \dfrac{r_+^2}{r_+ + r_-}\log(r - r_+)$

$\qquad - \dfrac{r_-^2}{r_+ - r_-}\log(r - r_-) = \text{constant}.$

18.9

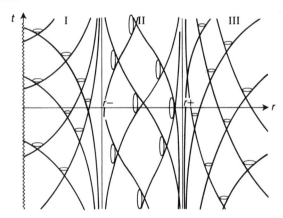

If $\varepsilon^2 = m^2$ there is no region II.

18.13 $\mathrm{d}s^2 = \left(1 - \dfrac{2m}{r} + \dfrac{\varepsilon^2}{r^2}\right)\mathrm{d}v^2$

$\qquad - 2\mathrm{d}v\,\mathrm{d}r - r^2(\mathrm{d}\theta^2 + \sin^2\theta\,\mathrm{d}\phi^2).$

18.14 $r = \varepsilon^2/m.$

$\qquad r = -m + (m^2 + \varepsilon^2)^{\frac{1}{2}},$

$\qquad -1.$

19.2

$$g_{ab} = \begin{pmatrix} 1 - 2m/r & -1 & 0 & 0 \\ -1 & 0 & 0 & 0 \\ 0 & 0 & -r^2 & 0 \\ 0 & 0 & 0 & -r^2\sin^2\theta \end{pmatrix},$$

19.14

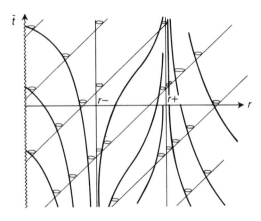

20.12

$\qquad u = U,$

$\qquad v = V - Y^2 f'/f - Z^2 g'/g,$

$\qquad y = Y/f,$

$\qquad z = Z/g,$

where $f = f(U)$ and $g = g(U)$.

21.1 A solution is cylindrically symmetric if it admits a symmetry axis and is invariant under both rotations about the axis and translations parallel to it.

$\qquad g_{ab} = g_{ab}(x^0, x^1)$ only.

$\qquad \mathrm{d}s^2 = g_{00}\mathrm{d}t^2 + 2g_{01}\mathrm{d}t\mathrm{d}\rho + g_{11}\mathrm{d}\rho^2.$

$\qquad g_{02} = g_{03} = g_{12} = g_{13} = 0.$

21.2 Invariant under $\phi \to -\phi$.
No cross term in $\mathrm{d}\phi\,\mathrm{d}z$, i.e. $\chi = 0$.

21.5 Non-zero independent components are:

$\Gamma^0_{00} = r^2 U^2 \mathrm{e}^{2\gamma - 2\beta}\gamma_{,1} + r^2 U \mathrm{e}^{2\gamma - 2\beta}U_{,1} + rU^2\mathrm{e}^{2\gamma - 2\beta}$

$\qquad + 2\beta_{,0} - r^{-1}V\beta_{,1} - \tfrac{1}{2}r^{-1}V_{,1} + \tfrac{1}{2}r^{-2}V,$

$\Gamma^0_{02} = -r^2 U\mathrm{e}^{2\gamma - 2\beta}\gamma_{,1} - \tfrac{1}{2}r^2\mathrm{e}^{2\gamma - 2\beta}U_{,1} - rU\mathrm{e}^{2\gamma - 2\beta} + \beta_{,2},$

$\Gamma^0_{22} = r^2 e^{2\gamma - 2\beta} \gamma_{,1} + r e^{2\gamma - 2\beta},$

$\Gamma^0_{33} = -r^2 e^{-(2\gamma + 2\beta)} \sin^2 \theta \gamma_{,1} + r e^{-(2\gamma + 2\beta)} \sin^2 \theta,$

$\Gamma^0_{00} = r^2 U^3 e^{2\gamma - 2\beta} \gamma_{,2} + r^2 U^2 e^{2\gamma - 2\beta} \gamma_{,0} + r^2 U^2 e^{2\gamma - 2\beta} U_{,2}$
$\qquad - r U^2 V e^{2\gamma - 2\beta} \gamma_{,1} - r U V e^{2\gamma - 2\beta} U_{,1}$
$\qquad - U^2 V e^{2\gamma - 2\beta} - r^{-1} U V \beta_{,2} - \frac{1}{2} r^{-1} U V_{,2}$
$\qquad - r^{-1} V \beta_{,0} + \frac{1}{2} r^{-1} V_{,0} + r^{-2} V^2 \beta_{,r}$
$\qquad + \frac{1}{2} r^{-2} V V_{,1} - \frac{1}{2} r^{-3} V^2,$

$\Gamma^1_{01} = -\frac{1}{2} r^2 U e^{2\gamma - 2\beta} U_{,1} - U \beta_2$
$\qquad + r^{-1} V \beta_{,1} + \frac{1}{2} r^{-1} V_{,r} - \frac{1}{2} r^{-2} V,$

$\Gamma^1_{02} = -r^2 U^2 e^{2\gamma - 2\beta} \gamma_{,2} - r^2 U e^{2\gamma - 2\beta} \gamma_{,0}$
$\qquad - r^2 U e^{2\gamma - 2\beta} U_{,2} + r U V e^{2\gamma - 2\beta} \gamma_{,1}$
$\qquad + \frac{1}{2} r V e^{2\gamma - 2\beta} U_{,1} + U V e^{2\gamma - 2\beta} + \frac{1}{2} r^{-1} V_{,2},$

$\Gamma^1_{11} = 2\beta_{,1},$

$\Gamma^1_{12} = \frac{1}{2} r^2 e^{2\gamma - 2\beta} U_{,1} + \beta_{,2},$

$\Gamma^1_{22} = r^2 U e^{2\gamma - 2\beta} \gamma_{,2} + r^2 e^{2\gamma - 2\beta} \gamma_{,0}$
$\qquad + r^2 e^{2\gamma - 2\beta} U_{,2} - r V e^{2\gamma - 2\beta} \gamma_{,1} - V e^{2\gamma - 2\beta},$

$\Gamma^1_{33} = r^2 U e^{-(2\gamma + 2\beta)} \cos \theta \sin \theta - r^2 U e^{-(2\gamma + 2\beta)} \sin^2 \theta \gamma_{,2}$
$\qquad - r^2 e^{-(2\gamma + 2\beta)} \sin^2 \theta \gamma_{,0}$
$\qquad + r V e^{-(2\gamma + 2\beta)} \sin^2 \theta \gamma_{,1} - V e^{-(2\gamma + 2\beta)} \sin^2 \theta,$

$\Gamma^2_{00} = r^2 U^3 e^{2\gamma - 2\beta} \gamma_{,1} + r^2 U^2 e^{2\gamma - 2\beta} U_{,1}$
$\qquad + r U^3 e^{2\gamma - 2\beta} - U^2 \gamma_{,2} + 2U \beta_{,0} - 2U \gamma_{,0} - U U_{,2}$
$\qquad - U_{,0} - r^{-1} U V \beta_{,1} - \frac{1}{2} r^{-1} U V_{,1} + \frac{1}{2} r^{-2} U V$
$\qquad + r^{-3} V e^{2\beta - 2\gamma} \beta_{,2} + \frac{1}{2} r^{-3} e^{2\beta - 2\gamma} V_{,2},$

$\Gamma^2_{01} = -U \gamma_{,1} - \frac{1}{2} U_{,1} - r^{-1} U + r^{-2} e^{2\beta - 2\gamma} \beta_{,2},$

$\Gamma^2_{02} = -r^2 U^2 e^{2\gamma - 2\beta} \gamma_{,1} - \frac{1}{2} r^2 U e^{2\gamma - 2\beta} U_{,1}$
$\qquad - r U^2 e^{2\gamma - 2\beta} + U \beta_{,2} + \gamma_{,0},$

$\Gamma^1_{12} = \frac{1}{2} r^2 e^{2\gamma - 2\beta} U_{,1} + \beta_{,2},$

$\Gamma^2_{22} = r^2 U e^{2\gamma - 2\beta} \gamma_{,1} + r U e^{2\gamma - 2\beta} + \gamma_{,2},$

$\Gamma^2_{33} = -r^2 U e^{-(2\gamma + 2\beta)} \sin^2 \theta \gamma_{,1} + r U e^{-(2\gamma + 2\beta)} \sin^2 \theta$
$\qquad - e^{-4\gamma} \cos \theta \sin \theta + e^{-4\gamma} \sin^2 \theta \gamma_{,2},$

$\Gamma^3_{03} = -\gamma_{,0},$

$\Gamma^3_{13} = -\gamma_{,1} + r^{-1},$

$\Gamma^3_{23} = \cot \theta - \gamma_{,2}.$

21.7

$R_{11} = -2\gamma^2_{,1} + 4r^{-1} \beta_{,1},$

$R_{12} = -r^2 e^{2\gamma - 2\beta} \beta_{,1} U_{,1} + r^2 e^{2\gamma - 2\beta} \gamma_{,1} U_{,1}$

$\qquad + \frac{1}{2} r^2 e^{2\gamma - 2\beta} U_{,11} + 2r e^{2\gamma - 2\beta} U_{,1} + 2 \cot \theta \gamma_{,1}$
$\qquad - \beta_{,12} - 2\gamma_{,1} \gamma_{,2} + \gamma_{,12} + 2r^{-1} \beta_{,2},$

$R_{22} = -\frac{1}{4} r^4 e^{4\gamma - 4\beta} U^2_{,1} + r^2 U e^{2\gamma - 2\beta} \cot \theta \gamma_{,1}$
$\qquad + 2r^2 U e^{2\gamma - 2\beta} \gamma_{,12} + r^2 e^{2\gamma - 2\beta} \gamma_{,1} U_{,2}$
$\qquad + r^2 e^{2\gamma - 2\beta} \gamma_{,2} U_{,1} + 2r^2 e^{2\gamma - 2\beta} \gamma_{,01}$
$\qquad + r^2 e^{2\gamma - 2\beta} U_{,12} + r U e^{2\gamma - 2\beta} \cot \theta + 2r U e^{2\gamma - 2\beta} \gamma_{,2}$
$\qquad - r V e^{2\gamma - 2\beta} \gamma_{,11} + 2r e^{2\gamma - 2\beta} \gamma_{,1}$
$\qquad - r e^{2\gamma - 2\beta} \gamma_{,1} V_{,1} + 3r e^{2\gamma - 2\beta} U_{,2} - V e^{2\gamma - 2\beta} \gamma_{,1}$
$\qquad + 3 \cot \theta \gamma_{,2} - 2\beta^2_{,2} + 2\beta_{,2} \gamma_{,2} - 2\beta_{,22}$
$\qquad - 2\gamma^2_{,2} + \gamma_{,22} + 1 - e^{2\gamma - 2\beta} V_{,1},$

$R_{33} = -r^2 U e^{-(2\gamma + 2\beta)} \cos \theta \sin \theta \gamma_{,1} - 2r^2 U e^{-(2\gamma + 2\beta)} \sin^2 \theta \gamma_{,12}$
$\qquad + r^2 e^{-(2\gamma + 2\beta)} \cos \theta \sin \theta U_{,1}$
$\qquad - r^2 e^{-(2\gamma + 2\beta)} \sin^2 \theta \gamma_{,1} U_{,2} - r^2 e^{-(2\gamma + 2\beta)} \sin^2 \theta \gamma_{,2} U_{,1}$
$\qquad - 2r^2 e^{-(2\gamma + 2\beta)} \sin^2 \theta \gamma_{,01}$
$\qquad + 3r U e^{-(2\gamma + 2\beta)} \cos \theta \sin \theta - 2r U e^{-(2\gamma + 2\beta)} \sin^2 \theta \gamma_{,2}$
$\qquad + r V e^{-(2\gamma + 2\beta)} \sin^2 \theta \gamma_{,11}$
$\qquad - 2r e^{-(2\gamma + 2\beta)} \sin^2 \theta \gamma_{,0} + r e^{-(2\gamma + 2\beta)} \sin^2 \theta \gamma_{,1} V_{,1}$
$\qquad + r e^{-(2\gamma + 2\beta)} \sin^2 \theta U_{,2}$
$\qquad + V e^{-(2\gamma + 2\beta)} \sin^2 \theta \gamma_{,1} - 2e^{-4\gamma} \cos \theta \sin \theta \beta_{,2}$
$\qquad + 3e^{-4\gamma} \cos \theta \sin \theta \gamma_{,2}$
$\qquad + 2e^{-4\gamma} \sin^2 \theta \beta_{,2} \gamma_{,2} - 2e^{-4\gamma} \sin^2 \theta \gamma^2_{,2}$
$\qquad + e^{-4\gamma} \sin^2 \theta \gamma_{,22} + e^{-4\gamma} \sin^2 \theta$
$\qquad - e^{-(2\gamma + 2\beta)} \sin^2 \theta V_{,1}.$

21.10 $l^a_{;b} l^b = \lambda l^a.$

21.11 $\theta = 2e^{-2\beta}/r.$

22.2 $V_{C_i} = -\dfrac{\Lambda}{6} m_i r^2_i.$

22.3 A is a half of the sum of the moments of inertia of the system about a set of orthogonal axes situated at the origin at epoch t_0 (i.e. a half of the trace of the inertia tensor).

22.5 c must be a constant.

$\qquad K = c/2.$

22.10
$\qquad (k = +1) \quad d\sigma^2 = R^2_0 \sin^2 \chi (d\theta^2 + \sin^2 \theta \, d\phi^2).$
$\qquad (k = -1) \quad d\sigma^2 = R^2_0 \sinh^2 \chi (d\theta^2 + \sinh^2 \theta \, d\phi^2).$

23.6 (ii) Oscillating model.

Further Reading

Here are a few suggestions for further reading. I have limited myself to books, and the full references can be found in the selected bibliography. The *Resource Letters* GR-1 and GI-1 in the *American Journal of Physics*, February 1968 and June 1982, list many books and articles on general relativity at an introductory and at more advanced levels. In addition, many of the texts in the selected bibliography have extensive reference sections, such as Schutz (1985) and Misner *et al.*

Chapter 1

It would be fun to have a look at Lieber. It is out of print, and there seem to be few copies in libraries so you will probably have to go through an inter-library loan service. It is certainly worth reading Einstein's autobiography in the Schilpp volume. I consider Pais to be the most authoritative biography.

Chapter 2

This chapter is based heavily on Bondi's article in the Brandeis volume (Trautmann *et al.*). I do not know any other treatment of the *k*-calculus in print.

Chapter 3

There are many fine texts around on special relativity, including two introductory ones written by an ex-Southampton university colleague, Les Marder (Marder 1968). Rindler is frequently cited as the most authoritative.

Chapter 4

See Rindler again, but also try Taylor and Wheeler, and Dixon.

Chapters 5, 6, and 7

As discussed in the book, we consider tensors via the index approach. The older texts adopt the same approach and one example of a classic text on differential geometry, which was a major source for this book, is Synge and Schild. Many of the modern books which utilize the index-free approach are, in my opinion, a bit sophisticated for a first course in general relativity. One exception, however, is the excellent book of Schutz (1985). This is written at about the same level as this book and, in part because it contains material not covered in this book, is used as a companion text to the course at Southampton. The earlier book of Schutz (1980) provides a more solid grounding in differential geometry, but is perhaps more of a graduate text. Wald is also excellent but yet more advanced. The most advanced and complete treatment of geometrical methods can be found in the two volumes of Penrose and Rindler. Our treatment has one important omission, and that is the topic of differential forms (which is omitted because we do not use it). Hughston and Tod is a book on general relativity that includes both a treatment and a subsequent application in discussing anisotropic cosmologies. The various sign conventions can be found on the inside cover of Misner *et al.* We use the timelike convention of Landau and Lifshitz (1971).

Chapter 8

For further consideration of the Lorentz group, see Carmeli. The axiomatic formulation of special relativity is that of Trautmann *et al.* (1964).

Chapters 9 and 10

The text which I used to recommend for a first course in general relativity is that of Adler *et al.* Unfortunately, I believe this is now out of print. There are many other excellent standard texts on general relativity which are currently available including Ohanion, Schutz, Stephani, and Hughston and Tod. We mostly avoid issues of units in this book by working in relativistic units, but for more on this topic see, for example, Schutz or Wald.

Chapter 11

My approach to the variational principle is based on the lovely little book of Schrödinger.

Chapter 12

Again, all the standard texts have a treatment of energy–momentum tensors. The dominant energy condition comes from Hawking and Ellis and the Newtonian limit from Trautmann (1964).

Chapter 13

The treatment of the Cauchy problem is taken from Adler *et al.* For a more advanced treatment see, for example, Smarr. I have written a review article on computer algebra in general relativity in the Einstein centenary volume edited by Held. Most of the known solutions of Einsteins equations can be found in the book of Kramer *et al.*

Chapter 14

There is another omission in this chapter in that I do not consider the interior Schwarzschild solution. For a simple treatment see Hughston and Tod.

Chapter 15

All standard texts include consideration of the experimental tests. A complete, but advanced treatment, can be found in Will. I would also recommend looking at the relevant sections of the text of Misner, Thorne, and Wheeler, known for short as 'MTW'. MTW is a rich resource and is certainly worth consulting for a whole string of topics. However, its style is not perhaps for everyone (I find it somewhat verbose in places and would not recommend it for a first course in relativity). MTW has a very extensive bibliography.

Chapters 16, 17, 18, and 19

The key source for black holes is, not surprisingly, Hawking and Ellis. For more on tidal forces see MTW. For a complete derivation of the Kerr solution see Adler *et al.* (second edition). A comprehensive, but advanced treatment, is contained in Chandrasekhar. See also the book of Thorne *et al.*

Chapters 20 and 21

The companion text of Schutz treats sources of gravitational radiation and their detection, which we do not. In particular, he derives the important quadrupole radiation formula and applies it to the binary pulsar. Ohanion is another source for these topics. A complete treatment of gravitational wave solutions can be found in Griffiths.

Chapter 22

The source for our approach to cosmology is the classic text of Bondi. Our brief look at Newtonian cosmology can be pursued further in Landsberg and Evans. Probably the most authoritative text on cosmology is that of Weinberg. In particular, he includes a comprehensive discussion of distance in cosmology. Tod and Hughston have a chapter on anisotropic cosmologies. Other important texts are Robertson and Noonan and MTW.

Chapter 23

The main source for this chapter is Hawking and Ellis, although the origins of the steady-state theory is considered in more detail in Bondi. The final topics of inflation and the anthropic principle can be found in the popular best seller of Hawking. A more mathematical treatment is contained in Barrow and Tipler.

Selected bibliography

A. Biographies of Einstein (1879–1955)

There are a large number of books and articles of a biographical nature on Einstein. Most of the post-1955 biographies are listed below, together with a few of the more important earlier biographies. The Schilpp volume is particularly significant because it includes Einstein's autobiography.

Bernstein, J. (1976). **Einstein**, Modern Masters Series. Penguin, London.

Born, M. (1971). **Born–Einstein letters, 1916–1955.** Macmillan, London.

Burke, T. F. (ed.) (1984). **Einstein: a portrait**. Pomegranate.

Cahn, W. (1955). **Einstein: A pictorial biography**. Citadel, New York.

Clark, R. W. (1973), **Einstein: the life and times**. Hodder and Stoughton, London.

Cuny, H. (1963). **Albert Einstein:** the man and his theories. Souvenir, London.

Dank, M. (1983). **Albert Einstein**, Impact Biography Series. Watts.

de Broglie, L., Armand, L., and Simon, P. (1979). **Einstein**. Peebles, New York.

Flückiger, M. (1974). **Albert Einstein in Bern**. Paul Hampt Verlag, Bern.

Frank, P. (1947). **Albert Einstein: his life and times**. Knopf, New York.

Hoffman, B. (1972). **Albert Einstein: creator and rebel**. Viking, New York.

Hunter, N. (1987). **Einstein**, Great Lives Series. Bookwright, Watts.

Infeld, L. (1950). **Albert Einstein: his work and his influence on the world**. Charles Scribner's Sons, New York.

Kirsten, C. and Treder, H. J. (eds.) (1979). **Albert Einstein in Berlin 1913–1933**. Akademie Verlag, Berlin.

Levinger, E. E. (1962). **Albert Einstein**. Dodson, London.

Pais, A. (1982). **'Subtle is the Lord . . .' the science and life of Albert Einstein**. Oxford University Press.

Reisner, A. (1930). **Albert Einstein, a biographical portrait**. A and C Boni, New York.

Schilpp, P. A. (ed.) (1949). **Albert Einstein: philosopher-scientist**, Volume VIII. The Library of Living Philosophers, Evanston, Illinois.

Seelig, C. (1960). **Albert Einstein**. Europa Verlag, Zürich.

Whitrow, G. J. (ed.) (1967). **Einstein: the man and his achievement**. Dover, New York.

B. Texts on Differential Geometry, Relativity, and Cosmology

The following texts are largely on relativity and cosmology, and consist of a selection of more recent as well as classic texts. They are supplemented by a few important references on differential geometry.

Adler, R., Bazin, M., and Schiffer, M. (1975). **Introduction to general relativity** (2nd edn). McGraw-Hill, New York.

Anderson, J. L. (1967). **Principles of relativity physics**, Academic Press, New York.

Barrow, J. D. and Tipler, F. J. (1986). **The anthropic cosmological principle**. Clarendon Press, Oxford.

Beem, J. K. and Ehrlich, P. E. (1981). **Global Lorentzian geometry**. Dekker, New York.

Bergmann, P. G. (1942). **Introduction to the theory of relativity**, Prentice-Hall, Englewood Cliffs, New Jersey.

Berry, M. (1976). **Principles of cosmology and gravitation**. Cambridge University Press.

Bishop, R. L. and Goldberg, S. I. (1968). **Tensor analysis on manifolds**. Macmillan, London.

Bohm, D. (1965). **The special theory of relativity**. Benjamin, New York.

Bondi, H. (1961). **Cosmology**. Cambridge University Press.

Bowler, M. G. (1976). **Gravitation and relativity**. Pergamon, Oxford.

Buchdahl, H. A. (1981). **Seventeen simple lectures on general relativity**. Wiley, New York.

Burke, W. L. (1980). **Spacetime, geometry, cosmology**. University Science Books, Mill Valley, California.

Carmeli, M. and Malin, S. (1976). **Representation of the rotation and Lorentz groups: an introduction**. Dekker, New York.

Chandrasekhar, S. (1957). **Stellar structure**. Dover, New York.

Chandrasekhar, S. (1983). **The mathematical theory of black holes**. Clarendon Press. Oxford.

Choquet-Bruhat, Y., De Witt-Morette, C., and Dillard-Bleick, M. (1977). **Analysis, manifolds and physics**. North-Holland, Amsterdam.

Clarke, C. (1979). **Elementary general relativity**. Arnold, London.

Davis, W. R. (1970). **Classical theory of particles and fields and the theory of relativity**. Gordon and Breach, New York.

Dixon, W. G. (1978). **Special relativity, the foundation of modern physics**. Cambridge University Press.

Eddington, A. (1923). **The mathematical theory of relativity**. Cambridge University Press.

Einstein, A. (1920). **Relativity: the special and general theory**. Methuen, London.

Einstein, A. (1951). **The meaning of relativity**. Methuen, London.

Eisenhart, L. P. (1926). **Riemannian geometry**. Princeton University Press.

Fock, V. (1959). **The theory of space, time and gravitation**. Pergamon, Oxford.

Fokker, A. D. (1965). **Time and space, weight and inertia**. Pergamon, Oxford.

Foster, J. and Nightingale, J. D. (1979). **A short course in general relativity**. Longman, London.

French, A. P. (1968). **Special relativity**. Norton, New York.

Geroch, R. (1978). **General relativity from A to B**. University of Chicago Press.

Griffiths, J. (1991). **Colliding waves in general relativity**. Oxford University Press.

Hawking, S. W. (1988). **A brief history of time**, Bantam Press, London.

Hawking, S. W. and Ellis, G. F. R. (1973). **The large scale structure of space-time**, Cambridge University Press.

Hawking, S. W. and Israel, W. (eds.) (1979). **An Einstein centenary survey**. Cambridge University Press.

Hawking, S. W. and Israel, W. (eds.) (1987). **300 years of gravitation**. Cambridge University Press.

Heidmann, J. (1980). **Relativistic cosmology**. Springer, Berlin.

Held, A. (ed.) (1980). **General relativity and gravitation (one hundred years after the birth of Albert Einstein)**, Vols. 1 and 2. Plenum, New York.

Hicks, N. J. (1965). **Notes in differential geometry**. Van Nostrand, New York.

Hughston, L. P. and Tod, K. P. (1990). **An introduction to general relativity**, Cambridge University Press.

Infeld, L. and Plebanski, J. **Motion and relativity**. Pergamon, New York.

Kobayashi, S. and Nomizu, K. (1963, 1969). **Foundations of differential geometry**. Vols. 1 and 2. Interscience, New York.

Kramer, D., Stephani, H., Herlt, E., and MacCallum, M. A. H. (1980). **Exact solutions of Einstein's field equations**, Cambridge University Press.

Landau, L. D. and Lifshitz, E. M. (1971). **The classical theory of fields**. Pergamon, Oxford.

Landsberg, P. and Evans, D. A. (1977). **Mathematical cosmology: an introduction**. Clarendon Press, Oxford.

Lawden, D. F. (1982). **An introduction to tensor calculus, relativity and cosmology**. Wiley, New York.

Lieber, L. R. (1949). **The Einstein theory of relativity**. Dennis Dobson, London.

Lightman, A. P., Press, W. H., Price, R. H., and Teukolsky, S. A. (1975). **Problem book in relativity and gravitation**. Princeton University Press, Princeton.

Marder, L. (1968). **An introduction to relativity**. Longman, London.

Marder, L. (1971). **Time and the space-traveller**. George Allen and Unwin, London.

McVittie, G. C. (1956). **General relativity and cosmology**. Chapman and Hall, London.

Milne, E. A. (1935). **Relativity, gravitation and world structure**. Oxford University Press.

Misner, C. W., Thorne, K. S., and Wheeler, J. A. (1973). **Gravitation**. Freeman, San Francisco.

Narlikar, J. V. (1983). **Introduction to cosmology**. Jones and Bartlett, Portola Valley, California.

Ohanion, H. C. (1976). **Gravitation and space-time**. Norton, New York.

O'Neil, B. (1983). **Semi-Riemannian geometry: with application to relativity**, Pure and Applied Mathematics Series, Academic Press, New York.

Papapetrou, A. (1974). **Lectures in general relativity**. Reidel, Dordrecht, The Netherlands.

Pauli, W. (1958). **Theory of relativity**, Pergamon, Oxford.

Peebles, P. J. E. (1971). **Physical cosmology**, Princeton University Press.

Peebles, P. J. E. (1980). **Large scale structure of the universe**. Princeton University Press.

Penrose, R. (1989). **The emperor's new mind**, Oxford University Press.

Penrose, R. and Rindler, W. (1986). **Spinors and space-time**. Vols 1 and 2. Cambridge University Press.

Rees, M., Ruffini, R., and Wheeler, J. A. (1974). **Black holes, gravitational waves and cosmology: an introduction to current research**. Gordon and Breach, New York.

Robertson, H. P. and Noonan, T. W. (1968). **Relativity and cosmology**, Saunders, Philadelphia.

Rindler, W. (1977). **Essential relativity**. Springer-Verlag, New York.

Rindler, W. (1991). **Introduction to special relativity** (2nd edn). Clarendon Press, Oxford.

Rucker, R. V. B. (1977). **Geometry, relativity and the fourth dimension**. Dover, New York.

Sachs, R. K. (1977). **General relativity for mathematicians**, Graduate Texts in Mathematics Series, Springer, Berlin.

Sciama, D. W. (1971). **Modern cosmology**, Cambridge University Press.

Schouten, J. A. (1954). **Ricci-calculus**, Springer, Berlin.

Schrödinger, E. (1950). **Space-time structure**. Cambridge University Press.

Schutz, B. F. (1980). **Geometrical methods in mathematical physics**. Cambridge University Press.

Schutz, B. F. (1985). **A first course in general relativity**. Cambridge University Press.

Smarr, L. (ed.) (1979). **Sources of gravitational radiation**. Cambridge University Press.

Spivak, M. (1965). **Calculus on manifolds**. Benjamin, New York.

Stephani, H. (1982). **General relativity: an introduction to the gravitational field**. Cambridge University Press.

Synge, J. L. (1960). **Relativity: the general theory**. North-Holland, Amsterdam.

Synge, J. L. (1965). **Relativity: the special theory**. North-Holland, Amsterdam.

Synge, J. L. and Schild, A. (1949). **Tensor calculus**, University of Toronto Press.

Taylor, E. F. and Wheeler, J. A. (1966). **Spacetime physics**. Freeman, San Francisco.

Thorne, K. S., Price, R. H., and MacDonald, D. A. (eds.) (1986). **Black holes: the membrane paradigm**. Yale University Press.

Tolman, R. C. (1934). **Relativity, thermodynamics and cosmology**. Clarendon Press, Oxford.

Trautmann, A., Pirani, F. A. E., and Bondi, H. (1964). **Lectures on general relativity**, Brandeis 1964 Summer Institute on Theoretical Physics, Vol. 1. Prentice-Hall, Englewood Cliffs, New Jersey.

Wald, R. M. (1984). **General relativity**. University of Chicago Press.

Weinberg, S. (1972). **Gravitation and cosmology**, Wiley, New York.

Weyl, S. (1952). **Space–time–matter**. Dover, New York.

Will, C. M. (1981). **Theory and experiment in gravitational physics**, Cambridge University Press.

Zeldovich, Ya. B. and Novikov, I. O. (1971, 1974). **Relativistic astrophysics**, Vols. 1 and 2. University of Chicago Press.

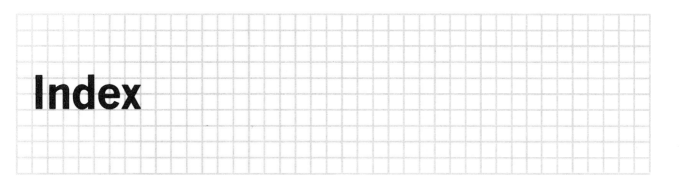

Index

The index is organised somewhat unconventionally in that it may be necessary to look under any of the words which occur in a multiple word entry (or which may be considered to be part of a multiple word entry) to locate it. For example, 'speed of sound' is located under 'speed' whereas 'speed of light' is located under 'light' (where it is grouped together with other related concepts, such as velocity of light). Again, references to particular solutions may occur under their name, under 'line element', 'Penrose diagram', 'space-time', or 'space-time diagram', depending on their actual usage in the text. Some commonly used terms (e.g. 'observer') have an initial reference only, whereas others (e.g. 'special relativity') have a comprehensive list of references.